Nitrogen Fixation

Nitrogen Fixation

Volume II: Symbiotic Associations and Cyanobacteria

Edited by

William E. Newton
Charles F. Kettering Research Laboratory
Yellow Springs, Ohio

and

William H. Orme-Johnson
University of Wisconsin-Madison

University Park Press
Baltimore

UNIVERSITY PARK PRESS
International Publishers in Science, Medicine, and Education
233 East Redwood Street
Baltimore, Maryland 21202

Typeset by American Graphic Arts Corporation.

Manufactured in the United States of America by
The Maple Press Company.

Library of Congress Cataloging in Publication Data

Kettering International Symposium on Nitrogen Fixation,
3d, Madison, Wis., 1978.
Nitrogen fixation.

Includes index.
1. Micro-organisms, Nitrogen-fixing—Congresses.
2. Nitrogen—Fixation—Congresses. I. Newton, William
Edward, 1938- II. Orme-Johnson, William H.
III. Title.
QR89.7.K47 1978 589 79-28285
ISBN 0-8391-1560-1 (v.1)
ISBN 0-8391-1561-X (v.2)

Contents

Contributors List vii
Preface xi

Section I **Leguminous Associations**

Host and *Rhizobium* Contributions to the Physiology of Legume Nodules *M. J. Dilworth* 3

Ammonia Assimilation in Nitrogen-fixing Legume Nodules *M. J. Boland, K. J. F. Farnden, and J. G. Robertson* 33

Utilization of Leghemoglobin-bound Oxygen by *Rhizobium* Bacteroids *J. B. Wittenberg* 53

Hydrogen Metabolism in the Legume-*Rhizobium* Symbiosis *H. J. Evans, D. W. Emerich, T. Ruiz-Argüeso, R. J. Maier, and S. L. Albrecht* 69

Control of Morphogenesis and Differentiation of Pea Root Nodules *W. Newcomb* 87

Factors Controlling the Legume-*Rhizobium* Symbiosis *J. M. Vincent* 103

Inoculation of Legumes with *Rhizobium* in Competition with Naturalized Strains *G. E. Ham* 131

Detection, Isolation, and Characterization of Large Plasmids in *Rhizobium* *R. K. Prakash, P. J. J. Hooykaas, A. M. Ledeboer, J. W. Kijne, R. A. Schilperoort, M. P. Nuti, A. A. Lepidi, F. Casse, C. Boucher, J. S. Julliot, and J. Dénarié* 139

Determinants of Host Specificity in the *Rhizobium*-Clover Symbiosis *F. B. Dazzo* 165

Host-Symbiont Interactions: Recognizing *Rhizobium* *C. Napoli, R. Sanders, R. Carlson, and P. Albersheim* 189

Role of Soybean Lectin in the Soybean–*Rhizobium japonicum* Symbiosis *W. D. Bauer* 205

Section II **Nonleguminous Associations**

On the Nature of the Endophyte Causing Root Nodulation in *Comptonia* *J. G. Torrey, D. Baker, D. Callaham, P. Del Tredici, W. Newcomb, R. L. Peterson, and J. D. Tjepkema* 217

Analysis of Factors Limiting Nitrogenase (C_2H_2) Activity in the Field *J. Balandreau and P. Ducerf* 229
Nitrogen Fixation Associated with Roots of Sorghum and Wheat *R. V. Klucas and W. Pedersen* 243

Section III Cyanobacteria and their Associations
Heterocyst Differentiation and Nitrogen Fixation in Cyanobacteria (Blue-Green Algae) *R. Haselkorn, B. Mazur, J. Orr, D. Rice, N. Wood, and R. Rippka* 259
Heterocysts, ^{13}N, and N_2-Fixing Plants *C. P. Wolk* 279
Azolla-Anabaena Association: Morphological and Physiological Studies *G. A. Peters, T. B. Ray, B. C. Mayne, and R. E. Toia, Jr.* 293
Azolla as a Nitrogen Source for Temperate Rice *S. N. Talley and D. W. Rains* 311

Index 321

Contributors List

Peter Albersheim
Department of Chemistry
University of Colorado
Boulder, Colorado 80309

Stephan L. Albrecht
Agronomy Department
University of Florida
Gainesville, Florida 32611

Dwight Baker
Department of Biology
Middlebury College
Middlebury, Vermont 05753

J. Balandreau
Centre de Pédologie Biologique
B. P. 5, 54500
Vandoeuvre-Les-Nancy, France

Wolfgang D. Bauer
Charles F. Kettering Research Laboratory
Yellow Springs, Ohio 45387

Michael J. Boland
Applied Biochemistry Division
DSIR
Palmerston North, New Zealand

C. Boucher
Laboratoire de Génétique des Microorganismes
I.N.R.A.
78000-Versailles, France

Dale Callaham
Department of Botany
University of Massachusetts
Amherst, Massachusetts 01003

Russell Carlson
Chemistry Department
Eastern Illinois University
Charleston, Illinois 61920

F. Casse
Laboratoire de Génétique des Microorganismes
I.N.R.A.
78000-Versailles, France

Frank B. Dazzo
Department of Microbiology and Public Health
Michigan State University
East Lansing, Michigan 48824

Peter Del Tredici
Arnold Arboretum
Harvard University
Jamaica Plain, Massachusetts 02130

J. Dénarié
Laboratoire de Génétique des Microorganismes
I.N.R.A.
78000-Versailles, France

M. J. Dilworth
School of Environmental and Life Sciences
Murdoch University
Murdoch, Western Australia

P. Ducerf
Centre de Pédologie Biologique
B. P. 5, 54500
Vandoeuvre-Les-Nancy, France

David W. Emerich
Laboratory for Nitrogen Fixation Research
Oregon State University
Corvallis, Oregon 97331

Harold J. Evans
Laboratory for Nitrogen Fixation Research
Oregon State University
Corvallis, Oregon 97331

Kevin J. F. Farnden
Department of Biochemistry
Otago University
Dunedin, New Zealand

G. E. Ham
Department of Soil Science
University of Minnesota
St. Paul, Minnesota 55108

Robert Haselkorn
Department of Biophysics and Theoretical Biology
University of Chicago
Chicago, Illinois 60637

P. J. J. Hooykaas
Biochemisch Laboratorium
Rijksuniversiteit Leiden
Leiden, The Netherlands

J. S. Julliot
Laboratoire de Génétique des Microorganismes
I.N.R.A.
78000-Versailles, France

J. W. Kijne
Biochemisch Laboratorium
Rijksuniversiteit Leiden
Leiden, The Netherlands

Robert V. Klucas
Laboratory of Agricultural Biochemistry
University of Nebraska
Lincoln, Nebraska 68583

A. M. Ledeboer
Biochemisch Laboratorium
Rijksuniversiteit Leiden
Leiden, The Netherlands

A. A. Lepidi
Istituto di Microbiologia agraria e tecnica
Università di Pisa
Centro di Studio per la Microbiologia del Suolo
C.N.R., Pisa, Italia

Robert J. Maier
Department of Biology
Mergenthaler Labs
Johns Hopkins University
Baltimore, Maryland 21218

B. C. Mayne
Charles F. Kettering Research Laboratory
Yellow Springs, Ohio 45387

Barbara Mazur
Department of Biophysics and Theoretical Biology
University of Chicago
Chicago, Illinois 60637

Carolyn Napoli
Department of Molecular, Cellular, and Developmental Biology
University of Colorado
Boulder, Colorado 80309

William Newcomb
Department of Biology
Queen's University
Kingston, Ontario
Canada K7L 3N6

M. P. Nuti
Instituto di Microbiologia agraria e tecnica
Università di Pisa
Centro di Studio per la Microbiologia del Suolo
C.N.R., Pisa, Italia

James Orr
Department of Biophysics and Theoretical Biology
University of Chicago
Chicago, Illinois 60637

Wayne Pederson
Department of Plant Pathology
Pennsylvania State University
University Park, Pennsylvania 16802

Gerald A. Peters
Charles F. Kettering Research Laboratory
Yellow Springs, Ohio 45387

R. L. Peterson
Department of Botany and Genetics
University of Guelph
Guelph, Ontario
Canada N1G 2W1

R. K. Prakash
Biochemisch Laboratorium
Rijksuniversiteit Leiden
Leiden, The Netherlands

D. William Rains
Plant Growth Laboratory
Department of Agronomy & Range Science
University of California
Davis, California 95616

T. B. Ray
Biochemicals Dept. 335/237
Experimental Station
E. I. Dupont de Nemours and Co.
Wilmington, Delaware 19898

Douglas Rice
Department of Biophysics and Theoretical Biology
University of Chicago
Chicago, Illinois 60637

Rosmarie Rippka
Unité de Physiologie Microbienne
Institut Pasteur
Paris, France

John G. Robertson
Applied Biochemistry Division
DSIR
Palmerston North, New Zealand

Tomás Ruiz-Argüeso
Departmento de Microbiologia
Escuela T. S. de Ingenieros Agronomos
Madrid -3, Spain

Richard Sanders
Department of Chemistry
University of Colorado
Boulder, Colorado 80309

R. A. Schilperoort
Biochemisch Laboratorium
Rijksuniversiteit Leiden
Leiden, The Netherlands

Steven N. Talley
Plant Growth Laboratory
Department of Agronomy & Range Science
University of California
Davis, California 95616

John D. Tjepkema
Harvard Forest
Harvard University
Petersham, Massachusetts 01366

R. E. Toia, Jr.
Charles F. Kettering Research Laboratory
Yellow Springs, Ohio 45387

John G. Torrey
Cabot Foundation
Harvard University
Petersham, Massachusetts 01366

J. M. Vincent
Department of Microbiology
University of Sydney
N.S.W. 2006, Australia

Jonathan B. Wittenberg
Department of Physiology
Albert Einstein College of Medicine
New York, New York 10461

C. Peter Wolk
MSU-DOE Plant Research Laboratory
Michigan State University
East Lansing, Michigan 48824

Nancy Wood
Department of Biophysics and Theoretical Biology
University of Chicago
Chicago, Illinois 60637

Preface

These volumes constitute the Proceedings of an international symposium that was held on June 12–16, 1978, in Madison, Wisconsin (USA). Three principal benefactors made this Symposium possible. The Steenbock Symposia Committee of the Department of Biochemistry, the University of Wisconsin–Madison, provided a grant as part of a regular program of annual symposia established in honor of the eminent biochemist, the late Professor Harry F. Steenbock, under the benefaction of Mrs. Evelyn Steenbock, whom we sincerely thank. The meeting was thus the Seventh Harry F. Steenbock Symposium. The Charles F. Kettering Foundation, Dayton, Ohio, and the Kettering Research Laboratory, Yellow Springs, Ohio, were established by the renowned scientist-inventor, Charles F. Kettering, and have organized several previous gatherings devoted to plant science. This meeting was the Third Kettering International Symposium on Nitrogen Fixation, the previous two being held in Pullman, Washington, in 1974 and Salamanca, Spain, in 1976. We thank Mr. Robert G. Chollar, President and Chairman of the Board of the C. F. Kettering Foundation, for his continued support of the concept of an interdisciplinary meeting in the area of nitrogen fixation. Finally, the Chemical Development Division of the Tennessee Valley Authority, under its programmatic interests in industrial nitrogen fixation and fertilizer economics, also provided support. Our thanks are extended to Mr. Charles H. Davis and his colleagues at TVA for their timely interest and generosity.

We are particularly pleased to note that three names widely associated with the application of science and technology to the betterment of mankind's condition have joined in this enterprise, which we hope has stimulated (and will continue to do so) interdisciplinary discussion on this vitally important contemporary problem.

The organization of this meeting was in the hands of a series of committees under our overall direction. We thank our colleagues on the International Program Committee, M. J. Dilworth (Murdoch University, Australia), R. W. F. Hardy (DuPont Company, USA), and J. R. Postgate (A.R.C. Unit of Nitrogen Fixation, UK), for their invaluable advice and counsel. R. H. Burris and W. J. Brill (University of Wisconsin-Madison) and H. J. Evans (Oregon State University) also contributed to this part of the endeavor, as did the many participants who wrote with programming suggestions. All are thanked.

The hard work of local arrangements and day-to-day organization fell in varying proportions to our colleagues on the Local Committee, Karen Davis, Vicki Hudson Newton, Bill Hamilton, Cynthia Touton, Nanette Orme-Johnson, Catherine Burris and Bob Burris, without whose efficiency and fidelity we would have been lost.

The interest in and importance of this area of research is demonstrated by the increasing demand for participation in these meetings. In 1974, about 200 scientists were in attendance, whereas in 1976 nearly 300 people participated. Attendance in 1978 was in excess of 400. This attendance trend is paralleled by the enormous progress made in the science in the four years since the Kettering-Pullman Symposium. These volumes speak very clearly to this point. We hope that this record of an outstanding symposium will benefit all persons interested in this area of research endeavor.

William E. Newton
Yellow Springs, Ohio

William H. Orme-Johnson
Madison, Wisconsin

Nitrogen Fixation

Section I
Leguminous Associations

Nitrogen Fixation, Volume II
Edited by W. E. Newton and W. H. Orme-Johnson

Host and *Rhizobium* Contributions to the Physiology of Legume Nodules

M. J. Dilworth

This paper concentrates on the following areas of the legume nodule symbiosis: 1) the biology of leghemoglobin; 2) the control of nitrogenase production and activity in *Rhizobium*, both free-living and symbiotic, and its relation to the enzymes of ammonia assimilation; 3) the evidence concerning DNA and RNA contents and stability in nodule bacteroids, and their implications for bacteroid viability; 4) the consequences of H_2 evolution and uptake by nodule bacteroids; and 5) the evidence regarding ammonia assimilation into amino acids in the bacteroid and plant fractions of nodules.

BIOLOGY OF LEGHEMOGLOBIN

Numerous earlier reports based on direct light microscopy (Bergersen, 1966), electron microscope autoradiography (Dilworth and Kidby, 1967) and cytochemical staining of leghemoglobin through its pseudoperoxidase activity (Truchet, 1972; Bergersen and Goodchild, 1973; Gourret and Fernandez-Arias, 1974) indicated that leghemoglobin was localized in the space between the peribacteroid membrane of plant origin and the bacteroid cell wall. Two recent reports throw doubt on this localization. When extensively washed, thin sections of fixed soybean nodules were treated with ferritin-coupled antibody to leghemoglobin, ferritin was found only in cytoplasmic locations and not in the space between bacteroid and plant cytoplasm (Verma and Bal, 1976). The failures by others to find leghemoglobin in the cytoplasm were attributed to inadequate fixation with consequent migration during processing. However, electron microscope autoradiography would have, but did not, detect migration of labeled leghemoglobin out of the tissue (Dilworth and Kidby, 1967). This reasoning does not explain the specific staining with diaminobenzidine seen by Bergersen and Goodchild (1973), Gourret and Fernandez-Arias (1974), and Truchet (1972). Furthermore, why should all leghemoglobin move from plant cytoplasm to membrane envelope? The result of Verma and Bal (1976) could be explained by the removal of surface-reactive leghemoglobin from all locations except where it was attached to plant ribosomes. Translation of globin messenger RNA undoubtedly occurs on plant cytoplasmic ribosomes (Verma, Nash, and Schulman, 1974; Verma and Bal, 1976).

In a second study (Robertson et al., 1978), membrane envelopes still containing bacteroids were isolated by sucrose density gradient centrifugation, after gentle rupture of nodules, and found to be free of leghemoglobin. To demonstrate that membranes had not ruptured during isolation and then resealed, ferritin was added to the isolation medium. Because ferritin was found inside only 1% of the membranes examined, their conclusion was that leghemoglobin must be located in the plant cytoplasm. However, the possibility of leakage through the peribacteroid membrane has not been excluded. In *Bacillus amyloliquefaciens*, simple cold shock to 16°–18°C causes complete leakage of ribonuclease inhibitor (molecular weight about 12,000) from the cells (Smeaton and Elliott, 1967), so leakage of leghemoglobin during isolation must be taken seriously. Simple techniques that do not involve fixation or tissue disruption are urgently needed to solve this localization problem.

The number of leghemoglobin (Lb) components found in any particular nodule type continues to increase. Better resolution techniques (Dilworth, 1969) applied to soybean Lb reveal that both Lb*c* and Lb*d* have two subcomponents (Appleby et al., 1975) and Lbc_2 has now been sequenced (Hurrell and Leach, 1977). Equal in length to Lb*a*, alignment required one COOH-terminal addition and two internal deletions, as well as a minimum of 14 substitutions, all conservative. The biological significance of these different Lb molecules remains to be determined. Thus far, only differences in O_2 affinity (Appleby, 1962) and in the pK for the conversion of CO-bound Lb from a neutral to an acid form (Fuchsman, unpublished data) have been reported. The changing ratio between different components during nodule development in soybean possibly suggests different physiological functions, different intracellular localization, or some combination of these (Fuchsman et al., 1976). No physiologically significant differences have been demonstrated, and when the leghemoglobin sequence variation between plants is considered (Dilworth and Appleby, 1977) for Lb with ostensibly the same function, genetic variability without physiological significance remains quite possible. Immunological studies on a variety of Lb (Hurrell et al., 1977) have produced a "family tree" of Lb that is, so far, parallel to the normal botanical relatedness of the plants, with the possible exception of serradella.

Loss of Lb has been implicated as a possible cause of the inhibitory effect of nitrate on N_2 (C_2H_2) fixation in pea nodules (Bisseling, van den Bos, and van Kammen, 1978). Pea nodule N_2 fixation was markedly inhibited by NH_4NO_3, $(NH_4)_2SO_4$, or NH_4Cl, but not by KNO_3, indicating that the effects were due to ammonium ion rather than nitrate, which contrasts with the results of Mahon (1977), where KNO_3 and NH_4NO_3 were equally inhibitory per unit of nitrogen. Bisseling, van den Bos, and van Kammen (1978) showed that, although activity declined, the amount of nitrogenase present (measured by the amount of $^{35}SO_4$ incorporation) remained constant. Since nodule heme content declined in parallel with nitrogenase activity, lowered nitrogenase activity was explained as Lb loss. In the pea, however, darkening is sufficient to cause Lb degradation (Roponen, 1970), probably due to decreased carbohydrate supply. Since nitrate will divert carbohydrate toward active growth sinks, low carbohydrate supply to pea nodules probably lowers their energy charge, thus lowering nitrogenase activity (Ching et al., 1975). Lb degradation and decline in nitrogenase activity would then have a common trigger rather than a causal relationship. Nitrite may also affect nitrogenase activity and Lb (Rigaud and Puppo, 1977). Nitrite inhibits nitrogenase activity in both bacteroids and extracts from soybean nodules (Rigaud et al., 1973; Kennedy, Rigaud, and Trinchant, 1975) and has now been shown to cause autoxidation of ferrous-oxyLb to the ferric form unable to transport O_2. However, the attempt to implicate nitrite as an intermediate in nitrate effects awaits the demonstration of significant nitrite concentrations in nodules.

Heme biosynthesis has been investigated, particularly in relation to the role of the bacteroid. Aminolaevulinate (ALA) synthetase activity is restricted to bacteroids in soybean nodules (Nadler and Avissar, 1977) and serradella nodules (Godfrey, Coventry, and Dilworth, 1975); however, ALA dehydratase activity occurs only in bacteroids in effective soybean nodules. Whereas Godfrey, Coventry, and Dilworth (1975) suggest that plant ALA dehydratase may function in heme synthesis for Lb, Nadler and Avissar (1977) consider that the time course of ALA dehydratase production and its absence in ineffective bacteroids indicate that heme synthesis for Lb is wholly a bacteroid property. This view must remain tentative until ALA synthesis and metabolism in the plant fraction can be established qualitatively and quantitatively. During microaerophilic growth of *Rhizobium japonicum*, a tenfold derepression of ALA synthetase and ALA dehydratase occurred and proto- and coproporphyrin were excreted into the media (Avissar and Nadler, 1978).

CONTROL OF NITROGENASE PRODUCTION AND ACTIVITY

With the discovery of rhizobial strains that would derepress nitrogenase in laboratory cultures, it became possible to examine the control mechanisms for nitrogenase. Until recently, the property appeared to be restricted to a few slow-growing *Rhizobium* strains (Pagan et al., 1975) and caution needed to be exercised in generalizing from such a few strains. Now, the regulation of nitrogenase production by O_2 concentration and by various nitrogen sources, notably ammonia, glutamate, and glutamine, are the key problems.

In liquid media, a very low O_2 concentration was clearly required for nitrogenase production (Tjepkema and Evans, 1975; Evans and Keister, 1976; Keister and Evans, 1976). On solid media, however, the highest nitrogenase activities were obtained with atmospheric O_2 concentrations (Pagan et al., 1975). This apparent contradiction proved to be a result of the assay time used to measure nitrogenase response to O_2. After 16-hr incubations with varying O_2 concentrations in the gas phase, nitrogenase derepression occurred and the optimal O_2 concentration was very low (Dilworth and McComb, 1975). In glutamine-limited continuous culture, nitrogenase activity appeared when dissolved oxygen concentration fell below 3 μM (Bergersen et al., 1976). If respiratory activity was lowered by stopping the supply of an easily respirable substrate (succinate), dissolved O_2 increased sharply and nitrogenase disappeared. Exposure to O_2 resulted in irreversible inactivation of nitrogenase and restoration under low O_2 concentration took much longer than inactivation. Addition of 5 mM ammonia to N_2-fixing continuous cultures resulted in partial loss of nitrogenase activity, not complete loss as found with other N_2-fixing bacteria (Postgate, 1974).

The response of respiration to O_2 concentration in such cultures showed an interesting contrast to nodule bacteroids. Bacteroids possess two distinct oxidase systems, one tightly coupled to ATP production and one more loosely coupled (Appleby, Turner, and Macnicol, 1975; Bergersen and Turner, 1975). The latter appears to be a flavoprotein oxidase located in the cell membrane (Appleby, unpublished data) that apparently offers "respiratory protection" for nitrogenase. Chemostat cultures of 32H1 grown at 30 μM or more O_2 had no nitrogenase but had both types of respiratory system. Cultures grown at around 1 μM O_2 had nitrogenase activity but not the flavoprotein oxidase. Tubb (1976) also observed in batch cultures ammonia-mediated decrease in nitrogenase activity, an effect that could be partially prevented by glutamate. Mutants of *Rhizobium trifolii* selected for the ability to derepress nitrogenase in the laboratory also showed this effect (O'Gara and Shanmugam, 1977).

Excretion of ammonia is a very important point of similarity between the rhizobia producing nitrogenase in culture and nodule bacteroids. With 32H1 or CB756 (cowpea strains), ammonia was excreted into the media when those strains fixed N_2 (O'Gara and Shanmugam, 1976a; Tubb, 1976; Bergersen and Turner, 1978). Soybean bacteroids (Bergersen and Turner, 1967) excrete ammonia virtually quantitatively in laboratory assays with $[^{15}N]N_2$. Thus far, N_2 fixation by *Rhizobium* in culture seems to be inadequate to provide fully for growth (Bergersen, 1977).

Known regulation of nitrogenase activity in gram-negative, N_2-fixing bacteria through glutamine synthetase and its state of adenylylation (Tubb, 1974; Ausubel, Margolskee, and Maizels, 1977) has focused considerable attention on this aspect with N_2-fixing laboratory cultures of rhizobia. Repression of glutamine synthetase occurred by addition of ammonia to laboratory cultures of *R. japonicum* but not by such additions to isolated bacteroids or to nodules before isolation (Bishop et al., 1976). In chemostat cultures of CB756 limited by O_2, and therefore high in nitrogenase, ammonia did not repress (Bergersen and Turner, 1976). When the growth-restricting O_2 level was increased, ammonia repression could be shown; the higher the ammonia concentration, the more stringent the oxygen limitation to growth for high nitrogenase activity. Increased adenylylation of glutamine synthetase was shown with lower nitrogenase activities (Bergersen and Turner, 1976). Bergersen and Turner (1978) showed a very strong correlation between nitrogenase activity and biosynthetic glutamine synthetase. Increasing the O_2 supply four to fivefold while only increasing the mean concentration twofold in cultures growing on 10 mM ammonia caused loss of nitrogenase, but, when ammonia became limiting at the higher O_2 supply rate, nitrogenase synthesis occurred, although not to the same extent as when O_2 was limiting. When the O_2 supply was decreased, glutamine synthetase relative adenylylation fell before nitrogenase activity rose. In general, then, O_2 concentration is far more important in regulating glutamine synthetase and nitrogenase activities than is ammonia. If there is another ammonia-mediated negative control on *nif* genes in *Klebsiella pneumoniae* besides glutamine synthetase (Ausubel, Margolskee, and Maizels, 1977), this ammonia control in rhizobia may be rather weak.

Regulation of nitrogenase by O_2 has been suggested for *K. pneumoniae* (St. John, Shah, and Brill, 1974). Further work (Eady et al., 1978) showed that 5% O_2 caused greater decrease in the rate of appearance of labeled nitrogenase proteins during derepression from ammonia than could be accounted for by destruction of nitrogenase components (stable for 3 hr). This indication of repression of nitrogenase synthesis by O_2 was reinforced by experiments with a strain (SK-24) that lacks glutamate synthase and is therefore constitutive for glutamine synthetase and nitrogenase. Twenty percent O_2 completely prevented nitrogenase derepression, although ammonia did not. Thus, for *Rhizobium*, which also only fixes N_2 microaerophilically, there may be regulation via O_2, via glutamine synthetase, and/or via ammonia.

Other evidence comes from glutamate/glutamine metabolism in *Rhizobium* mutants. A glutamine auxotroph of 32H1 with low activity in the transferase assay for glutamine synthetase failed to produce nitrogenase under inducing conditions in laboratory culture (Ludwig and Signer, 1977). Its nodules on siratro were completely devoid of nitrogenase and isolates from such nodules contained gln^+ revertants as well as the mutant. Nodulation was very severely delayed by glutamine in the rooting medium, but the ineffective nodules eventually formed also contained glutamine-independent revertants. Laboratory revertants recovered the ability to synthesize glutamine, but only two also recovered nitrogenase production in nodules. Thus, two strains recovered both catalytic and regulatory roles for glutamine synthetase, and the others regained only catalytic activity. The involvement of glutamine synthetase with nitrogenase production in laboratory culture was strongly suggested. Since the activities of the other ammonia-assimilatory enzymes were not reported, the lesion in the original mutant could be similar to that in *Salmonella typhimurium*, which requires a gene product additional to glutamine synthetase protein for transcription of glutamine synthetase DNA (Bloom, Streicher, and Tyler, 1977; Garcia et al., 1977).

Mutants of *R. trifolii* have been isolated on the criteria of resistance to methionine sulfoximine or slow growth on ammonia as sole source of nitrogen following mutagenesis (O'Gara and Shanmugam, 1977). Such mutants formed nitrogenase in laboratory culture under low O_2 concentration, but nitrogenase synthesis was lowered 50% when ammonia was added to glutamate as the nitrogen source. The lesion in these mutants has complex effects: lowered glutamate synthase (both NADH and NADPH) and glutamate dehydrogenase; lowered ribitol dehydrogenase and β-galactosidase; and a threefold increase in generation time. Although mutants formed either white ineffective nodules or none at all, revertants regained all their laboratory culture characteristics as well as nodulation. These artificial mutants apparently have a different reason for their ability to fix N_2 in culture because strains like 32H1, 61A76, or CB756 retain full effectiveness on their hosts.

In studies of transformation of *nif*$^-$ *Azotobacter vinelandii* strains with DNA from various *Rhizobium* species (Page, 1978), only the two strains 32H1 and 41A1 that were capable of laboratory nitrogenase production would transform the I^aII^a phenotype thought to be due to a defect in the system regulating component I and II transcription (Shah et al., 1974). This genetic evidence of difference in the control system in these strains parallels the obvious difference in nitrogenase production capability.

Ammonia Assimilation and Its Regulation in *Rhizobium*

There are probably three major reasons, two of them amenable to experimental treatment, for the lack of a consensus in this area. The first is the wide difference in the rhizobial strains being used and the difficulty in studying sufficient numbers of strains to produce acceptable generalizations. The second lies in the virtually complete lack of data concerning the actual intracellular pools of the various nitrogen compounds under study. For example, although it is known that glutamate prevents ammonia utilization (O'Gara and Shanmugam, 1976b), this does not say whether ammonium levels in the cell are or are not changed by glutamate addition. Furthermore, glutamate synthase in *Rhizobium* is extremely sensitive to inhibition by amino acids (Figures 1 and 2) and accordingly even measured activities of this enzyme may mean very little in terms of the cell's capability for ammonia assimilation. The third reason lies in the difficulties of interpreting the effects of growth in a specific medium on enzyme activities in batch culture measured at one particular time. Activities of glutamate synthase are particularly awkward in this respect. Data for *Rhizobium leguminosarum* illustrate the large variations to be expected over the growth cycle (Figure 3). Similar effects (O'Gara and Shanmugan, 1976b) on the specific activity of glutamate synthase from *R. trifolii* are found even during log phase. The only solution is the wider use of controlled chemostat cultures where a constant physiological state can be guaranteed. These reservations should be kept in mind during the following examination of the recently published work on ammonia assimilation by *Rhizobium*.

O'Gara and Shanmugam (1976a, 1976b) have studied regulation of ammonia uptake and production in *R. trifolii*, *R. japonicum*, *R. leguminosarum*, and cowpea *Rhizobium*, with principal emphasis on the first. They showed that several strains were unable to grow on ammonia as sole source of nitrogen, although many can (Brown and Dilworth, 1975). More interesting was the indication that glutamate prevented utilization of exogenous ammonia, as had been shown for endogenous ammonia produced by N_2 fixation (Tubb, 1976). Mutants able to grow on ammonia as sole source of nitrogen lost the effect of glutamate on ammonia uptake (O'Gara and Shanmugam, 1976b). Ammonia liberation into the medium was also noted (O'Gara and Shanmugam, 1976a).

There is general agreement that ammonia represses glutamine synthetase activity in laboratory cultures whether they are N_2-fixing or not (Brown and Dilworth, 1975; Bergersen and Turner, 1976, 1978; Bishop et al., 1976; O'Gara and Shanmugam, 1976a, 1976b, 1977; Upchurch and Elkan, 1978b), although the size of the effect on activity cannot be estimated where only transferase activities were measured (O'Gara and Shanmugam, 1976a, 1976b, 1977; Upchurch and Elkan, 1978b). In some

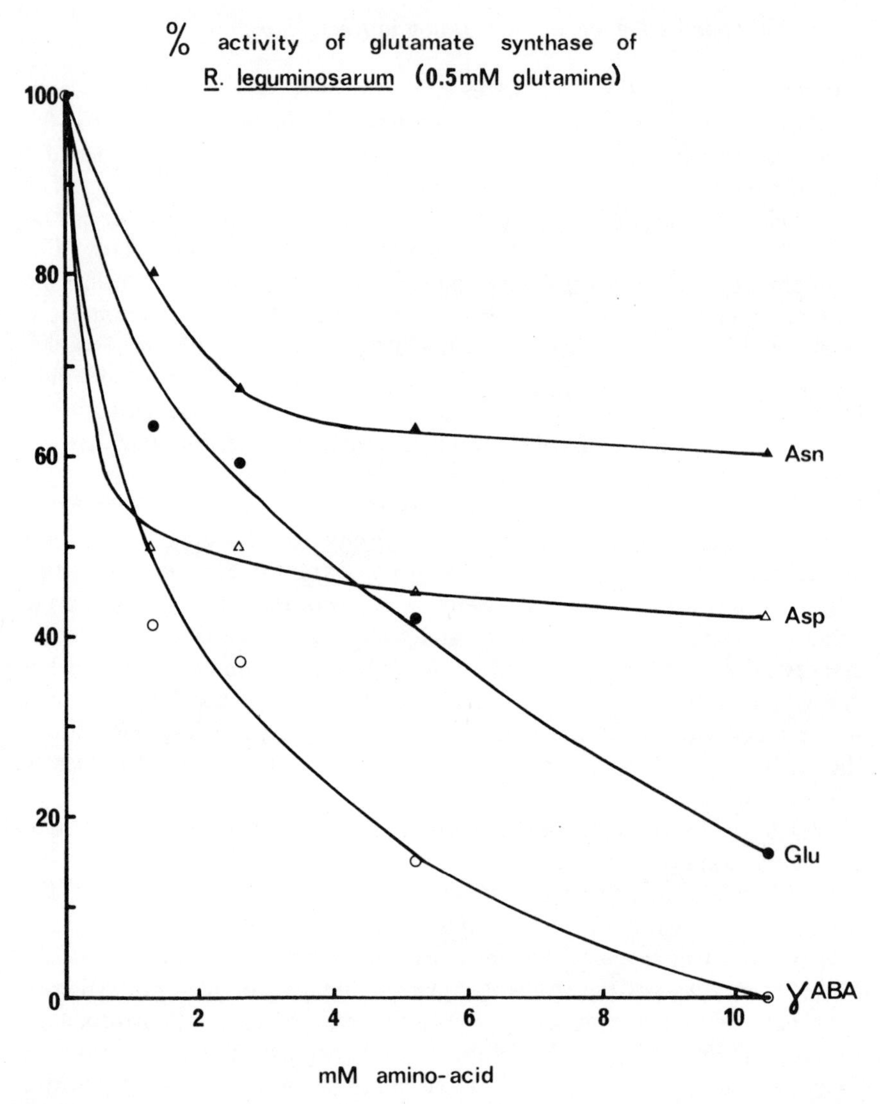

Figure 1. Inhibition of a partially (× 10) purified fraction of *R. leguminosarum* glutamate synthase by some important nodule amino acids.

strains of *R. japonicum*, glutamate also repressed glutamine synthetase in batch culture (Upchurch and Elkan, 1977). Whereas chemostat cultures of *R. japonicum* were not affected, *R. trifolii* continuous cultures did show glutamate repression (Brown and Dilworth, 1975). Ammonia (Bergersen and Turner, 1976, 1978; Bishop et al., 1976) and glutamate (Upchurch and

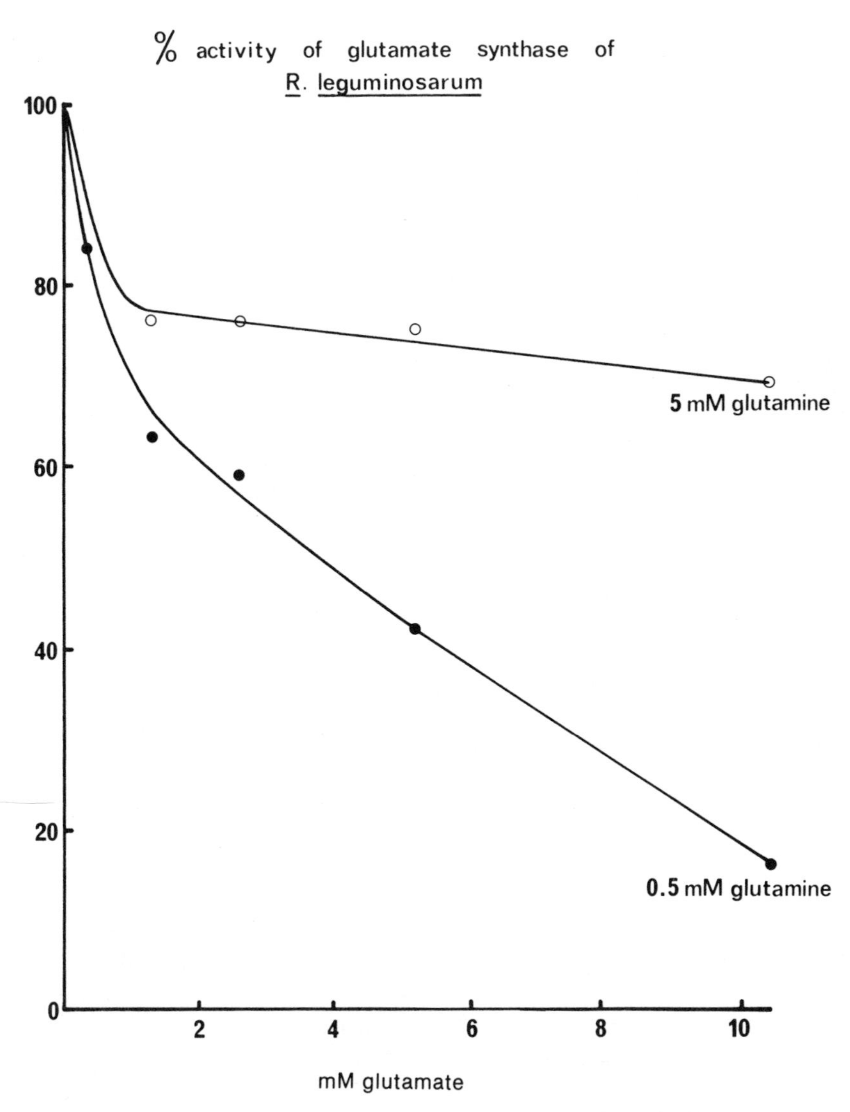

Figure 2. Inhibition of partially purified glutamate synthase from *R. leguminosarum* by glutamate in the presence of 0.5 or 5 mM glutamine in the assay medium.

Elkan, 1978a) increase the average adenylylation state of glutamine synthetase. Cyclic 3′,5′-adenosine monophosphate (cAMP) at 1 mM appeared to repress glutamine synthetase production by *R. japonicum* (Upchurch and Elkan, 1978a) in the presence of ammonia, quite contrary to its effect with *Escherichia coli* (Prusiner et al., 1972).

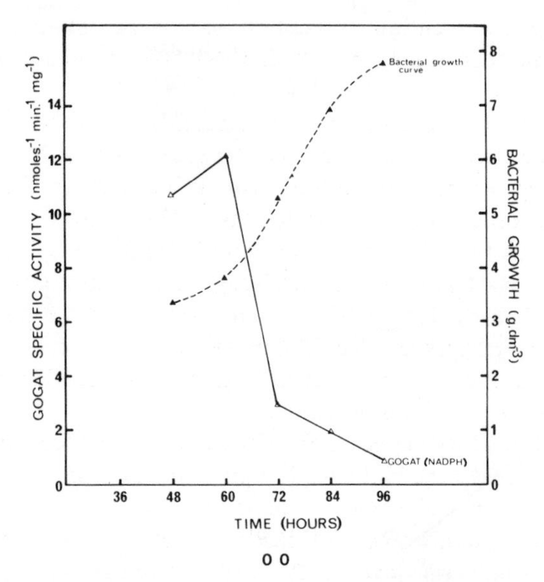

Figure 3. Activity of glutamate synthase as a function of growth in 1-liter batch cultures of *R. leguminosarum* WU47.

Glutamate synthase activities in *R. japonicum* have been variously reported as unresponsive to ammonia or glutamate in continuous culture (Brown and Dilworth, 1975) or strongly repressed by glutamate but not by ammonia in batch cultures (Upchurch and Elkan, 1978a). Microcolony derivatives of *R. japonicum* with higher symbiotic and asymbiotic nitrogenase activities appeared to have lower activities for both glutamate synthase and glutamine synthetase than larger colony types during aerobic growth or in the bacteroid forms (Upchurch and Elkan, 1978b), although both types were low in glutamate synthase activity as compared to other effective soybean strains (Dunn and Klucas, 1973; Brown and Dilworth, 1975). Microaerophilic N_2-fixing *R. japonicum* strains showed very low activities of glutamine synthetase, glutamate synthase, or glutamate dehydrogenase (Upchurch and Elkan, 1978b). By contrast, chemostat cultures of CB756 (cowpea) rhizobia have normal activities of glutamine synthetase and glutamate synthase, although they lack glutamate dehydrogenase (Bergersen and Turner, 1978). For *R. trifolii*, glutamate synthase activities were very low in glutamate-grown cells (Brown and Dilworth, 1975). Those strains, apparently unable to grow on ammonium as sole source of nitrogen, produced little glutamate synthase, but aspartate addition restored both the ability to use ammonium and glutamate synthase activity (O'Gara and Shanmugam, 1976b). Strains able to use ammonium as sole source of nitrogen had normal glutamate synthase activities. Mutant strains able to derepress nitrogenase in culture showed, in one case, no glutamate synthase activity and, in another, a moderate level of NADH-linked activity (O'Gara and Shanmugam, 1977).

Some resolution is required on the nucleotide specificity of glutamate synthase in *Rhizobium*. Thus, activity dependent on either NADH or NADPH has been reported for *R. japonicum* (Upchurch and Elkan, 1978a, 1978b), whereas others (Dunn and Klucas, 1973; Brown and Dilworth, 1975) found only NADH-linked activity. Similarly, *R. trifolii* is reported to have either both NADH- and NADPH-dependent activities (O'Gara and Shanmugam, 1976a, 1976b, 1977) or only an NADPH-linked activity (Brown and Dilworth, 1975). Resolution of this point should impact on whether the same enzyme can interact with both nucleotides or whether there are, in fact, two enzymes present.

Bacteroid studies will indicate the situation with respect to ammonia-assimilatory enzymes in the biologically significant environment. Glutamine synthetase activities in the bacteroid fraction are low (Brown and Dilworth, 1975; Kurz, Rokosh, and LaRue, 1975; Planqué et al., 1977; Upchurch and Elkan, 1978b), usually too low for the flux of fixed nitrogen through the nodule to be assimilated via the bacteroid. Glutamate synthase activities may be high or low depending on the legume in question (Brown and Dilworth, 1975; Kurz, Rokosh, and LaRue, 1975; Upchurch and Elkan, 1978b). Glutamate dehydrogenase, both NADH and NADPH dependent, shows no consistent pattern of activity. Localization studies on bacteroids suggest that nitrogenase and glutamate dehydrogenase occur in the bacteroid cytoplasm, glutamate synthase is weakly membrane bound, and glutamine synthetase partly cytoplasmic and partly membrane bound (Planqué et al., 1977). O'Gara and Shanmugam (1976a) have proposed that ammonia-assimilatory enzyme activities are suppressed during N_2 fixation in the bacteroid state, which leads to ammonia export rather than assimilation. The only clearly consistent effect on the ammonia-assimilatory enzymes is with glutamine synthetase, although glutamate synthase may also be involved in some strains. The experiments with the small and large colony variants (Upchurch and Elkan, 1978b) support the concept that low ammonia assimilatory enzyme activities are involved with both laboratory culture and nodular N_2 fixation.

Ammonium transport systems may also be important in regulation of internal ammonium concentration, nitrogenase, and ammonia-assimilatory enzymes. Some interesting observations are: 1) ammonia concentrations outside bacteroids in the nodule are higher than inside (Upchurch and Elkan, 1978b), although leakage could have occurred from bacteroid to cytosol during isolation, as occurs with amino acids (Kennedy, Parker, and Kidby, 1966) (if true, however, it would indicate a need for an active transport system out of the bacteroid); 2) in *A. vinelandii*, inward ammonium transport is suppressed by anaerobiosis (Kleiner, 1975) and the interior of nodules is very close to being anaerobic (Appleby, 1969); 3) new permeases can be induced in *A. vinelandii* to facilitate ammonia uptake (Kleiner, 1975); 4) mutants in *R. trifolii* with the ability to derepress nitrogenase in the laboratory are very slow growing as compared to their parent strain (O'Gara and Shanmugam, 1977) and could be limited for ammonia intake; and 5) glutamate quite specifically antagonizes ammonium uptake, whereas equal concentrations of nitrogen as aspartate do not (O'Gara and Shanmugam, 1976a, 1976b; Tubb, 1976), suggesting a possible interaction on a carrier system.

If the small NH_3 pool that saturates with ^{15}N after 3–4 min (Kennedy, 1966a, 1966b) is assumed to be the bacteroid pool, the intrabacteroid ammonia concentration is only about 0.3 mM. Total nodule ammonia is 2–4 mM (Kennedy, 1966a), which thus demands an active transport system for NH_3. Calculations based on data (Bergersen and Turner, 1967) for isolated bacteroids do not, however, suggest any concentration gradient between bacteroids and medium, but isolated bacteroids are clearly often leaky (Kennedy, Parker, and Kidby, 1966). Therefore, the whole question of ammonium import-export in rhizobia, and its interactions with uptake of amino acids, nitrogenase induction, and ammonia-assimilatory enzyme levels should be considered.

NUCLEIC ACIDS IN NODULE BACTEROIDS

Since Almon (1933) demonstrated low viability for single isolated bacteroids, most bacteroids are considered nonviable (Bergersen, 1974; Nutman, 1975). Investigations into DNA content and properties in bacteroids as compared to laboratory cultures have therefore attempted to explain this apparent nonviability. The DNA concentration in bacteroids falls during nodule maturation in *Lupinus luteus* (Dilworth and Williams, 1967). In *Lotus*, bacteroids of fast-growing strains had less DNA per cell than corresponding broth cultures, while slow-growing strains had the same amounts (Sutton, 1974). Subsequent work, using techniques of cytofluorimetry on individual bacteroids, has indicated that, in all systems except cowpea CB756, amounts of DNA per cell increased by a factor of 1.5 to 7.8 (Reijnders et al., 1975; Bisseling et al., 1977; Paau, Lee, and Cowles, 1977). The final DNA per cell reported by Dilworth and Williams (1967) is also consistent with values for *R. lupini* bacteroids determined by fluorimetry (Bisseling et al., 1977).

Some conflict exists about DNA properties in bacteroids. Although Agarwal and Mehta (1974) reported changes in the DNA buoyant density and melting behavior for *Cicer* and *Phaseolus* bacteroids, no such changes were detected in buoyant density, melting temperature, molecular weight, or kinetic complexity in *Rhizobium lupini* from *Lotus* (Sutton, 1974) or in buoyant density or RNA-DNA or DNA-DNA hybridization characteristics for DNA from pea bacteroids (Reijnders et al., 1975). Plasmid DNA has been frequently detected (Sutton, 1974; Tshitenge et al., 1975; Zurkowski and Lorkiewicz, 1976; Nuti et al., 1977), although not invariably (Reijnders et al., 1975), but the precise functions of such plasmids are unknown. The Ti plasmid from *Agrobacterium tumefaciens* changed the host specificity pattern to that of *R. trifolii* (Hooykaas et al., 1977). DNA polymerase activity was 10–18 times higher in laboratory-grown *Rhizobium meliloti* than in bacteroids from alfalfa, but no such difference was found for *R. japonicum* (Paau and Cowles, 1975). Ribonucleotide reductase activity was lower in alfalfa bacteroids than in laboratory-grown *R. meliloti* (Cowles, Evans, and Russell, 1969). RNA content in bacteroids fell during nodule development in *L. luteus* (Dilworth and Williams, 1967) and such bacteroids had a low ratio of intact rRNA to DNA (Sutton and Robertson, 1974). Changes in polynucleotide phosphorylase activity in *R. meliloti* do not account for the loss (Hunt and Cowles, 1977). Apparent protein synthetic activity is also lower in bacteroids than in laboratory cultures, although how far oxygen and osmotic shock contribute to low activities in most experiments has not been fully assessed.

Viability of *R. lupini* bacteroids was markedly increased by isolation and cultivation in the presence of 0.3 M mannitol (Sutton, Jepsen, and Shaw, 1977). Protein synthetic ability of bacteroids was also improved by high osmotic pressure media, although laboratory cultures were inhibited by them. Even though mannitol increased colony-forming ability at all nodule ages, there was still evidence that viability declined steadily as nodules aged (Sutton, Jepsen, and Shaw, 1977). Clover bacteroids isolated from protoplasts produced in high osmotic pressure media (0.5 M mannitol) and diluted in media containing at least 0.2 M mannitol, Mg^{2+}, and Ca^{2+} showed viabilities of 0%–90%, even though 70%–90% of the cells were branched (Gresshoff et al., 1977). With *R. japonicum* bacteroids, viabilities of about 80% have been reported with no decrease in viability even in 17-week-old nodules, and no special protective treatments were used (Tsien, Cain, and Schmidt, 1977).

In some legumes, such as soybean and serradella, division of rhizobia occurs intracellularly after their enclosure in membrane envelopes, but unrestricted multiplication obviously does not. Some mechanism must therefore limit bacterial cell division and cause the decrease in transcriptional activity for rRNA and the decrease in protein synthetic capability. These changes now seem unlikely to be mediated by destruction of any part of the genome, but rather to be associated with the frequently reported changes in the cell envelope (wall and membrane).

Broad bean apices exposed to *R. leguminosarum* contain sequences in their cells capable of hybridizing with iodinated rhizobial DNA (Lepidi et al., 1976). This implies that rhizobial DNA can be transferred to plant cells.

Further confirmation of such transfers and their significance in nodulation would be of obvious importance.

H_2 EVOLUTION AND UPTAKE BY NODULE BACTEROIDS

Some legume root nodules evolve H_2 (Hoch, Schneider, and Burris, 1960; Bergersen, 1963), take up H_2 (Dixon, 1967, 1968, 1972), or do both (see Evans et al., 1977 for a review). It was established (Dixon, 1967, 1968, 1972) that: 1) H_2 was evolved by pea nodules during N_2 fixation and O_2 was required for both processes; 2) D_2 was simultaneously taken up by pea nodules, and that uptake also required O_2; 3) pea bacteroids could catalyze H_2 oxidation by O_2, with a stoichiometry of 2:1; 4) H_2 uptake could be coupled to ATP formation with an ATP:H_2 ratio of about 2, and respiratory CO_2 output was decreased by H_2; and 5) hydrogenase activity was a bacteroid character, but its expression was affected by the host plant.

Three suggestions of possible hydrogenase function were advanced (Dixon, 1972): 1) alleviation of feedback inhibition of nitrogenase by the H_2 it produced; 2) protection of nitrogenase from O_2 by reduction of O_2 with H_2; and 3) recycling of electrons from nitrogenase, either to serve as reductant for the enzyme or to produce ATP via respiration. In blue-green algae, reutilization of H_2 evolved from nitrogenase has been demonstrated, together with the uptake hydrogenase system (Bothe et al., 1977). Hydrogen can protect nitrogenase from O_2 inactivation, enter the respiratory chain and lead to ATP synthesis, and act as an electron donor for nitrogenase (Benemann and Weare, 1974; Bothe, Tennigkeit, and Eisbrenner, 1977). Since H_2 evolution from nitrogenase involves hydrolysis of ATP, H_2 evolution from legume nodules represents an apparently unproductive use of photosynthate (Schubert and Evans, 1976) and raises the possibility of making legume systems more efficient if it can be prevented.

The concept of "relative efficiency" compares the H_2 wastage from different symbiotic associations (Schubert and Evans, 1976).

$$\text{Relative efficiency} = 1 - \frac{H_2 \text{ production in air}}{C_2H_2 \text{ reduction}} \text{ or } 1 - \frac{H_2 \text{ evolution in air}}{H_2 \text{ evolution in Ar:}O_2\text{:}CO_2}$$

It attempts to relate the electron (and therefore energy) wastage by H^+ reduction to the total electron flux (or total energy use) through nitrogenase. One of the assumptions used is that the electron flux measured by C_2H_2 reduction is equal to that measured by H_2 evolution under argon:oxygen:CO_2 and represents total electron supply to nitrogenase. In at least one report (Ruiz-Argueso, Hanus, and Evans, 1978), this condition is not met, even with 0.1 atm C_2H_2 in the assays. Relative efficiencies varied widely in legume nodules, whereas nonlegume root nodules gave higher values (Schubert and Evans, 1976, 1977). Thirty to sixty percent of the energy used by nitrogenase in legume nodules was involved with H_2 evolution. Nodules that failed to evolve H_2 (relative efficiency close to 1.0) contained uptake hydrogenase activity, including nonleguminous nodules (Schubert and Evans, 1977). H_2 uptake activity was also present in laboratory cultures of the rhizobial strains whose nodules took up H_2. Addition of H_2 to nodules with uptake hydrogenase led to an increase in O_2 uptake, indicating that electrons derived from H_2 participated in respiration (Schubert and Evans, 1977).

From studies with soybeans and cowpeas (Schubert et al., 1977; Carter et al., 1978), uptake hydrogenase was clearly a rhizobial characteristic. A decline in C_2H_2 reduction rate with time of exposure was evident after 15 min for all cowpea and soybean treatments, but was faster for nodules capable of H_2 uptake. Determining whether those rhizobial strains not showing uptake hydrogenase activity are genetically hydrogenase negative requires further investigation in cultures or in alternative hosts because one plant may repress hydrogenase production and another permit it (Dixon, 1972). The properties of the uptake hydrogenase in *R. japonicum* bacteroids are summarized in Table 1. Although no number is available for H_2 concentration in nodules, free O_2 concentration can be inferred from the degree of oxygenation of Lb (Appleby, 1969). A 20% oxygenation of Lb corresponds to about 0.09 μM free O_2; uptake hydrogenase would be very inactive if its K_m is as high as 1.6 μM. Assays of bacteroids in Lb solutions (Appleby et al., 1975; Bergersen and Turner, 1975) could be useful to assess whether or not O_2 transport via Lb is involved with hydrogenase activity. A sparing effect of H_2 uptake on respirable substrate was evident in bacteroids since CO_2 evolution was lowered to 40% when H_2 was present (McCrae, Hanus, and Evans, 1978). In pea nodules with relative efficiencies from 0.55 to 0.8, hydrogenase activities were found in 6 of 15 strains but none had sufficient activity to catalyze the oxidation of all evolved H_2 (Ruiz-Argueso, Hanus, and Evans, 1978). The reason suggested for the incomplete oxidation of H_2 was an inadequate amount of hydrogenase or electron transport components, but another possibility is O_2 deficiency, as suggested above.

A fundamental corollary of relative efficiency being important in N_2 fixation is a positive correlation between relative efficiency and plant growth or nitrogen content. An inverse correlation would also be expected between

Table 1. Uptake hydrogenase properties in bacteroids of *R. japonicum*

Property	
K_m for H_2	2.8 μM[a]
K_m for O_2	1.3 μM
H_2 uptake/O_2 uptake	1.9 ± 0.1
H_2 uptake (O_2)/H_2 uptake (Methylene blue)	18

From McCrae, Hanus, and Evans (1978).

[a] 4 μM in pea bacteroids (Ruiz-Argueso, Hanus, and Evans, 1978).

relative efficiency and nodule CO_2 production (estimated by Minchin and Pate [1973], Mahon [1977], or Herridge and Pate [1977]). A comparison of soybeans or cowpeas inoculated with one H_2-evolving strain and a nonevolving strain showed the latter to give 11% higher dry weight and 14% higher total nitrogen for cowpeas, with 24% and 31% higher values in soybeans (Evans et al., 1977). Peas (Ruiz-Argueso, Hanus, and Evans, 1978) with a wider selection of rhizobial strains gave no significant correlation between dry weight and relative efficiency. Gibson (unpublished data) found significant negative correlations between relative efficiency and either dry weight or plant nitrogen content for subterranean clover and a range of *R. trifolii* strains of different effectiveness (Table 2). Similar negative correlations were found for four strains of *R. trifolii* on white clover, 10 strains of *R. leguminosarum* on pea and four strains on *Vicia faba*, and 10 strains of *R. lupini* on *Lupinus angustifolius*. No correlation was apparent between relative efficiency and plant weight or nitrogen content for soybean–*R. japonicum* combinations or for *Vigna unguicaluata* or *Terramus uncinatum*.

In attempting to resolve this difference between theory and performance, Gibson (unpublished data) calculated the theoretical cost of H_2 evolution not in relation to total N_2 fixation, but in relation to total plant carbon assimilation. His two assumptions were: that H_2 evolution was steady over day and night (not so for clover nodules, which are lower at night); and that glucose oxidation by bacteroids yields 36 ATP/glucose. Although substrate oxidation by bacteroids is clearly aerobic, the ATP yield is not known exactly. Table 3 summarizes these calculations and, if correct, shows that H_2 evolution would consume such a small percentage of photosynthate as to be swamped by other interstrain variables in determining N_2 fixation or plant yield. Physiological variation in relative efficiency has been shown for varying light intensity (Bethlenfalvay and Phillips, 1977), shading, or nitrate addition (Gibson, unpublished data). Although each of these factors lowered nitrogenase activity, they increased relative efficiency. Accordingly, H_2 evolution, although it consumes a significant fraction of the energy used by nitrogenase, may not be a major growth-limiting factor in legumes.

Table 2. Correlation coefficients (over strain means) between relative efficiency and plant dry weight or total nitrogen (after Gibson, personal communication)

Time	Dry weight	Total N_2
Day 14	-0.69^a	-0.85^b
Day 21	-0.63^a	-0.69^a
Day 29	-0.67^a	-0.67^a
Day 35	-0.92^c	-0.84^b

[a] $p < 0.05$.
[b] $p < 0.01$.
[c] $p < 0.001$.

Table 3. Calculation of energy loss implicit in H_2 evolution in nodules of soybeans and clover (from Gibson, personal communication)

	Soybean	Clover
H_2 evolution ($\mu mol \cdot hr^{-1}$)	2.2	0.34
H_2 evolution ($\mu mol \cdot day^{-1}$)	52.8	8.2
At 3 ATP/H_2, μmol of ATP $\cdot day^{-1}$	158	24.5
At 36 ATP/glucose, mg of glucose $\cdot day^{-1}$	0.79	0.12
Dry weight increase, mg $\cdot day^{-1}$	185	8.0
Percent photosynthate for H_2 evolution	0.43	1.5

AMMONIA ASSIMILATION IN NODULES

Assays of nodule bacteroids for various ammonia-assimilatory enzymes show that bacteroid glutamine synthetase is inadequate to cope with NH_3 from N_2 fixation (Dunn and Klucas, 1973; Brown and Dilworth, 1975; Kurz, Rokosh, and LaRue, 1975; Planqué et al., 1977). Activities of glutamine synthetase in the plant cytosol are more than sufficient to assimilate all ammonia produced. Glutamine synthetase activity was shown to be about 500 times higher in lupin nodules than in roots and to increase during nodule development roughly parallel to nitrogenase and leghemoglobin (Robertson et al., 1975). A nodule glutamate synthase coupled to NADH has been demonstrated in lupin (*Lupinus angustifolius*) nodule cytosol; its activity increased with the onset of N_2 fixation, while the bacteroid enzyme activity remained constant (Robertson, Warburton, and Farnden, 1975). Glutamate dehydrogenase activity was too low to have any significant role in ammonia assimilation (Robertson, Warburton, and Farnden, 1975).

Since the main exported amino acid from lupin nodules is asparagine, Scott, Farnden, and Robertson (1976) investigated asparagine synthetase activity in nodules. They found an abrupt induction of asparagine synthetase at day 13, following the disappearance of the plant cytosol asparaginase. The enzyme required ATP, aspartate, glutamine, and Mg^{2+}, as reported for the yellow lupin cotyledon enzyme (Rognes, 1975). There were requirements for an —SH group protector and for glutamine or 5-diazo-4-oxo-L-norvaline to stabilize the enzyme. Ammonia would substitute for glutamine, but only poorly, and the K_m values indicated that ammonia would be preferentially incorporated into glutamine and not directly into asparagine. The demonstrated asparagine synthetase activities accounted for about 40% of the asparagine synthesis demonstrated by xylem bleeding experiments (Scott, Farnden, and Robertson 1976).

Accordingly, Scott, Farnden, and Robertson (1976) and Mifflin and Lea (1976) both proposed the following scheme for nodular ammonia incorporation (Figure 4). NH_3 from fixation reaches the plant cytosol, is converted into glutamine via glutamine synthetase, and thence into glutamate via glutamate synthase, with part of the glutamate nitrogen being used to

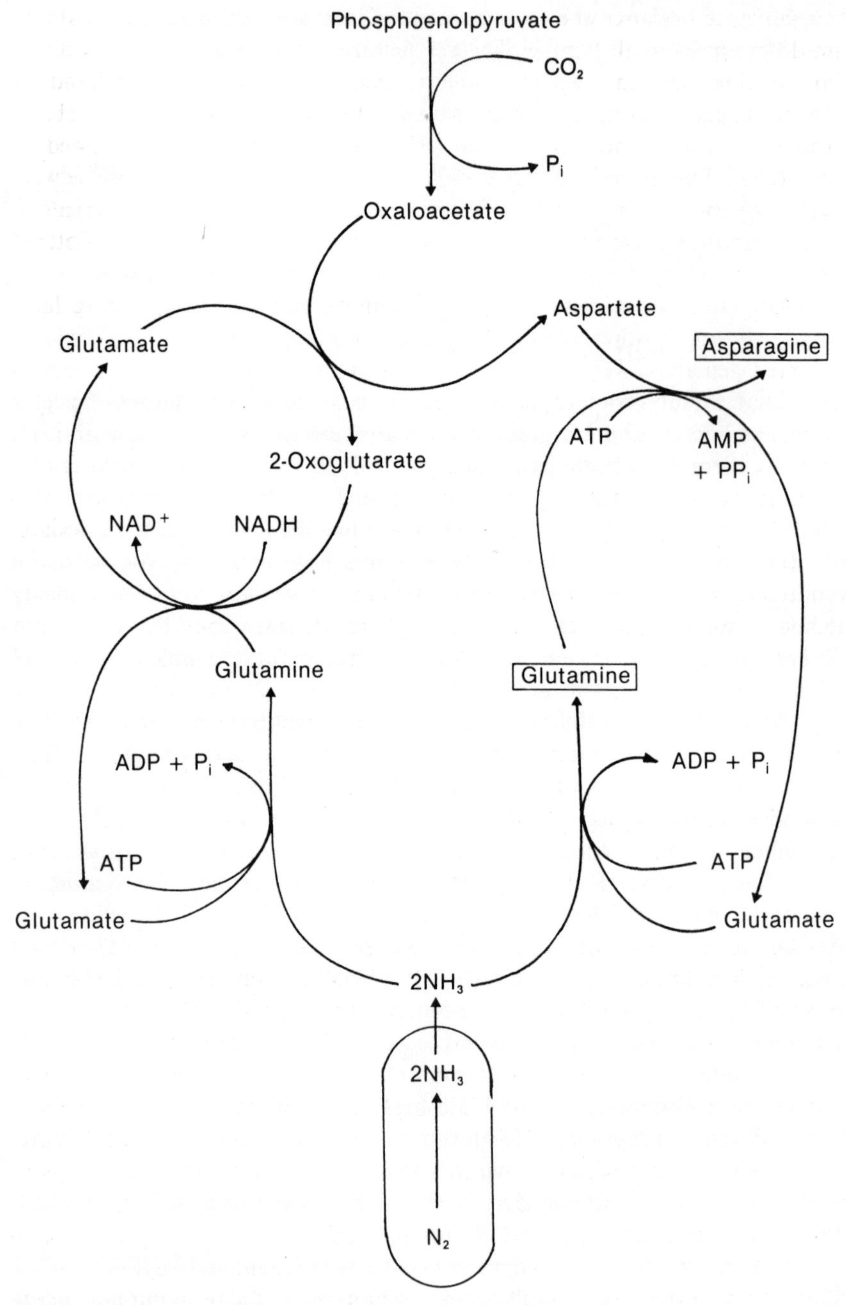

Figure 4. Scheme for ammonia assimilation in lupin root nodules (modified from Scott, Farden, and Robertson, 1976).

transaminate oxaloacetate to aspartate for formation of asparagine with the amido group from glutamine. The net requirements for asparagine synthesis from oxaloacetate and two ammonia molecules would be three ATP and one NADH (Scott, Farnden, and Robertson, 1976). Oxaloacetate is probably produced by phosphoenolpyruvate (PEP) carboxylation, as suggested by Christeller, Laing, and Sutton (1977), from the strong correlation between C_2H_2-reduction activity and PEP carboxylase activity. Glutamate-oxaloacetate aminotransferase activity is also required (Bodley, Ryan, and Fottrell, 1970).

Although the evidence for the glutamine pathway operating in lupin nodules is good, other nodule types need investigation before it could be accepted generally. We have found glutamine pathway enzymes in nodules from lucerne, subterranean clover, pea, broad bean, yellow lupin, serradella, vetch, *Albizzia lophantha*, and winged bean (Ling and Dilworth, unpublished data). A correlation between the activities of the two enzymes might also be expected in view of their likely common function; such a correlation does exist (Figure 5). The correlation coefficient for the different legume nodules was 0.83 ($p < 0.001$). Accordingly, ammonia incorporation via glutamine synthetase–glutamate synthase seems likely to be general in legume nodules, although subsequent metabolism may lead to the export of glutamine, asparagine, allantoin, allantoic acid, γ-aminobutyrate, and a variety of other compounds (Pate, 1976).

Representative nodule cytosol enzymes have been purified and their properties examined. Glutamine synthetase from soybean nodule cytosol (McPharland et al., 1976) accounted for about 2% of cytosol protein, was O_2 sensitive during concentration, and appeared to be an octamer of molecular weight 376,000. Regulation by amino acids was slight, but activity was strongly affected by energy charge with 5 mM ADP inhibiting 61% and 5 mM AMP, 29%.

Glutamate synthase from lupin nodules (Boland and Benny, 1977) is a single polypeptide of molecular weight 235,000 with a flavin prosthetic group (FMN). Our results (Ling and Dilworth, unpublished data) show that it is metal free, unlike the bacterial enzymes (Miller and Stadtman, 1972; Adachi and Suzuki, 1977). The lupin cytosol enzyme has K_m values for 2-oxoglutarate, glutamine, and NADH of 39 μM, 400 μM, and 1.3 μM, respectively (Boland and Benny, 1977). The enzyme is inhibited by glutamate, oxaloacetate, aspartate, and asparagine, all acting competitively with respect to 2-oxoglutarate, and by NAD^+, which is competitive with NADH. The enzyme is completely specific for its pyridine nucleotide, like the two bacterial enzymes (Miller and Stadtman, 1972; Adachi and Suzuki, 1977). The physiological importance of the inhibition of the enzyme by nodule amino acids remains to be assessed quantitatively. Whether plant glutamate synthase can utilize either NADH or NADPH or there are two separate

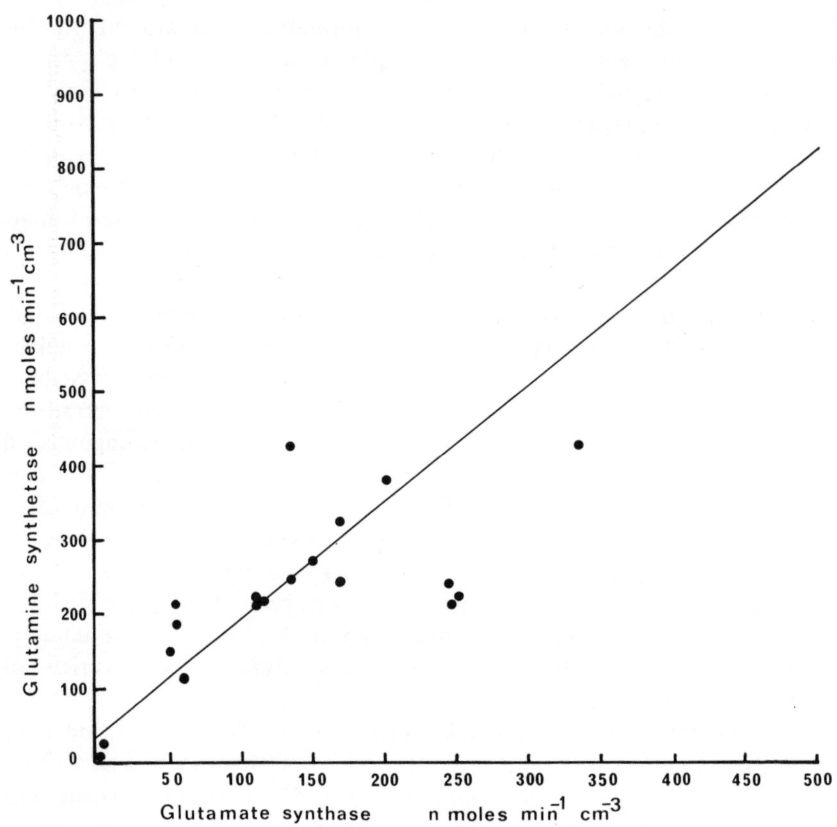

Figure 5. Relationship between plant cytosol glutamine synthetase and glutamate synthase activities from a variety of root nodules. Correlation coefficient (r) = 0.83 with 20 d.f. ($p < 0.001$).

enzymes with different coenzyme specificities remains to be established. In lupin nodules, only the NADH-linked activity occurs. In pea nodules (Ling and Dilworth, unpublished data), roots, or cotyledons, and carrot callus cells (Dougall, 1974; Beevers and Storey, 1976; Lea and Mifflin, 1975), activity dependent on NADH or NADPH can be demonstrated.

With the demonstration that cytosol glutamine synthetase and glutamate synthase are induced in nodules at a particular stage (Robertson et al., 1975a; Robertson, Warburton, and Farnden, 1975), the question arises as to what the inducing signal may be. The actual control of ammonia-assimilatory enzymes in plants remains to be defined. Studies with *Lemna* indicate that, with a low ammonia supply, high activities of glutamate synthase and glutamine synthetase are found (Rhodes, Rendon, and Stewart, 1976). Increasing ammonia availability lowered glutamine

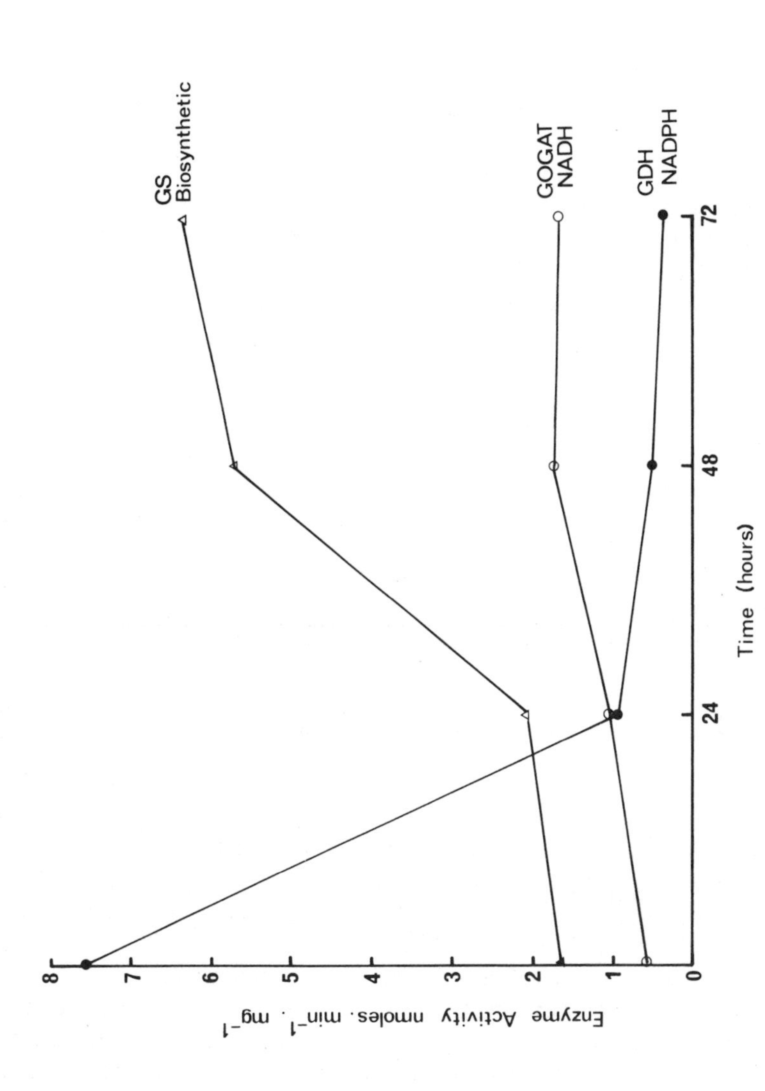

Figure 6. Activities of ammonia-assimilatory enzymes in lucerne callus cultures as a function of time after transfer to a nitrogen-free "starvation" medium.

synthetase and glutamate synthase while increasing glutamate dehydrogenase. Glutamine rather than ammonia appeared to be the intracellular regulator, decreasing glutamine synthetase and increasing glutamate dehydrogenase (Rhodes, Rendon, and Stewart, 1976). Studies in our laboratory with lucerne (alfalfa) callus grown on solid media indicate that ammonium level in the medium varies the activities of the ammonia-assimilatory enzymes. Callus grown with ammonium ion as nitrogen source transferred to "starvation" medium free of ammonium showed a rapid decline in glutamate dehydrogenase and corresponding increases in glutamate synthase and glutamine synthetase (Figure 6). The disappearance of asparaginase noted by Scott, Farnden, and Robertson (1976) in lupin nodules led them to propose that asparagine supply from cotyledons to nodule had ceased at this time. Nitrogen starvation could briefly occur before ammonia from N_2 fixation became available and be responsible for induction of the necessary cytosol enzymes for ammonia assimilation.

CONCLUDING REMARKS

Some important problems that still require attention are:

1. Intracellular localization of Lb in nodule cells.
2. Control and induction of Lb synthesis, both apoprotein and heme.
3. Induction and repression of nitrogenase and its relationship to O_2 and the enzymes of ammonia assimilation.
4. Ammonia assimilation and its regulation in both laboratory cultures and bacteroids.
5. The significance of ammonia transport in rhizobia, particularly in N_2-fixing forms in the laboratory culture or nodule.
6. The agronomic and biochemical consequences of H_2 evolution and uptake and the role of the legume host in modifying the rhizobial activity.
7. The mechanism of induction and control of plant cytosol enzymes from ammonia assimilation.

REFERENCES

Adachi, K., and I. Suzuki. 1977. Purification and properties of glutamate synthase from *Thiobacillus thioparus*. J. Bacteriol. 129:1173–1183.

Agarwal, A. K., and S. L. Mehta. 1974. Characterization of DNA from *Rhizobium* cells and their bacteroids from root nodules. Biochem. Biophys. Res. Commun. 60:257–265.

Almon, L. 1933. Concerning the reproduction of bacteroids. Zentralbl. Parasitenkd. Infektionskr. 87:289–297.

Appleby, C. A. 1962. Oxygen equilibrium of leghemoglobin. Biochim. Biophys. Acta 60:226–235.

Appleby, C. A. 1969. Properties of leghaemoglobin *in vivo* and its isolation as ferrous oxyleghaemoglobin. Biochim. Biophys. Acta 188:222–229.

Appleby, C. A., N. A. Nicola, J. G. R. Hurrell, and S. J. Leach. 1975. Characterization and improved separation of soybean leghemoglobins. Biochemistry 14:4444–4450.

Appleby, C. A., G. L. Turner, and P. K. Macnicol. 1975. Involvement of oxyleghaemoglobin and cytochrome P-450 in an efficient oxidative phosphorylation pathway which supports nitrogen fixation in *Rhizobium*. Biochim. Biophys. Acta 387:461–474.

Ausubel, F. M., R. F. Margolskee, and N. Maizels. 1977. Mutants of *Klebsiella pneumoniae* in which expression of nitrogenase is independent of glutamine synthetase control. In: W. Newton, J. R. Postgate, and C. Rodriguez-Barrueco (eds.), Recent Developments in Nitrogen Fixation, pp. 347–356. Academic Press, London.

Avissar, Y. J., and K. D. Nadler. 1978. Stimulation of tetrapyrrole formation in *Rhizobium japonicum* by restricted aeration. J. Bacteriol. 135:782–789.

Beevers, L., and R. Storey. 1976. Glutamate synthetase in developing cotyledons of *Pisum sativum*. Plant Physiol. 57:862–866.

Benemann, J. R., and N. M. Weare. 1974. Nitrogen fixation by *Anabaena cylindrica*. III. Hydrogen-supported nitrogenase activity. Arch. Microbiol. 101: 401–408.

Bergersen, F. J. 1963. The relationship between hydrogen evolution, hydrogen exchange, nitrogen fixation and applied O_2 tension in soybean root nodules. Aust. J. Biol. Sci. 16:669–680.

Bergersen, F. J. 1966. Nitrogen fixation in the legume root nodule. In: Proceedings of the 9th International Congress of Microbiology, Moscow, pp. 97–101. Pergamon Press, London.

Bergersen, F. J. 1974. Formation and function of bacteroids. In: A. Quispel (ed.), The Biology of Nitrogen Fixation, pp. 476–498. North-Holland Publishing Company, Amsterdam.

Bergersen, F. J. 1977. Nitrogenase in chemostat cultures of rhizobia. In: W. Newton, J. R. Postgate, and C. Rodriguez-Barrueco (eds.), Recent Developments in Nitrogen Fixation, pp. 309–320. Academic Press, London.

Bergersen, F. J., and D. J. Goodchild. 1973. Cellular location and concentration of leghaemoglobin in soybean root nodules. Aust. J. Biol. Sci. 26:741–756.

Bergersen, F. J., and G. L. Turner. 1967. Nitrogen fixation by the bacteroid fraction of breis of soybean root nodules. Biochim. Biophys. Acta 141:507–515.

Bergersen, F. J., and G. L. Turner. 1975. Leghaemoglobin and the supply of O_2 to nitrogen-fixing root nodule bacteroids: Presence of two oxidase systems and ATP production at low free O_2 concentration. J. Gen. Microbiol. 91:345–354.

Bergersen, F. J., and G. L. Turner. 1976. The role of O_2-limitation in control of nitrogenase in continuous cultures of *Rhizobium* ssp. Biochem. Biophys. Res. Commun. 73:524–531.

Bergersen, F. J., and G. L. Turner. 1978. Activity of nitrogenase and glutamine synthetase in relation to availability of oxygen in continuous cultures of a strain of cowpea *Rhizobium* sp. supplied with excess ammonium. Biochim. Biophys. Acta 538:406–416.

Bergersen, F. J., G. L. Turner, A. H. Gibson, and W. H. Dudman. 1976. Nitrogenase activity and respiration of cultures of *Rhizobium* spp. with special reference to concentration of dissolved oxygen. Biochim. Biophys. Acta 444:164–174.

Bethlenfalvay, G. J., and D. A. Phillips. 1977. Effect of light intensity on efficiency of carbon dioxide and nitrogen reduction in *Pisum sativum* L. Plant Physiol. 60:868–871.

Bishop, P. E., J. G. Guevara, J. A. Engelke, and H. J. Evans. 1976. Relation between glutamine synthetase and nitrogenase activities in the symbiotic association between *Rhizobium japonicum* and *Glycine max*. Plant Physiol. 57:542–546.

Bisseling, T., R. C. Van den Bos, and A. Van Kammen. 1978. The effect of ammonium nitrate on the synthesis of nitrogenase and the concentration of leghemoglobin in pea root nodules induced by *Rhizobium leguminosarum*. Biochim. Biophys. Acta 539:1–11.

Bisseling, T., R. C. Van den Bos, A. Van Kammen, M. Van der Ploeg, P. van Duijn, and A. Houwers. 1977. Cytofluorometric determination of the DNA contents of bacteroids and corresponding broth-cultured *Rhizobium* bacteria. J. Gen. Microbiol. 101:79–84.

Bloom, F. R., S. L. Streicher, and B. Tyler. 1977. Regulation of enzyme synthesis by the glutamine synthetase of *Salmonella typhimurium*: A factor in addition to glutamine synthetase is required for activation of enzyme formation. J. Bacteriol. 130:983–990.

Bodley, F., E. Ryan, and P. F. Fottrell. (1970) Purification of aspartate amino transferases from soybean root nodules and *Rhizobium japonicum*. Biochem. J. 119:7P.

Boland, M. J., and A. G. Benny. 1977. Enzymes of nitrogen metabolism in legume nodules. Purification and properties of NADH-dependent glutamate synthase from lupin nodules. Eur. J. Biochem. 79:355–362.

Bothe, H., J. Tennigkeit, and G. Eisbrenner. 1977. The utilization of molecular hydrogen by the blue-green alga *Anabaena cylindrica*. Arch. Microbiol. 114:43–49.

Bothe, H., J. Tennigkeit, G. Eisbrenner, and M. G. Yates. 1977. The hydrogenase-nitrogenase relationship in the blue-green alga *Anabaena cylindrica*. Planta 133:237–242.

Brown, C. M., and M. J. Dilworth. 1975. Ammonia assimilation by *Rhizobium* cultures and bacteroids. J. Gen. Microbiol. 86:39–48.

Carter, K. R., N. T. Jennings, J. Hanus, and H. J. Evans. 1978. Hydrogen evolution and uptake by nodules of soybeans inoculated with different strains of *Rhizobium japonicum*. Can. J. Microbiol. 24:307–311.

Ching, T. M., S. A. Hedtke, S. A. Russell, and H. J. Evans. 1975. Energy state and dinitrogen fixation in soybean nodules of dark-grown plants. Plant Physiol. 55:796–798.

Christeller, J. T., W. A. Laing, and W. D. Sutton. 1977. Carbon dioxide fixation by lupin root nodules. I. Characterization, association with phosphoenolpyruvate carboxylase and correlation with nitrogen fixation during nodule development. Plant Physiol. 60:47–50.

Cowles, J. R., H. J. Evans, and S. A. Russell. 1969. B_{12}-coenzyme dependent ribonucleotide reductase in *Rhizobium* species and the effects of cobalt deficiency on the activity of the enzyme. J. Bacteriol. 97:1460–1465.

Dilworth, M. J. 1969. The plant as the genetic determinant for leghemoglobin production in the legume root nodule. Biochim. Biophys. Acta 184:432–441.

Dilworth, M. J., and C. A. Appleby. 1977. Leghemoglobin and *Rhizobium* hemoproteins. In: R. C. Burns (ed.), A Treatise on Dinitrogen Fixation, pp. 691–764. John Wiley & Sons, Inc., New York.

Dilworth, M. J., and D. K. Kidby. 1967. Localization of iron and leghaemoglobin in

the legume nodule by electron microscope autoradiography. Exp. Cell Res. 49:148–159.
Dilworth, M. J., and J. A. McComb. 1977. Recent advances in tissue-cultures studies of the legume *Rhizobium* symbiosis. In: A. Ayanaba and P. J. Dart (eds.), Nitrogen Fixation in Farming Systems of the Humid Tropics, pp. 135–149. John Wiley & Sons, Ltd., London.
Dilworth, M. J., and D. C. Williams. 1967. Nucleic acid changes in bacteroids of *Rhizobium lupini* during nodule development. J. Gen. Microbiol. 48:31–36.
Dixon, R. O. D. 1967. Hydrogenase uptake and exchange by pea root nodules Ann. Bot. 31:179–188.
Dixon, R. O. D. 1968. Hydrogen in pea root nodule bacteroids. Arch. Mikrobiol. 62:272–283.
Dixon, R. O. D. 1972. Hydrogenase in legume root nodule bacteroids: Occurrence and properties. Arch. Mikrobiol. 85:193–201.
Dougall, D. K. 1974. Evidence for the presence of glutamate synthase in extracts of carrot cell cultures. Biochem. Biophys. Res. Commun. 58:639–646.
Dunn, S. D., and R. V. Klucas. 1973. Studies on possible routes of ammonium assimilation in soybean root nodule bacteroids. Can. J. Microbiol. 19:1493–1499.
Eady, R. R., R. Issack, C. Kennedy, J. R. Postgate, and H. D. Ratcliffe. 1978. Nitrogenase synthesis in *Klebsiella pneumoniae:* Comparison of ammonium and oxygen regulation. J. Gen. Microbiol. 104:277–285.
Evans, H. J., T. Ruiz-Argueso, N. Jennings, and J. Hanus. 1977. Energy coupling efficiency of symbiotic nitrogen fixation. In: A. Hollaender (ed.), Genetic Engineering for Nitrogen Fixation, pp. 333–354. Plenum Publishing Corp., New York.
Evans, W. R., and D. L. Keister. 1976. Reduction of acetylene by stationary cultures of free-living *Rhizobium* spp. under atmospheric oxygen levels. Can. J. Microbiol. 22:949–952.
Fuchsman, W. H., C. R. Barton, M. M. Stein, J. T. Thompson, and R. M. Willett. 1976. Leghemoglobin: Different roles for different components? Biochem. Biophys. Res. Commun. 68:387–392.
Garcia E., S. Bancroft, S. G. Rhee, and S. Kustu. 1977. The product of a newly identified gene, *glnF*, is required for synthesis of glutamine synthetase in *Salmonella*. Proc. Natl. Acad. Sci. USA 74:1662–1666.
Godfrey, C. A., D. R. Coventry, and M. J. Dilworth. 1975. Some aspects of leghaemoglobin biosynthesis. In: W. D. P. Stewart (ed.), Nitrogen Fixation by Free-Living Microorganisms, pp. 311–332. Cambridge University Press, New York.
Gourret, J. P., and H. Fernandez-Arias. 1974. Étude ultrastructurale et cytochimique de la differéntiation des bactéroidesde *Rhizobium trifolii* Dangeard dans les nodules de *Trifolium repens* L. Can. J. Microbiol. 20:1169–1181.
Gresshoff, P. M., M. L. Skotnicki, J. F. Eadie, and B. G. Rolfe. 1977. Viability of *Rhizobium trifolii* bacteroids from clover root nodules. Plant Sci. Lett. 10:299–304.
Herridge, D. F., and J. S. Pate. 1977. Utilization of net photosynthate for nitrogen fixation and protein production in an annual legume. Plant Physiol. 60:759–764.
Hoch, G. E., K. C. Schneider, and R. H. Burris. 1960. Hydrogen evolution and exchange and conversion of N_2O to N_2 by soybean root nodules. Biochim. Biophys. Acta 37:273–279.
Hooykaas, P. J. J., P. M. Klapwijk, M. P. Nuti, R. A. Schilperoort, and A. Rörsch. 1977. Transfer of the *Agrobacterium tumefaciens* TI plasmid to avirulent agrobacteria and to *Rhizobium ex planta*. J. Gen. Microbiol. 98:477–484.

Hunt, R. E., and J. R. Cowles. 1977. Physiological levels and properties of polynucleotide phosphorylase of *Rhizobium meliloti*. J. Gen. Microbiol. 102:403–411.

Hurrell, J. G. R., and S. J. Leach. 1977. The amino acid sequence of soybean leghaemoglobin c_2. FEBS Lett. 80:23–26.

Hurrell, J. G. R., K. R. Thulborn, W. J. Broughton, M. J. Dilworth, and S. J. Leach. 1977. Leghemoglobins: Immunochemistry and phylogenetic relationships. FEBS Lett. 84:244–246.

Keister, D. L., and W. R. Evans. 1976. Oxygen requirement for acetylene reduction by pure cultures of rhizobia. J. Bacteriol. 127:149–153.

Kennedy, I. R. 1966a. Primary products of symbiotic nitrogen fixation. I. Short-term exposures of serradella nodules to $^{15}N_2$. Biochim. Biophys. Acta 130;285–294.

Kennedy, I. R. 1966b. Primary products of symbiotic nitrogen fixation II. Pulse labelling of serradella nodules with $^{15}N_2$. Biochim. Biophys. Acta 130:295–303.

Kennedy, I. R., C. A. Parker, and D. K. Kidby. 1966. The probable site of nitrogen fixation in root nodules of *Ornithopus sativus*. Biochim. Biophys. Acta 130:517–519.

Kennedy, I. R., J. Rigaud, and J. C. Trinchant. 1975. Nitrate reductase from bacteroids of *Rhizobium japonicum:* Enzyme characteristics and possible interaction with nitrogen fixation. Biochim. Biophys. Acta 397:24–35.

Kleiner, D. 1975. Ammonium uptake by nitrogen fixing bacteria. I. *Azotobacter vinelandii*. Arch. Microbiol. 104:163–169.

Kurz, W. G. W., D. A. Rokosh, and T. A. LaRue. 1975. Enzymes of ammonia assimilation in *Rhizobium leguminosarum*. Can. J. Microbiol. 21:1009–1012.

Lea, P. J., and B. J. Mifflin. 1975. Glutamine and asparagine as nitrogen donors for reductant-dependent glutamate synthesis in pea roots. Biochem. J. 149:403–409.

Lepidi, A. A., M. P. Nuti, G. Bernacchi, and R. Neglia. 1976. Physiology of the *Rhizobium*-legume association: I—The presence of bacterial DNA within the plant root. Plant Soil 45:555–564.

Ludwig, R. A., and E. R. Signer. 1977. Glutamine synthetase and control of nitrogen fixation in *Rhizobium*. Nature 267:245–248.

Mahon, J. D. 1977. Root and nodule respiration in relation to acetylene reduction in intact nodulated peas. Plant Physiol. 60:812–816.

McCrae, R. E., J. Hanus, and H. J. Evans. 1978. Properties of the hydrogenase system in *Rhizobium japonicum* bacteroids. Biochem. Biophys. Res. Commun. 80:384–390.

McPharland, R. H., J. G. Guevara, R. R. Becker, and H. J. Evans. 1976. The purification and properties of the glutamine synthetase of soybean root nodules. Biochem. J. 153:597–606.

Mifflin, B. J., and P. J. Lea. 1976. The path of ammonia assimilation in the plant kingdom. Trends Biochem. Sci. 1:103–106.

Miller, R. E., and E. R. Stadtman. 1972. Glutamate synthase from *Escherichia coli*. J. Biol. Chem. 247:7407–7419.

Minchin, F. R., and J. S. Pate. 1973. The carbon balance of a legume and the functional economy of its nodules. J. Exp. Bot. 24:259–271.

Nadler, K. D., and Y. J. Avissar. 1977. Heme synthesis in soybean root nodules. I. On the role of bacteroid δ-amino-levulinic acid synthetase and δ-amino-levulinic acid dehydrase in the synthesis of the heme of leghemoglobin. Plant Physiol. 60:433–436.

Nuti, M. P., A. M. Ledeboer, A. M. Lepidi, and R. A. Schilperoort. 1977. Large plasmids in different *Rhizobium* species. J. Gen. Microbiol. 100:241–248.

Nutman, P. S. 1975. *Rhizobium* in the soil. In: N. Walker (ed.), Soil Microbiology, pp. 111–131. Butterworths, London.
O'Gara, F., and K. T. Shanmugam. 1976a. Regulation of nitrogen fixation by Rhizobia. Export of fixed N_2 and NH_4^+. Biochim. Biophys. Acta 437:313–321.
O'Gara, F., and K. T. Shanmugam. 1976b. Control of symbiotic nitrogen fixation in Rhizobia. Regulation of NH_4^+ assimilation. Biochim. Biophys. Acta 451:342–352.
O'Gara, F., and K. T. Shanmugam, 1977. Regulation of nitrogen fixation in *Rhizobium* spp. Isolation of mutants of *Rhizobium trifolii* which induce nitrogenase activity. Biochim. Biophys. Acta 500:277–290.
Paau, A., and J. R. Cowles, 1975. Comparison of DNA polymerase of *Rhizobium meliloti* and alfalfa bacteroids. Plant Physiol. 56:526–528.
Paau, A. S., D. Lee, and J. R. Cowles, 1977. Comparison of nucleic acid content in populations of free-living and symbiotic *Rhizobium meliloti* by flow microfluorometry. J. Bacteriol. 129:1156–1158.
Pagan, J. D., W. R. Scowcroft, J. J. Child, and A. H. Gibson. 1975. Nitrogen fixation by *Rhizobium* cultured on a defined medium. Nature 256:406–407.
Page, W. J. 1978. Transformation of *Azotobacter vinelandii* strains unable to fix nitrogen with *Rhizobium* spp. DNA. Can. J. Microbiol. 24:209–214.
Pate, J. S. 1976. Nutrient mobilization and cycling: Case studies for carbon and nitrogen in organs of a legume. In: I. Wordlaw and J. Passioura (eds.), Transport and Transfer Processes in Plants, pp. 447–462. Academic Press, Inc., New York.
Planqué, K., I. R. Kennedy, G. E. de Vries, A. Quispel, and A. A. N. van Brussel. 1977. Location of nitrogenase and ammonia-assimilatory enzymes in bacteroids of *Rhizobium leguminosarum* and *Rhizobium lupini*. J. Gen. Microbiol. 102:95–104.
Postgate, J. R. 1974. Pre-requisites for biological nitrogen fixation in free-living heterotrophic bacteria. In: A. Quispel (ed.), The Biology of Nitrogen Fixation, pp. 663–686. North-Holland Publishing Co., Amsterdam.
Prusiner, S., R. E. Miller, and R. C. Valentine. 1972. Adenosine 3′, 5′-cyclic monophosphate control of the enzymes of glutamine metabolism in *Escherichia coli*. Proc. Natl. Acad, Sci. USA 69:2922–2926.
Reijnders, L., L. Visser, A. M. J. Aaalbers, A. van Kammen, and A. Houwers. 1975. A comparison of DNA from free living and endosymbiotic *Rhizobium leguminosarum* (strain PRE). Biochim. Biophys. Acta 414:206–216.
Rhodes, D., G. A. Rendon, and G. R. Stewart. 1976. The regulation of ammonia assimilating enzymes in *Lemna minor*. Planta 129:203–210.
Rigaud, J., F. J. Bergersen, G. L. Turner, and R. M. Daniel. 1973. Nitrate-dependent anaerobic acetylene-reduction and nitrogen-fixation by soybean bacteroids. J. Gen. Microbiol. 77:137–144.
Rigaud, J., and A. Puppo. 1977. Effect of nitrite upon leghemoglobin and interaction with nitrogen fixation. Biochim. Biophys. Acta 497:702–706.
Robertson, J. G., K. J. F. Farnden, M. P. Warburton, and J. A. M. Banks. 1975. Induction of glutamine synthetase during nodule development in lupin. Aust. J. Plant Physiol. 2:265–272.
Robertson, J. G., M. P. Warburton, and K. J. F. Farnden. 1975. Induction of glutamate synthase during nodule development in lupin. FEBS Lett. 55:33–37.
Robertson, J. G., M. P. Warburton, P. Lyttleton, A. M. Fordyce, and S. Bullivant. 1978. Membranes in lupin root nodules. II. Preparation and properties of peribacteroid membranes and bacteroid envelope inner membranes from developing lupin nodules. J. Cell Sci. 30:151–174.
Rognes, S. E. 1975. Glutamine-dependent asparagine synthetase from *Lupinus luteus*. Phytochemistry 14:1975–1984.

Roponen, I. 1970. The effect of darkness on the leghaemoglobin content and amino-acid levels in the root nodules of pea plants. Physiol. Plant. 23:452–460.

Ruiz-Argueso, T., J. Hanus, and H. J. Evans. 1978. Hydrogen production and uptake by pea nodules as affected by strains of *Rhizobium leguminosarum*. Arch. Microbiol. 116:113–118.

St. John, R. T., V. K. Shaw, and W. J. Brill. 1974. Regulation of nitrogenase synthesis by oxygen in *Klebsiella pneumoniae*. J. Bacteriol. 119:266–269.

Schubert, K. R., J. A. Engelke, S. A. Russell, and H. J. Evans. 1977. Hydrogen reactions of nodulated leguminous plants. I. Effect of rhizobial strain and plant age. Plant Physiol. 60:651–654.

Schubert, K., and H. J. Evans. 1976. Hydrogen evolution: A major factor affecting the efficiency of nitrogen fixation in nodulated symbionts. Proc. Natl. Acad. Sci. USA 73:1207–1211.

Schubert, K. R., and H. J. Evans. 1977. The relation of hydrogen reactions to nitrogen fixation in nodulated symbionts. In: W. E. Newton, J. R. Postgate, and C. Rodriguez-Barrueco (eds.), Recent Developments in Nitrogen Fixation, pp. 469–485. Academic Press, London.

Scott, D. B., K. J. F. Farnden, and J. G. Robertson. 1976. Ammonia assimilation in lupin nodules. Nature 263:703–704.

Shah, V. K., L. C. Davis, M. Steighorst, and W. J. Brill. 1974. Mutant of *Azotobacter vinelandii* that hyperproduces nitrogenase component II. J. Bacteriol. 117:917–919.

Smeaton, J. R., and W. H. Elliott. 1967. Selective release of ribonuclease-inhibitor from *Bacillus subtilis* cells by cold-shock treatment. Biochem. Biophys. Res. Commun. 26:75–81.

Sutton, W. D. 1974. Some features of the DNA of *Rhizobium* bacteroids and bacteria. Biochim. Biophys. Acta 366:1–10.

Sutton, W. D., N. M. Jepsen, and B. D. Shaw. 1977. Changes in the number, viability and amino-acid-incorporating activity of *Rhizobium* bacteroids during lupin nodule development. Plant Physiol. 59:741–744.

Sutton, W. D., and W. D. Robertson. 1974. Control of gene action in nitrogen-fixing bacteroids. In: R. L. Bieleski, A. R. Ferguson, and M. M. Cresswell (eds.), Mechanisms of Regulation of Plant Growth, pp. 23–30. Royal Society of New Zealand Bulletin No. 12.

Tjepkema, J., and H. J. Evans. 1975. Nitrogen fixation by free-living *Rhizobium* in a defined liquid medium. Biochem. Biophys. Res. Commun. 65:625–628.

Truchet, G. 1972. Mise en évidence de l'activité peroxidasique dans les différentes zones des nodules radiculaires de pois (*Pisum sativum* L.) Localization de la leghémoglobine. C. R. Acad. Sci. [D] (Paris) 274:1290–1293.

Tshitenge, G., N. Luyindula, P. F. Lurquin, and L. Ledoux. 1975. Plasmid deoxyribonucleic acid in *Rhizobium vigna* and *Rhizobium trifolii*. Biochim. Biophys. Acta 414:357–361.

Tsien, H. C., P. S. Cain, and E. L. Schmidt. 1977. Viability of *Rhizobium* bacteroids. Appl. Envir. Microbiol. 34:854–856.

Tubb, R. S. 1974. Glutamine synthetase and ammonium regulation of nitrogenase synthesis in *Klebsiella*. Nature 251:481–485.

Tubb, R. S. 1976. Regulation of nitrogen fixation in *Rhizobium* sp. Appl. Envir. Microbiol. 32:483–488.

Upchurch R. G., and G. H. Elkan. 1977. Comparison of colony morphology, salt tolerance and effectiveness in *Rhizobium japonicum*. Can. J. Microbiol. 23:1118–1122.

Upchurch, R. G., and G. H. Elkan. 1978a. The role of ammonia, L-glutamate and cyclic adenosine-3′,5′-monophosphate in the regulation of ammonia assimilation in *Rhizobium japonicum*. Biochim. Biophys. Acta 538:244–248.

Upchurch, R. G., and G. H. Elkan. 1978b. Ammonia assimilation in *Rhizobium japonicum* colonial derivatives differing in nitrogen-fixing efficiency. J. Gen. Microbiol. 104:219–225.

Verma, D. P. S., and A. K. Bal. 1976. Intracellular site of synthesis and localization of leghemoglobin in root nodules. Proc. Natl. Acad. Sci. USA 73:3843–3847.

Verma, D. P. S., D. T. Nash, and H. M. Schulman. 1974. Isolation and *in vitro* translation of soybean leghaemoglobin mRNA. Nature 251:74–77.

Zurkowski, W., and Z. Lorkiewicz. 1976. Plasmid deoxyribonucleic acid in *Rhizobium trifolii*. J. Bacteriol. 128:481–484.

Nitrogen Fixation, Volume II
Edited by W. E. Newton and W. H. Orme-Johnson
Copyright 1980 University Park Press Baltimore

Ammonia Assimilation in Nitrogen-fixing Legume Nodules

M. J. Boland, K. J. F. Farnden,
and J. G. Robertson

The study of symbiotic nitrogen fixation has traditionally centered on the mechanism of the reaction catalyzed by nitrogenase. Assimilation of the ammonia produced is an integral part of the overall biological nitrogen fixation process and is equally as important as the nitrogenase reaction. Both are necessary if the organism is to benefit from nitrogen fixation.

Ammonia can be assimilated into two distinctly different types of chemical groups. The first of these, the amides, are formed by an ATP-linked condensation of ammonia with a carboxyl group. The commonest example is the formation of glutamine catalyzed by glutamine synthetase.

$$l\text{-glutamate} + NH_3 + ATP \rightarrow l\text{-glutamine} + ADP + P_i$$

The second type of group is the α amino group. Formation of α amino groups involves reductive amination of an oxo acid and requires a reducing agent, usually NAD(P)H or reduced ferredoxin. An outline of the metabolism of the two types of groups is presented in Figure 1.

Because of its role in protein biosynthesis, formation of the α amino group is of considerable importance. Amino group biosynthesis can be carried out by two different enzymes, glutamate dehydrogenase and glutamate synthase. The reactions catalyzed are glutamate dehydrogenase:

$$\text{2-oxoglutarate} + NH_4^+ + NAD(P)H + H^+ \rightleftharpoons \text{glutamate} + NAD(P)^+ + H_2O$$

and glutamate synthase:

$$\text{2-oxoglutarate} + \text{glutamine} + NAD(P)H + H^+ \rightarrow 2\ \text{glutamate} + NAD^+$$

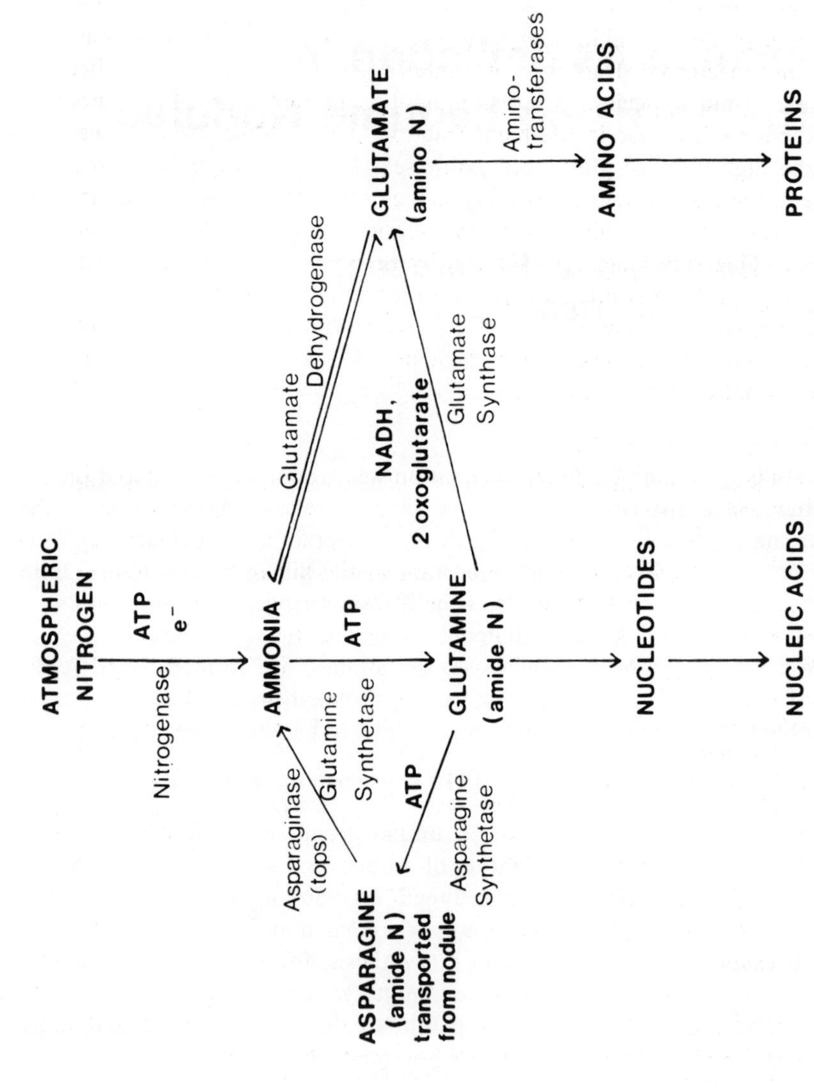

Figure 1. Simplified scheme showing metabolism and ultimate fate of fixed nitrogen in legumes.

Prior to 1970, the glutamate synthase reaction was unknown and, because most nodules contain insufficient glutamate dehydrogenase to account for known rates of amino acid biosynthesis, ammonia assimilation was not easily explained. Glutamate synthase was first discovered in *Aerobacter aerogenes* by Tempest, Meers, and Brown in 1970 and has subsequently been found in nodules of many legumes. Table 1 lists specific activities of glutamine synthetase, glutamate synthase, and glutamate dehydrogenase in plant and bacteroid fractions of nodules from a number of different legumes. In most species, the predominant ammonia-metabolizing enzyme is glutamine synthetase in the plant fraction of the nodule, which, together with the high level of glutamate synthase relative to glutamate dehydrogenase in the plant fraction, strongly suggests that most ammonia assimilation is via the plant glutamine synthetase reaction, with the α amino nitrogen being subsequently produced by the glutamate synthase reaction. Glutamate synthase is more suited to a biosynthetic role than the dehydrogenase because the reaction is effectively irreversible in the direction of ammonia assimilation (Boland and Benny, 1977) and the overall amino group formation takes advantage of the very low K_m of glutamine syn-

Table 1. Specific activities of ammonia-assimilating enzymes in legume nodule cytoplasm and bacteroids[a]

Species of legume	Glutamine synthetase[b] (nmol/min/mg of protein)		Glutamate synthase[c] (nmol/min/mg of protein)		Glutamate dehydrogenase[c] (nmol/min/mg of protein)	
	Plant	Bacteroid	Plant	Bacteroid	Plant	Bacteroid
Lupinus angustifolius	102	10	20	15	<0.2	—
Galega officinalis	143	—	49	—	17	—
Phaseolus vulgaris	116	2.6	8	1.2	4.5	1.6
Glycine max	89	6.4	10	56	4	20
Pisum sativum	80	0.4	24	1.4	16	33
Trifolium repens	91	—	22	—	138	—
Medicago sativa	26	4.0	26	ND[d]	4	16.7
Lotus pedunculatus	88	—	12	—	6	—
Ornithopus compressus	38	—	14	—	3	—

[a] Specific activities of enzymes in the plant fraction are from Boland, Fordyce, and Greenwood (1978) and those of bacteroid enzymes are from Brown and Dilworth (1975).

[b] Biosynthesis of γ-glutamyl hydroxamate from glutamate, hydroxylamine, and ATP.

[c] Either NADH or NADPH dependent. Where either cofactor is used, the higher value is shown.

[d] ND = none detectable.

thetase for ammonia. The glutamine synthetase–glutamate synthase pathway for α amino group formation has the same overall stoichiometry as the glutamate dehydrogenase reaction but with the additional hydrolysis of an ATP molecule. The latter presumably provides energy to drive the assimilation of ammonia.

AMMONIA ASSIMILATION AND EXCRETION BY BACTEROIDS AND FREE-LIVING NITROGEN-FIXING RHIZOBIA

Incorporation of $[^{15}N]N_2$ into nonprotein nitrogen in detached soybean nodules was first reported by Aprison and Burris in 1952. Following that report, ammonia was shown to be the initial product of nitrogen reduction in intact nodules and also to be rapidly incorporated into ammonia acids (Bergersen, 1965). Subsequently, Bergersen and Turner (1967) showed that fixation of $[^{15}N]N_2$ occurred in isolated preparations of bacteroids, with 94%–95% being excreted into the medium as ammonia and the remainder being incorporated into bacteroid nonprotein nitrogen. The acetylene reduction technique confirmed the site of nitrogen fixation as the bacteroids (Koch, Evans, and Russell, 1967) and led to the purification of bacteroid nitrogenase (Klucas et al., 1968; Bergersen and Turner, 1970).

Bergersen (1977) has reviewed the efforts to determine whether ammonia is assimilated in the bacteroids or in the plant cytosol of legume nodules. Studies of incorporation of $[^{15}N]N_2$ into serradella nodules (Kennedy, Parker, and Kidby, 1966) suggested the bacteroids as the primary site of ammonia assimilation, since glutamic acid from this fraction contained the highest enrichment of ^{15}N. However, studies with soybean nodules (Bergersen, 1960; Klucas and Burris, 1966) suggested that the ^{15}N accumulated in the plant cytosol fraction and that the bacteroids were virtually unlabeled. These apparently conflicting reports have led to the conclusion that the bacteroid forms of different species or strains of rhizobia may differ in the degree to which they assimilate or excrete ammonia (Bergersen, 1977).

Excretion of ammonia by nitrogen-fixing bacteroids (Bergersen and Turner, 1967) is not an artifact caused by some alteration in their metabolic state during isolation because free-living, nitrogen-fixing rhizobia also excrete ammonia (O'Gara and Shanmugam, 1976). These organisms, like bacteroids, vary in the relative amounts of ammonia assimilated and excreted by different strains. Bergersen and Turner (1978) have shown that, in continuous cultures of cowpea strain CB756, newly formed ammonia is in equilibrium with the medium from which it is assimilated by the glutamine synthetase–glutamate synthase pathway. Upchurch and Elkan (1978) have proposed that derivatives of *Rhizobium japonicum* that fix greater amounts of nitrogen assimilate less of the fixed nitrogen.

Factors causing variations in the amount of ammonia assimilated or excreted by different strains of free-living nitrogen-fixing rhizobia or by bacteroids have not been determined. Possibilities include: differences in the levels of ammonia-assimilatory enzymes; their inhibition by metabolites; their accessibility to newly formed ammonia; or leakiness of rhizobium strains with respect to ammonia. The activities of glutamine synthetase, glutamate synthase, and glutamate dehydrogenase in extracts of bacteroids from legume nodules are too low to assimilate all of the ammonia produced by nitrogen fixation in most (Brown and Dilworth, 1975; Kurz, Rokosh, and LaRue, 1975; Robertson, Warburton, and Farnden, 1975; Planqué et al., 1977), but not all (Dunn and Klucas, 1973; Ryan and Fottrell, 1974) cases. These enzyme levels should, however, be interpreted with caution because both bacteroid preparations (Ching, Hedtke, and Newcomb, 1977) and pure cultures of free-living nitrogen-fixing rhizobia (W. Evans and D. Crist, unpublished data) consist of populations of cells of different densities that may also have different complements of enzymes.

The question of why rhizobia excrete the ammonia formed by nitrogen fixation is both highly intriguing and of considerable agronomic importance. It has been suggested that this mechanism may have been evolved to maximize the benefits of symbiosis. It is also possible that ammonia excretion is a reflection of some metabolic imbalance in free-living, nitrogen-fixing rhizobia and that some plants have evolved to take advantage of it.

INDUCTION OF AMMONIA-ASSIMILATING ENZYMES IN LEGUME NODULES

That ammonia assimilation is primarily due to plant cytosol as opposed to the bacteroids has been strongly suggested by examination of specific activities of ammonia-assimilating enzymes during nodule development (Robertson et al., 1975; Robertson, Warburton, and Farnden, 1975). The results are summarized in Figure 2. *Lupinus angustifolius* plants grown under controlled conditions show a dramatic rise in nitrogenase (acetylene reduction) activity and leghemoglobin concentration in the nodules from the twelfth to about the twentieth day after planting out and inoculation. Over this same period, a similar increase is seen in the specific activities of both glutamine synthetase and glutamate synthase in the plant cytoplasm fraction of the nodule. Simultaneously, in the bacteroid, glutamine synthetase remains constant at a very low level and glutamate synthase, which has a relatively high specific activity (three-quarters of the final value reached by the plant enzyme), also remains constant. These results imply a causal relationship between nitrogen fixation and plant cytoplasm glutamine synthetase and glutamate synthase levels, thus providing a strong argument

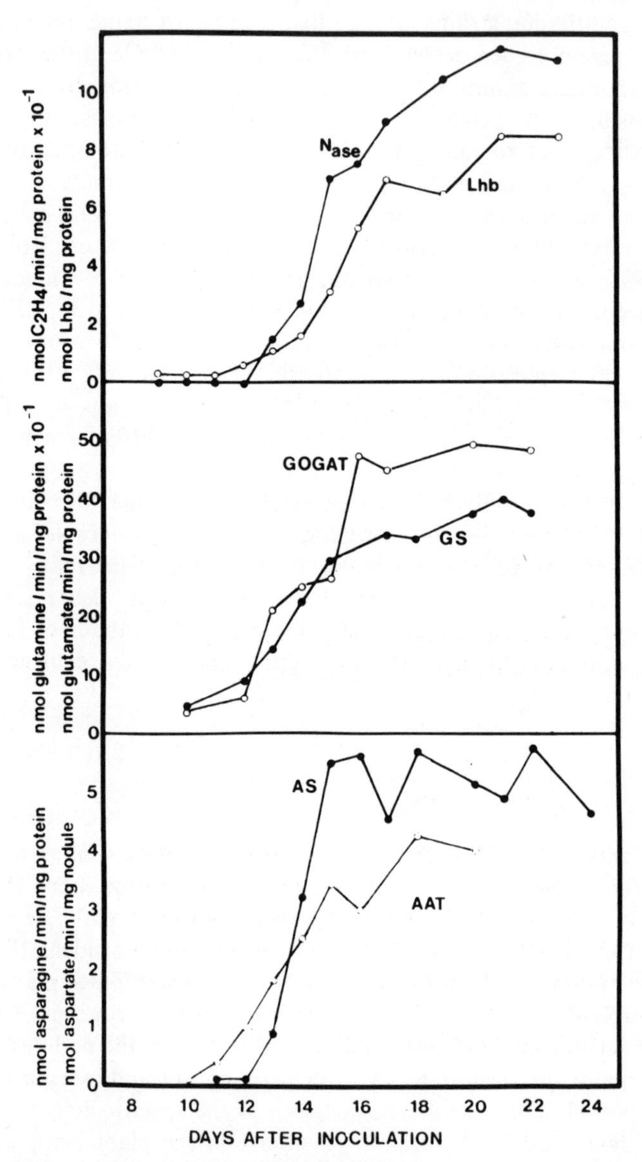

Figure 2. Induction of enzymes of nitrogen fixation and ammonia metabolism in nodules from *Lupinus angustifolius* plants. All activities other than nitrogenase refer to plant cytoplasm enzymes. Nase, nitrogenase; Lhb, leghemoglobin; GOGAT, glutamate synthase; GS, glutamine synthetase; AAT, aspartate aminotransferase; AS, asparagine synthetase. Values shown are derived from Robertson et al. (1975); Robertson, Warburton, and Farnden, (1975); Scott, Farnden, and Robertson (1976); and Reynolds and Farnden (unpublished data).

Table 2. Amino acid content of xylem sap from different legume species (all values not given are less than 0.1 mM)

Amino acid	*L. angustifolius*[a] (mM)	*Vicia faba*[b] (mM)	*Glycine max*[c] (mM)
Asparagine	2.35	5.91	4.14
Aspartic acid	0.54	2.12	1.22
Glutamine	0.41	0.60	0.34
Glutamic acid	0.36	0.24	0.16
Threonine	0.40	—	—
Histidine	—	—	0.24
Arginine	—	—	0.14
Lysine	—	—	0.21
Valine	—	—	0.16

[a] For 18-day plants in growth cabinet (Scott, Farnden, and Robertson, 1976).

[b] Field-grown plants, age not known (Pate, Gunning, and Briarty, 1969).

[c] Calculated from data for 21-day field-grown plants (Streeter, 1972).

that ammonia assimilation is carried out in the plant cytoplasm. This suggestion is consistent with the observed excretion of ammonia by both free-living nitrogen-fixing rhizobia and bacteroids fixing nitrogen in vitro.

GLUTAMATE METABOLISM AND ASPARAGINE SYNTHESIS IN LEGUME NODULES

Isotope labeling studies using $[^{15}N]N_2$ (Kennedy, 1966a, 1966b) have indicated that the first stable organic products from nitrogen fixation in nodules are glutamate and glutamine and that asparagine is an end point for nitrogen in the nodule. This result correlates with high levels of asparagine found in the xylem sap of several legume species (Pate, Gunning, and Briarty, 1969; Streeter, 1972; Scott, Farden, and Robertson, 1976), as shown in Table 2. Asparagine is a logical choice for nitrogen transport because it has a high nitrogen:carbon ratio and is relatively easily synthesized. Arginine has a higher nitrogen:carbon ratio but the biosynthesis is considerably more complicated. Synthesis of asparagine from oxaloacetate using glutamate and glutamine as nitrogen donors involves two enzymes: aspartate aminotransferase

$$\text{oxaloacetate} + l\text{-glutamate} \rightleftarrows l\text{-aspartate} + \text{2-oxoglutarate}$$

and asparagine synthetase

$$l\text{-aspartate} + l\text{-glutamine} + \text{ATP} \rightarrow l\text{-asparagine} + l\text{-glutamate} + \text{AMP} + \text{PP}_i$$

Specific activities of these enzymes in the plant cytoplasm of the nodule have been found to increase concurrently with the onset of nitrogen fixation (Scott, Farnden, and Robertson, 1976; Reynolds and Farnden, unpublished data). These results are shown in Figure 2. Asparagine synthetase was not detected in bacteroids or in cell-free extracts of free-living rhizobia, even in the presence of asparaginase inhibitors. Ammonia can replace glutamine in the asparagine synthetase reaction. However, ammonia at 4 times the glutamine concentration gives only half the rate of synthesis observed with glutamine (Scott, Farnden, and Robertson, 1976).

PROPERTIES OF ENZYMES INVOLVED IN AMMONIA ASSIMILATION AND ASPARAGINE BIOSYNTHESIS

To understand the behavior of the ammonia-assimilating and asparagine-synthesizing system in the nodule, it is necessary to isolate and study each individual enzyme. Here we present data on these enzymes, particularly from *Lupinus angustifolius*.

Glutamine Synthetase

High levels of glutamine synthetase in the nodules of many legume species have been reported (Sloger, 1973; Dunn and Klucas, 1973; Robertson et al., 1975; Boland, Fordyce and Greenwood, 1978). Glutamine synthetase from nodule cytoplasm of *Glycine max* has been purified 57-fold and some physical characteristics have been determined (Guevara, McParland, and Evans, 1975; McParland et al., 1976). In contrast to bacteroid glutamine synthetase, the plant enzyme is not adenylylated. The enzyme molecule is described as consisting of eight subunits arranged in two planar tetramers, giving an overall cubic configuration with dimensions of 10 nm on each side. Molecular weights of 376,000 for the native enzyme and 47,300 for the subunit were determined. Kinetic constants were not reported. We have purified glutamine synthetase 88-fold to >95% homogeneity from the plant cytoplasm of *L. angustifolius* nodules (McCormack, Farnden, and Boland, unpublished data). Kinetic constants were determined for the formation of glutamine by measuring the rate of hydrolysis of ATP, using a linked assay with pyruvate kinase and lactic dehydrogenase. The reaction appears to be sequential, involving a tetramolecular intermediate. By measuring the dependence of the reaction rate on pairs of substrates with the third held at an effectively saturating concentration, saturating K_m values for each substrate were determined. At pH 7.5 and 24°C, these were 4 mM for *l*-glutamate, 0.15 mM for ATP, and 0.12 mM for $[NH_3 + NH_4^+]$. From pH dependence studies, it seems likely that the true substrate is NH_3, with a K_m of about 1 μM. A more detailed analysis of the interdependence of the rate of reaction on the concentrations of the three substrates is being attempted in our laboratory according to the method of Dalzeil (1969) with rate determination using the conversion of [^{14}C]glutamate to glutamine according to Prusiner and Milner (1970). Results so far indicate that the kinetics is complex, with the concentration of each substrate having a profound effect on the apparent K_m of each of the others.

Glutamate Synthase

The presence of glutamate synthase in the plant cytosol and bacteroid fractions of nodules has been reported for *Glycine max* (Sloger, 1973) and for *L. angustifolius* (Robertson et al., 1975). Glutamate synthase has been purified 500-fold to >95% homogeneity from the plant fraction of *L. angustifolius* nodules (Boland and Benny, 1977). This enzyme is extremely susceptible to oxidation, requiring high levels of reducing agents for protection, and lacks stability in Tris buffers. In contrast to bacterial glutamate synthases that contain two different types of subunits (Miller and Stadtman, 1972), the lupin enzyme is monomeric and has a molecular weight of 235,000. The enzyme shows an absolute specificity for NADH (cf. Stewart and Rhodes, 1977). It contains two flavin groups, probably FMN, that are presumably involved in the electron transfer process. K_m values for the *L. angustifolius* glutamate synthase–catalyzed reaction are shown in Table 3, together with some binding constants for inhibitor amino acids involved in the ammonia assimilation system. The low K_m for NADH is typical of many dehydrogenases. The V_{max} is independent of pH between 6.5 and 9.5. K_m values for both 2-oxoglutarate and glutamine show optima at 8.5, although the pH dependence is not great, except for the glutamine K_m above pH 8.5, where a first order pH effect is seen with a pK of about 9.1.

Inhibition by other amino acids is purely competitive with respect to 2-oxoglutarate, in view of which the comparatively low K_m for the oxo acid could be very important. The most interesting and important inhibitor from a physiological as well as a kinetics viewpoint is glutamate, 2 molecules of which are produced by the reaction. The competitive pattern of inhibition with respect to 2-oxoglutarate is shown in Figure 3. Inhibition shows only first order dependence on glutamate concentration, which implies an irre-

Table 3. Kinetics constants of glutamate synthase from lupin at pH 8.5

Constant		
K_m 2-oxoglutarate	39 μM	
K_m glutamine	0.4 mM	
K_m NADH	1.3 μM	
K_i NAD^+	0.7 mM (competitive with NADH)	
K_i glutamate	0.7 mM	(all competitive with 2-oxoglutarate)
K_i aspartate	2.7 mM	
K_i asparagine	14 mM	
Specific activity	18.2 μmol min^{-1} mg of $enzyme^{-1}$	

Data from Boland and Benny (1977).

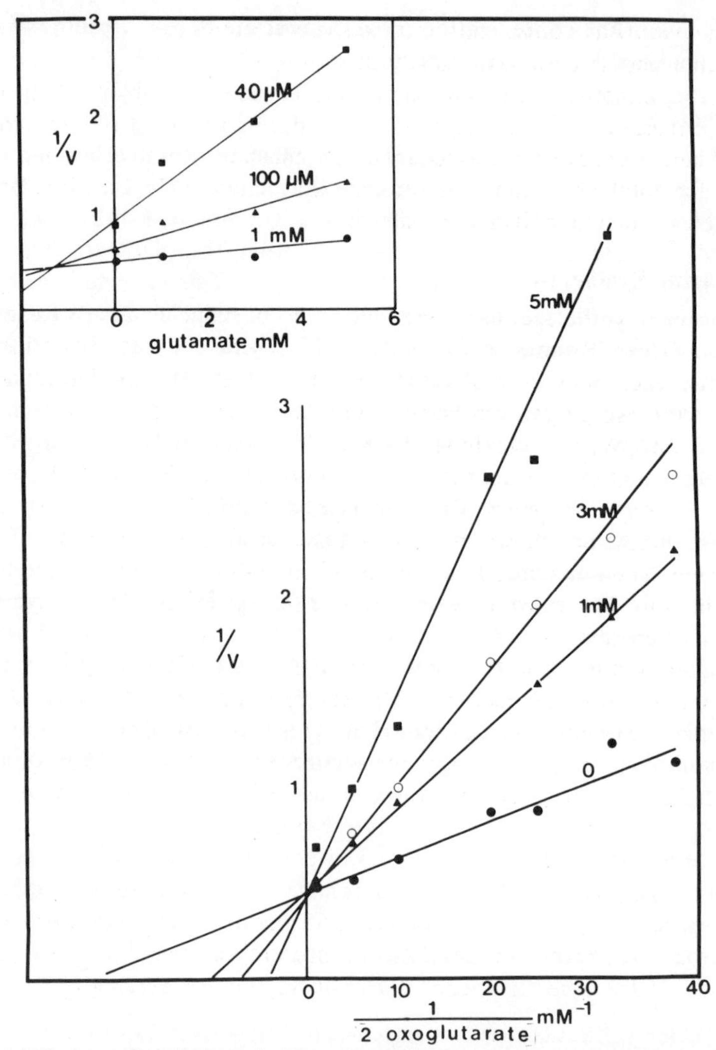

Figure 3. Pattern of inhibition of glutamate synthase by glutamate (Boland, unpublished data). Strict competitive inhibition occurs with respect to 2-oxoglutarate. The linearity of lines in the Dixon plot (inset) shows that only 1 glutamate molecule is causing inhibition, although 2 are produced in the reaction.

versible step in the reaction mechanism between the release of the first and second glutamate molecules.

Aspartate Aminotransferase

Aspartate aminotransferase has been detected in the plant fraction of nodules from *Lupinus angustifolius* (Reynolds and Farnden, unpublished data),

Glycine max, *Vicia faba*, and *Pisum sativum* (Grimes and Masterson, 1971). The lupin nodule cytoplasm has been shown to contain two isoenzymes of aspartate aminotransferase (Reynolds and Farnden, unpublished data). One of these increased dramatically in activity during the onset of nitrogen fixation, whereas the other remained relatively constant. Results shown in Figure 2 are for total aspartate aminotransferase activity. No kinetics data are available yet on the individual isoenzymes.

Asparagine Synthetase

Asparagine synthetase has been partially purified from cotyledons of *Lupinus luteus* (Rognes, 1970, 1975). This enzyme consists of two identical subunits, each with a molecular weight of 160,000. We have partially purified the asparagine synthetase from *L. angustifolius* nodule cytoplasm (Boland and Wong, unpublished data). This enzyme shows considerable instability, and if it is to be kept for more than a few hours freezing at −70°C becomes necessary. Preliminary results obtained using this partially purified enzyme are shown in Table 4, together with similar results of the *L. luteus* cotyledon enzyme. It would be of interest to know if the differences arise because the enzymes came from different tissues or if they reflect species differences.

Kinetics studies on the enzyme from *L. luteus* (Rognes, 1975) suggest a mechanism with up to three different forms of enzyme. Since ATP is cleaved to AMP plus pyrophosphate in asparagine synthesis and the enzyme catalyzes a vigorous ATP-pyrophosphate exchange, it is likely that one intermediate is an adenylylated form of the enzyme.

Glutamate Dehydrogenase

It seems likely (see Table 1) that glutamate dehydrogenase plays little part in ammonia metabolism in nodules of most legumes. The low level of activity in most species is probably of mitochondrial origin (e.g., Davies and Teixeira, 1975). The significance of relatively large amounts in one or two

Table 4. Properties of asparagine synthetases from *Lupinus angustifolius* nodules and *Lupinus luteus* cotyledons

	L. angustifolius[a] (mM)	*L. luteus*[b] (mM)
K_m aspartate	3.6	1.3
K_m glutamine	0.26	0.16
K_m MgATP	0.14	0.14

[a] Boland and Wong, unpublished data; Scott, Farnden, and Robertson, 1976.

[b] Rognes, 1975.

species, particularly white clover, is unknown. Approximate K_m values determined on crude plant cytoplasm extracts from which small molecules had been removed (Boland et al., 1978) are shown in Table 5. It seems that quite high concentrations of ammonia must be present for efficient glutamate dehydrogenase function. Such an ammonia level could not build up unless the supply of ammonia exceeded the capacity of glutamine synthetase to metabolize it. Under those circumstances, glutamate dehydrogenase might form a backup ammonia-metabolizing system, although ammonia-dependent activity of asparagine synthetase would also become significant (Scott, Farnden, and Robertson, 1976). Inhibition or repression of nitrogenase might, however, be expected to occur once ammonia levels exceeded 1 mM (Gibson, Scowcroft, and Pagan, 1977). Another possibility is that glutamate dehydrogenase is functioning in the reverse (ammonia-releasing) direction, although the usefulness of such a reaction is hard to imagine.

SCHEME FOR AMMONIA ASSIMILATION AND ASPARAGINE BIOSYNTHESIS IN LEGUME NODULES

A scheme showing the interrelationships of the four main enzymes of ammonia assimilation and asparagine biosynthesis is shown in Figure 4 (Scott, Farnden, and Robertson, 1976). The net reaction of the system is:

$$2NH_3 + \text{oxaloacetate} + NADH + H^+ + 3ATP \rightarrow$$
$$l\text{-asparagine} + NAD^+ + 2ADP + AMP + 2P_i + PP_i$$

Although we do not yet have sufficient kinetics data to describe the behavior of the system in detail, some observations can be made, using the *L. angustifolius* nodule as a model system. Specific activities of the relevant enzymes in nodule cytoplasm extracts are given in Table 6. Inevitably, some enzyme activity will have been lost by denaturation in the extraction procedure, with the greatest losses expected in glutamate synthase and,

Table 5. Approximate K_m values for ammonia-assimilating enzymes (determined on crude extracts of plant cytoplasm after passage through Sephadex)

Species	Glutamine synthetase K_m for hydroxylamine[a] at pH 7.5 (mM)	Glutamate dehydrogenase K_m for ammonia at pH 7.9 (mM)
Pea	0.02	91
White clover	<0.2	36

Boland, Fordyce, and Greenwood (1978).

[a] Biosynthetic assay.

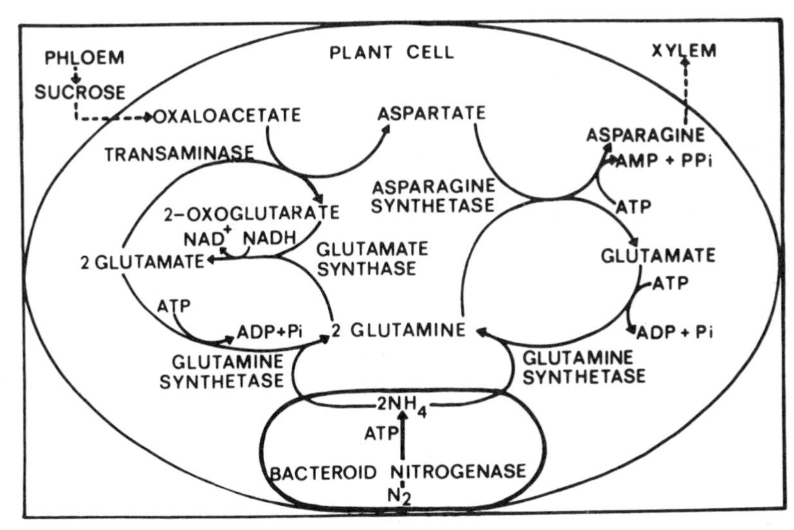

Figure 4. Proposed pathway for the assimilation of ammonia in legume root nodules (after Scott, Farnden, and Robertson, 1976).

more particularly, asparagine synthetase activities because of their known instabilities (Rognes, 1975; Scott, Farnden, and Robertson, 1976; Boland and Benny, 1977; Boland and Wong, unpublished data). Even so, it is likely that specific activities of these two enzymes are considerably lower than those of other enzymes. Activities of ammonia-assimilating enzymes in the nodule tissue will be governed not only by specific activities, but also by concentrations of substrate molecules relative to their respective K_m values. Some K_m values of these enzymes for amino acid substrates are compared with levels of these substrates in the nodule tissue and in xylem sap (Table 7). All enzymes are between 40% and 83% saturated with their respective substrates and thus there are adequate concentrations of substrate for the enzymes to function fairly efficiently.

Table 6. Specific activities of ammonia-assimilating enzymes in nitrogen-fixing lupin nodules (18 days after inoculation, grown in 12-hr photoperiod at 24°–21° C; nmol/min/mg of protein at 25°C)

Enzyme	S.A.	Reference
Glutamine synthetase	530[a]	McCormack et al. (unpublished data)
Glutamate synthase	37	Boland and Benny (1977)
Aspartate aminotransferase	1200	Reynolds and Farnden (unpublished data)
Asparagine synthetase	5	Scott, Farnden, and Robertson (1976)
(PEP Carboxylase)	(350)	Christeller, Laing, and Sutton (1977)

[a] Hydroxamate biosynthesis.

Table 7. Comparison of levels of amino acids in 18-day lupin nodule tissue and xylem sap with K_m values of enzymes utilizing them

Amino acid	Level in xylem sap[a] (mM)	Level in nodule tissue[b] (mM)	K_m (mM)	Enzyme
Glutamate	0.36	2.86	4.0[b]	Glutamine synthetase
Glutamine	0.41	2.04	0.4[c]	Glutamate synthase
			0.26[d]	Asparagine synthetase
Aspartate	0.54	3.21	3.6[a,d]	Asparagine synthetase
Asparagine	2.35	18.39	—	

[a] Scott, Farnden, and Robertson (1976).
[b] McCormack et al. (unpublished data).
[c] Boland and Benny (1977).
[d] Boland and Wong (unpublished data).

SUPPLY OF ENERGY, REDUCING POWER, AND OXALOACETATE FOR AMMONIA ASSIMILATION AND ASPARAGINE BIOSYNTHESIS

As shown above, the biosynthesis of an asparagine molecule requires 3 ATP, 1 NADH and 1 oxaloacetate molecule. The source of these is an important consideration.

For nitrogen fixation to occur in isolated bacteroids or free-living rhizobia, the organism must be supplied with a carboxylic acid, such as succinate or fumarate (Bergersen and Turner, 1967; Gibson et al., 1976). Because the main supply of photosynthate to the nodule will be as sugars, the supply of carboxylic acids to the bacteroids implies a suitable rate of glycolysis in the plant cytoplasm. Glycolysis will provide a source of NADH from the glyceraldehyde-3-phosphate dehydrogenase–catalyzed reaction and will also supply ATP. In view of the high level of phosphoenolpyruvate carboxylase in the nodule cytoplasm (Christeller, Laing, and Sutton, 1977; Table 6), it seems probable that an adequate supply of oxaloacetate will be maintained by carboxylation of the glycolytic intermediate phosphoenolpyruvate (PEP). Whether or not there is a sufficient rate of ATP production to saturate the ammonia-assimilating system is not known; however, it seems likely that the plant will put its own needs before those of the bacteroids, and one would expect an adequate supply of ATP for all nodule cytoplasm functions. The supply of NADH is almost certain to be adequate because 1 NADH molecule will be generated in the production of every PEP molecule used in oxaloacetate synthesis.

OVERALL RATES OF FLOW OF NITROGEN THROUGH THE NODULE

Rates of transport of the main amino acids found in xylem sap of 18-day *L. angustifolius* plants are compared with levels of various enzymes, calculated on a per plant basis, in Table 8. The level of nitrogenase activity is adequate to account for the total rate of nitrogen transported out of the nodule. The activities of two enzymes, glutamate synthase and asparagine synthetase, are likely to be limiting in the overall rate of flow of nitrogen through the nodule. The measured activity of asparagine synthetase is insufficient to account for the required rate of asparagine biosynthesis. This situation is due partly to the denaturation problems mentioned previously and partly to incomplete recovery in the extraction procedure. It is probable that asparagine synthetase activity in the nodule is as high as 2 μmol/hr/plant, thus accounting for all the asparagine exported from the nodule. The reaction of asparagine synthetase will, however, be rate limiting in asparagine biosynthesis since all the other enzyme activities are well in excess of this value. This argument is supported by the observation that appreciable levels of aspartate, up to a quarter of those of asparagine, are found in the xylem sap.

The rate of reaction of glutamate synthase will be limiting for total amino acid biosynthesis provided that glutamine synthetase is adequately supplied with ATP and ammonia. The indicated level of nitrogenase activity will ensure an adequate supply of ammonia, provided that sufficient ATP and reducing power are available. The close agreement between measured glutamate synthase activity and rate of export of amino acids in xylem sap (Table 8) is fortuitous and arises because of two factors: first, the level of glutamate synthase in the nodule extract will be lower than the level in the nodule (see above); and second, the rate of reaction of the enzyme in the nodule will be lower than that found under (optimized) assay conditions because of nonsaturation with substrates and presence of inhibitors. From values in Table 7, it can be calculated that glutamate synthase will be 83% saturated with glutamine under conditions found in the nodule and that the effective K_m for 2-oxoglutarate will be about 0.9 mM because of competitive

Table 8. Rates of transport of amino acids in xylem sap of 18-day *Lupinus angustifolius* plants compared with levels of nodule enzymes

Rate of transport[a] (μmol/hr/plant)		Enzyme level (μmol/hr/plant)	
Asparagine	2	Asparagine synthetase	0.8[a]
Aspartate	0.5		
Glutamine	0.4		
Glutamate	0.3		
Threonine	0.1		
Total nitrogen	5.7	Nitrogenase	7.7[b]
Total α amino group	3.3	Glutamate synthase	3.3[c]

[a] Scott, Farnden, and Robertson (1976).
[b] Calculated from Robertson et al. (1975).
[c] Calculated from Boland and Benny (1977).

inhibition, mainly by glutamate. We have not yet measured the 2-oxoglutarate concentration in the nodule, but since the glutamate and aspartate concentrations are 2–3 mM, and there is a high level of aspartate aminotransferase, it seems possible that the level of the oxo acid might be similar.

If the supply of ammonia becomes limiting for the ammonia assimilation system, then the rate of the glutamine synthetase–catalyzed reaction must become partially rate limiting. Although the level of glutamine synthetase activity shown in Table 6 is apparently fifteenfold greater than that of glutamate synthase, the glutamate concentration in nodule tissue is only sufficient for 40% saturation of the enzyme. Furthermore, the glutamine synthetase–catalyzed reaction must occur twice for every molecule of asparagine and glutamine synthesized, and these constitute most of the exported nitrogen. The K_m of glutamine synthetase for $[NH_3 + NH_4^+]$ at pH 7.5 is about 10^{-4} M and, when the steady state concentration drops much below this, glutamine biosynthesis will become the rate-limiting step in the ammonia-assimilating system. As a consequence, there will be competition between asparagine synthetase and glutamate synthase for glutamine. Because asparagine synthetase has the slightly lower K_m, this situation may result in a higher proportion of asparagine exported from the nodule. The level of glutamine in xylem sap suggests a slight excess of glutamine synthetase activity.

Pool sizes of intermediates in ammonia assimilation and asparagine biosynthesis are indicated in Table 9. These pool sizes have been calculated from concentrations in cytoplasm of nitrogen-fixing nodule tissue and weight of total nodule tissue per plant, with a correction for the volume of cytoplasm per unit weight of nodule tissue. Therefore, they probably err on the high side because lower concentrations of amino acids will be found in the cortical and vascular tissue of the nodule. They will, however, represent an upper limit. In the table, the time required for the rate-limiting enzyme

Table 9. Estimated pool size and replenishment time for intermediates in ammonia assimilation and asparagine biosynthesis in nodule cytosol of 18-day *Lupinus angustifolius* plants

Intermediate	Pool size (μmol/plant)	Enzyme likely to be rate limiting[a]	Replenishment time[b] (min)
Glutamine	0.16	Glutamate synthase	3.0
Glutamate	0.23	Glutamate synthase	4.0
Aspartate	0.25	Glutamate synthase	4.1
Asparagine	1.50	Asparagine synthetase	108
Ammonia	0.48	Nitrogenase	4

[a] Assuming adequate levels of ATP, NADH, and oxo acids.

[b] Replenishment time is the time required to synthesize the entire pool of amino acid from a non-amino acid source, or ammonia from N_2.

to entirely replenish the pool has been estimated. No great significance attaches to the times indicated; however, they do indicate that turnover of all intermediates other than asparagine is very rapid, i.e., pool sizes are small compared with rates of metabolism. The rate of turnover of asparagine is far slower, however, due to both a large pool size and a relatively slow rate of synthesis. The large pool size may be necessary to generate a concentration gradient for the export of asparagine (cf. Pate, Gunning, and Briarty, 1969) and is consistent with asparagine being the end point of overall ammonia assimilation in the nodule (Kennedy, 1966a, 1966b).

LIMITING FACTORS IN LEGUMINOUS NITROGEN FIXATION

From the preceding discussion, it is not obvious which step in the N_2-to-xylem amino acids pathway is limiting. If an adequate supply of ATP is maintained for all processes in the cytoplasm, it seems likely that the glutamate synthase-catalyzed reaction will be rate limiting. The high level of ammonia found in the nodule cytoplasm (Table 9) tends to support this suggestion. Although the level of glutamate in the nodule is higher than that of glutamine, glutamine is present at a concentration much greater than K_m for either glutamate synthase or asparagine synthetase. Glutamate, however, is at a concentration lower than K_m for the glutamine synthetase. These data also indicate that glutamate synthase activity is limiting. The question then arises as to whether the plant synthesizes just enough enzyme to metabolize the ammonia being produced or whether nitrogenase activity is limited to produce only as much ammonia as the plant can cope with. Recent work has shown that the rate of nitrogen fixation in the nodules can be enhanced by raising the partial pressure of carbon dioxide around the plant (Hardy, Criswell, and Havelka, 1977) and it was indicated that the limiting factor for nitrogen fixation was supply of photosynthate to the nodules. Therefore, it is reasonable to speculate that nitrogenase activity in the bacteroid will be regulated by the supply of energy and reductant and the plant, in turn, will regulate levels of ammonia-assimilating enzymes to those just sufficient to cope with ammonia produced by the bacteroid.

It is our opinion that future attempts to increase productivity in legumes should include matching of studies of N_2-fixing efficiency of various strains of rhizobia, with experiments to enhance the ability of the host plant to supply photosynthate and to assimilate the ammonia produced.

ACKNOWLEDGMENTS

We wish to thank Wendy Ulyatt for growing the plants, Derek McCormack, Paul Reynolds, and Mee Wong for unpublished data, Glenda Lover for typing the paper, and members of the staff of the C. F. Kettering Research Laboratory for many helpful discussions.

REFERENCES

Aprison, M. H., and R. H. Burris. 1952. Time course of fixation of N_2 by excised soybean nodules. Science 115:264–265.

Bergersen, F. J. 1960. Incorporation of $^{15}N_2$ into various fractions of soybean root nodules. J. Gen. Microbiol. 22:671–677.

Bergersen, F. J. 1965. Ammonia—an early stable product of nitrogen fixation by soybean root nodules. Aust. J. Biol. Sci. 18:1–9.

Bergersen, F. J. 1977. Physiological chemistry of dinitrogen fixation by legumes. In: R. W. F. Hardy and W. S. Silver (eds.), A Treatise on Dinitrogen Fixation, pp. 519–555. John Wiley & Sons, Inc., New York.

Bergersen, F. J., and G. L. Turner. 1967. Nitrogen fixation by the bacteroid fraction of breis of soybean root nodules. Biochim. Biophys. Acta 141:507–515.

Bergersen, F. J., and G. L. Turner. 1970. Gel filtration of nitrogenase from soybean root-nodule bacteroids. Biochim. Biophys. Acta 214:28–36.

Bergersen, F. J., and G. L. Turner. 1978. Activity of nitrogenase and glutamine synthetase in relation to availabiltiy of oxygen in continuous cultures of a strain of cowpea *Rhizobium* sp. supplied with excess ammonium. Biochim. Biophys. Acta 538:406–416.

Boland, M. J., and A. G. Benny. 1977. Enzymes of nitrogen metabolism in legume nodules: Purification and properties of NADH-dependent glutamate synthase from lupin nodules. Eur. J. Biochem. 79:355–362.

Boland, M. J., A. M. Fordyce, and R. M. Greenwood. 1978. Enzymes of nitrogen metabolism in legume nodules: A comparative study. Aust. J. Plant Physiol. 5:553–559.

Brown, C. M., and M. J. Dilworth. 1975. Ammonia assimilation by rhizobium cultures and bacteroids. J. Gen. Microbiol. 86:39–48.

Ching, T. M., S. Hedtke, and W. Newcomb. 1977. Isolation of bacteria, transforming bacteria, and bacteroids from soybean nodules. Plant Physiol. 60:771–774.

Christeller, J. T., W. A. Laing, and W. D. Sutton. 1977. Carbon dioxide fixation by lupin root nodules. Plant Physiol. 60:47–50.

Dalzeil, K. 1969. The interpretation of kinetic data for enzyme-catalysed reactions involving three substrates. Biochem. J. 114:547–556.

Davies, D. D., and A. N. Teixeira. 1975. The synthesis of glutamate and the control of glutamate dehydrogenase in pea mitochondria. Phytochemistry 14:647–656.

Dunn, S. D., and R. V. Klucas. 1973. Studies on possible routes of ammonium assimilation in soybean root nodule bacteroids. Can. J. Microbiol. 19:1493–1499.

Gibson, A. H., W. R. Scowcroft, J. J. Child, and J. D. Pagan. 1976. Nitrogenase activity in cultures of rhizobia sp. strain 32H1: Nutritional and physical considerations. Arch. Microbiol. 108:45–54.

Gibson, A. H., W. R. Scowcroft, and J. D. Pagan. 1977. Nitrogen fixation in plants: An expanding horizon? In: W. Newton, J. R. Postgate, and C. Rodriguez-Barrueco (eds.), Recent Developments in Nitrogen Fixation, pp. 388–417. Academic Press, London.

Grimes, H., and C. L. Masterson. 1971. Isoenzyme patterns of nodules from specific legume-*Rhizobium* combinations. Plant Soil 35:289–297.

Guevara, J. G., R. H. McParland, and H. J. Evans. 1975. The purification of glutamine synthetase from soybean root nodule bacteroids. Plant Physiol. 56:34S.

Hardy, R. W. F., J. G. Criswell, and U. D. Havelka. 1977. Investigations of possible limitations of nitrogen fixation by legumes: (1) Methodology, (2) Identification and (3) Assessment of significance. In: W. Newton, J. R. Postgate, and C.

Rodriguez-Barrueco (eds.), Recent Developments in Nitrogen Fixation, pp. 451–467. Academic Press, London.

Kennedy, I. R. 1966a. Primary products of symbiotic nitrogen fixation I. Short-term exposures of serradella nodules to $^{15}N_2$. Biochim. Biophys. Acta 130:285–294.

Kennedy, I. R. 1966b. Primary products of symbiotic nitrogen fixation II. Pulse-labelling of serradella nodules with $^{15}N_2$. Biochim. Biophys. Acta 130:295–303.

Kennedy, I. R., C. A. Parker, and D. K. Kidby. 1966. The probable site of nitrogen fixation in root nodules of *Ornithopus sativus*. Biochim. Biophys. Acta 130: 517–519.

Klucas, R. V., and R. H. Burris. 1966. Locus of nitrogen fixation in soybean nodules. Fixation by crushed nodules. Biochim. Biophys. Acta 136:399–401.

Klucas, R. V., B. Koch, S. A. Russell, and H. J. Evans. 1968. Purification and some properties of the nitrogenase from soybean (*Glycine max* Merr.) nodules. Plant Physiol. 43:1906–1912.

Koch, B., H. J. Evans, and S. Russell. 1967. Reduction of acetylene and nitrogen gas by breis and cell-free extracts of soybean root nodules. Plant Physiol. 42:466–468.

Kurz, W. G. W., D. A. Rokosh, and T. A. LaRue. 1975. Enzymes of ammonia assimilation in *Rhizobium leguminosarum* bacteroids. Can. J. Microbiol. 21:1009–1012.

McParland, R. H., J. G. Guevara, R. R. Becker, and H. J. Evans. 1976. The purification and properties of the glutamine synthetase from the cytosol of soybean root nodules. Biochem. J. 153:597–606.

Miller, R. E., and E. R. Stadtman. 1972. Glutamate synthase from *Escherichia coli*. An iron sulfide flavoprotein. J. Biol. Chem. 247:7407–7419.

O'Gara, F., and K. T. Shanmugam. 1976. Regulation of nitrogen fixation by *Rhizobia;* export of fixed N_2 as NH_4^+. Biochim. Biophys. Acta 437:313–321.

Pate, J. S., B. E. S. Gunning, and L. G. Briarty. 1969. Ultrastructure and functioning of the transport system of the leguminous root nodule. Planta 85:11–34.

Planqué, K., I. R. Kennedy, G. E. de Vries, A. Quispel, and A. A. N. van Brussel. 1977. Location of nitrogenase and ammonia assimilatory enzymes in bacteroids of *Rhizobium leguminosarum* and *R. lupini*. J. Gen. Microbiol. 102:95–104.

Prusiner, S., and L. Milner. 1970. A rapid radioactive assay for glutamine synthetase, glutaminase, asparagine synthetase and asparaginase. Anal. Biochem. 37:429–438.

Robertson, J. G., K. J. F. Farnden, M. P. Warburton, and J. Banks. 1975. Induction of glutamine synthetase during nodule development in lupin. Aust. J. Plant Physiol. 2:265–272.

Robertson, J. G., M. P. Warburton, and K. J. F. Farnden. 1975. Induction of glutamate synthase during nodule development in lupin. FEBS Lett. 55:33–37.

Rognes, S. E. 1970. A glutamine-dependent asparagine synthetase from yellow lupin seedlings. FEBS Lett. 10:62.

Rognes, S. E. 1975. Glutamine-dependent asparagine synthetase from *Lupinus luteus*. Phytochemistry 14:1975–1982.

Ryan, E., and P. F. Fottrell. 1974. Subcellular location of enzymes involved in the assimilation of ammonia by soybean root nodules. Phytochemistry 13:2647–2652.

Scott, D. B., K. J. F. Farnden, and J. G. Robertson. 1976. Ammonia assimilation in lupin nodules. Nature 263:703–705.

Sloger, C. 1973. Assimilation of ammonia by glutamine synthetase and glutamate synthase in N_2-fixing bacteroids from soybean nodules. Plant Physiol. 51:34S.

Stewart, G. R., and D. Rhodes. 1977. Nitrogen metabolism of halophytes III. Enzymes of ammonia assimilation. New Phytol. 80:307–316.

Streeter, J. G. 1972. Nitrogen nutrition of field-grown soybean plants: 1. Seasonal variations in soil nitrogen and nitrogen composition of stem exudate. Agron. J. 64:311–314.

Tempest, D. W., J. L. Meers, and C. M. Brown. 1970. Synthesis of glutamate in *Aerobacter aerogenes* by a hitherto unknown route. Biochem. J. 117:405–407.

Upchurch, R. G., and G. H. Elkan. 1978. Ammonia assimilation in *Rhizobium japonicum* colonial derivatives differing in nitrogen fixing efficiency. J. Gen. Microbiol. 104:219–225.

Nitrogen Fixation, Volume II
Edited by W. E. Newton and W. H. Orme-Johnson

Utilization of Leghemoglobin-bound Oxygen by *Rhizobium* Bacteroids

J. B. Wittenberg

The role of leghemoglobin in assuring the delivery of oxygen to bacteroids of legume root nodules has been reviewed previously (Appleby et. al., 1976; Wittenberg, 1976). This paper shows that the greatest part of the oxygen used by the bacteroids is transported through the plant cytoplasm by leghemoglobin and addresses the question: How is oxygen transferred to the bacteroid terminal oxidase?

LEGHEMOGLOBIN IN NONLEGUME SYMBIOSES

Leghemoglobin serves an essential function in all effective *Rhizobium*-legume associations. However, in the single instance of a symbiosis between *Rhizobium* and a nonlegume, *Parasponia* (Trinick, 1973; Trinick and Galbraith, 1976), no leghemoglobin is detectable in the nodules (Coventry, Trinick, and Appleby, 1976) even though nitrogen fixation is rapid (Trinick, 1973). The bacterium is competent to elicit leghemoglobin synthesis, and nodules produced on roots of several legumes infected with *Rhizobium* from *Parasponia* contain leghemoglobin (Trinick, 1973) and are effective in nitrogen fixation. Akkermans, Abdulkadir, and Trinick (1978) have identified the host plant as *Parasponia parviflora* Miq. (Ulmaceae); previously it was believed to be a *Trema* sp.

A particle-bound, but otherwise typical, hemoglobin has been isolated from nodules of *Casuarina cunninghamiana* (Davenport, 1960). Although

---

This research was supported by National Science Foundation grants 76-07572 and 78-06073. The author is a Research Career Program Awardee 5K06 HL00733 of the U.S. Public Health Service, National Heart, Lung, and Blood Institute.

hemoglobin was abundant in the nodules of plants grown under these particular conditions and the hemoglobin nature of the pigment was established beyond doubt, hemoglobin has never been detected in other *Casuarina* nodules or in nodules produced by actinomycetes on any other woody dicot.

OXYGEN PRESSURE IN THE SOYBEAN ROOT NODULE AND LEGHEMOGLOBIN-MEDIATED OXYGEN TRANSPORT

Oxygen entry into the soybean root nodule is limited by diffusion across the nodule cortex (Tjepkema and Yocum, 1973, 1974). The rate of entry is about that calculated for diffusion across a layer of water of the same thickness, although gas exchange may be by way of lenticels (Pankhurst and Sprent, 1975). Vigorous oxygen consumption, for which the bacteroids are mainly responsible (Bergersen, 1962), keeps the internal oxygen pressure low. Nitrogen fixation in the intact nodule is oxygen limited and increases immediately in response to increased ambient oxygen pressure (Burris, Magee, and Bach, 1955; Bergersen, 1962, 1970); this effect is dependent on leghemoglobin function (Bergersen, Turner, and Appleby, 1973). However, after several hours of continuous exposure, in this case to very low rhizosphere oxygen concentration, the system adapts and recovers its initial rate of nitrogen fixation (Criswell et al., 1976; Hardy, Criswell, and Havelka, 1977). The authors suggest that the adaptation results from return to an optimal internal oxygen pressure, possibly by adjustment of the protective respiratory rate. If so, the optimum falls in the range where oxygen supply limits nitrogen fixation.

Within the central tissue of the nodule a system of air passages (Sprent, 1972; Bergersen and Goodchild, 1973a) probably maintains a uniform oxygen pressure at the surface of the cells. The role of leghemoglobin, accordingly, is to assure the flow of oxygen within each individual cell.

Appleby (1969c) found the fractional oxygenation of leghemoglobin in the intact soybean nodule surrounded by air to be about 20%, which corresponds to an intracellular oxygen pressure of about 0.006 mm Hg or 10 nM oxygen. The effective leghemoglobin concentration in the bacteroid-containing cells of the soybean nodule is estimated to be about 1 mM (Bergersen and Goodchild, 1973b) or about 100,000 times the concentration of dissolved oxygen.

Thus, the conditions within the nodule, in particular the high leghemoglobin concentration, favor a dominant role of leghemoglobin-facilitated oxygen diffusion in oxygen translocation (Wittenberg, 1970). However, the rate of facilitated diffusion is proportional to the translational diffusion coefficient of the protein and disappears when that becomes zero (Riveros-

Moreno and Wittenberg, 1972), and we have no assurance that leghemoglobin is free to diffuse in the cytoplasm.

LEGHEMOGLOBIN-DEPENDENT NITROGEN FIXATION

Nitrogen fixation by bacteroids parallels the presence of leghemoglobin in the nodule (Evans and Russell, 1971). Abolition of the oxygen-binding ability of leghemoglobin by carbon monoxide decreases nitrogenase activity (as H_2 evolution) in the intact nodule to one-tenth or one-twentieth of its previous value, although oxygen consumption was scarcely affected (Smith, 1949; Bergersen, Turner, and Appleby, 1973). Conversely, addition of leghemoglobin to dense suspensions of bacteroids increased nitrogenase activity twentyfold with somewhat less than a doubling of oxygen uptake (Bergersen, Turner, and Appleby, 1973). These experiments clearly differentiate leghemoglobin-supported nitrogenase activity from bacteroid oxygen consumption per se. Wittenberg et al. (1974) and subsequently Melik-Sarkisyan et al. (1976) extended these observations to very dilute suspensions of bacteroids in the hope that events at or near the bacteroid surface would dominate the results. Nitrogenase activity was very small in the absence of leghemoglobin and increased monotonically with increasing concentration of leghemoglobin (Figure 1). Oxygen consumption, in contrast, was large in the absence of leghemoglobin. An incremental, leghemoglobin-dependent oxygen uptake was superimposed on the leghemoglobin-independent oxygen uptake (Figure 1). These results led the authors to postulate two distinct terminal oxidase systems in bacteroids: an oxidase accepting leghemoglobin-bound oxygen that is *effective* in supporting nitrogenase activity, and an oxidase accepting free, dissolved oxygen that is *ineffective* in supporting nitrogen fixation.

The reality of these two oxidase systems was demonstrated by their differential inhibition (see below) and in experiments by Bergersen and Turner (1975a, 1975b), who eliminated the gas phase that had served as an oxygen reservoir in earlier experiments. The actual oxygen pressure in the solution could now be measured. They found that the leghemoglobin-dependent oxidase system operates only at very low oxygen pressure with an optimum near 40 nM oxygen, coincident with half-saturation of leghemoglobin with oxygen. Oxygen consumption at more than about 1 μM oxygen is independent of leghemoglobin and supports nitrogenase activity only weakly.

OXYGEN TRANSFER FROM LEGHEMOGLOBIN TO BACTEROIDS

In experiments in which respiring isolated bacteroids receive oxygen from oxyleghemoglobin, the oxygen pressure at the very surface of the bacteroid

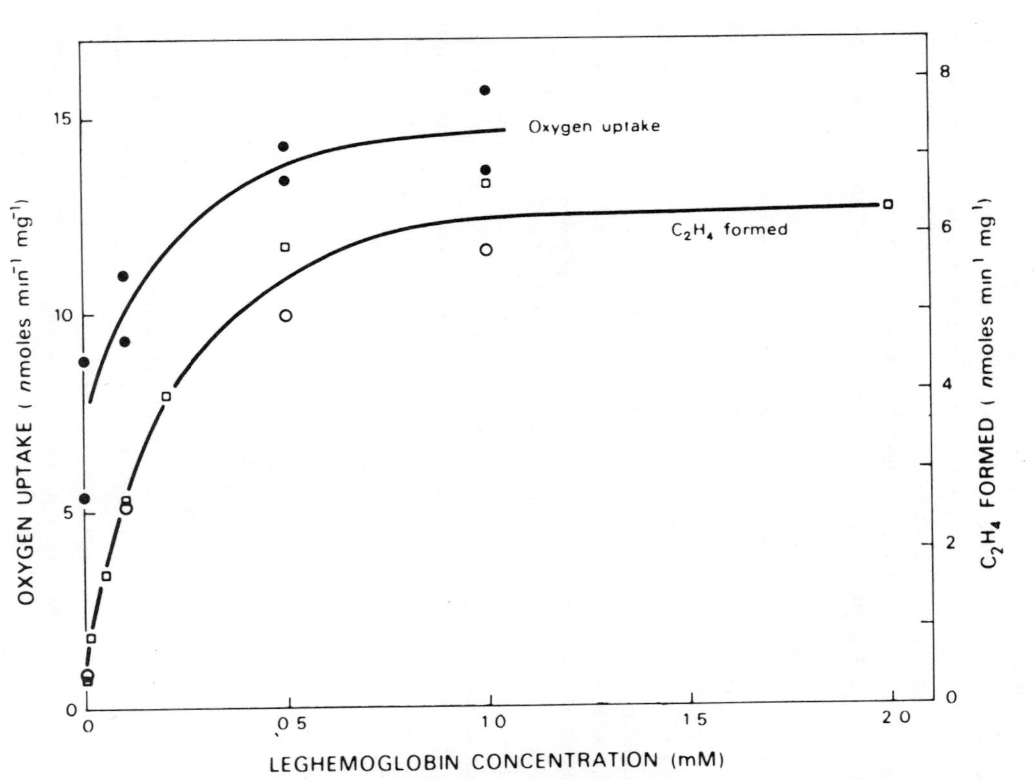

Figure 1. Oxygen uptake (solid symbols) and acetylene reduction (open symbols) by bacteroid suspensions as functions of leghemoglobin concentration. (From Wittenberg et al., 1974.)

will be larger in the presence than in the absence of leghemoglobin. This effect results from dissociation of leghemoglobin-bound oxygen brought to the bacteroid surface by translational diffusion of oxyleghemoglobin molecules and has been variously described as facilitated diffusion (Wittenberg et al., 1974), oxygen buffering (Appleby et al., 1976) or "stabilization" of oxygen pressure (Stokes, 1975). The question arises as to whether or not this is the only function of oxyleghemoglobin in oxygen delivery.

Oxygen consumption and nitrogenase activity of dilute suspensions of bacteroids were unexpectedly insensitive to leghemoglobin concentration (Bergersen and Turner, 1975a, 1975b) and the leghemoglobin concentration required (no more than 50 μM) is at least an order of magnitude less than that believed optimal for facilitated diffusion (about 500 to 7000 μM) (Wittenberg, 1970). In these experiments, oxygen consumption, the rate of ATP generation (as nitrogenase activity), and ATP levels were maximal near the very low oxygen pressure at which leghemoglobin is half-saturated with oxygen. In experiments in which oxygen consumption depleted the oxygen stores, nitrogenase activity was initially low and the onset of enhanced nitrogenase activity coincided with partial desaturation of the oxygen carrier protein, whether it was leghemoglobin (half saturated at 70 nM O_2) or myoglobin (half saturated at 1000 nM O_2). This effect is seen clearly in an experiment using nitrogen-fixing rhizobia grown in continuous culture (Bergersen, 1977), presented in Figure 2. Bergersen's experiments were not done in the steady state and are open to the objection that bacteroid ATP levels take time to adjust. Veeger (this volume) showed that optimal steady state nitrogenase activity of bacteroids approximately coincides with half-saturation of myoglobin or leghemoglobin. Finally, Bergersen (1978) carried out a new series of experiments with a longer time course and again found the same result. These experiments suggest a role for deoxyleghemoglobin in oxygen transfer.

In an attempt to understand the role of the oxygen carrier, Wittenberg et al. (1974) tested an array of oxygen-binding proteins. All supported nitrogenase activity to a greater or lesser extent. Among these were *Ascaris* body wall and perienteric hemoglobins with oxygen dissociation rate constants of 0.23 sec^{-1} and 0.0041 sec^{-1}, respectively. These rates may not be sufficient to supply free oxygen to the respiring bacteroid surface and the possibility must be considered that electrons are somehow transferred from the bacteroid surface to the oxygenated proteins.

LEGHEMOGLOBIN-DEPENDENT, EFFECTIVE BACTEROID TERMINAL OXIDASE

Leghemoglobin-delivered oxygen supports an efficient formation of ATP in bacteroids. This ATP supports nitrogenase activity at a rate proportional to

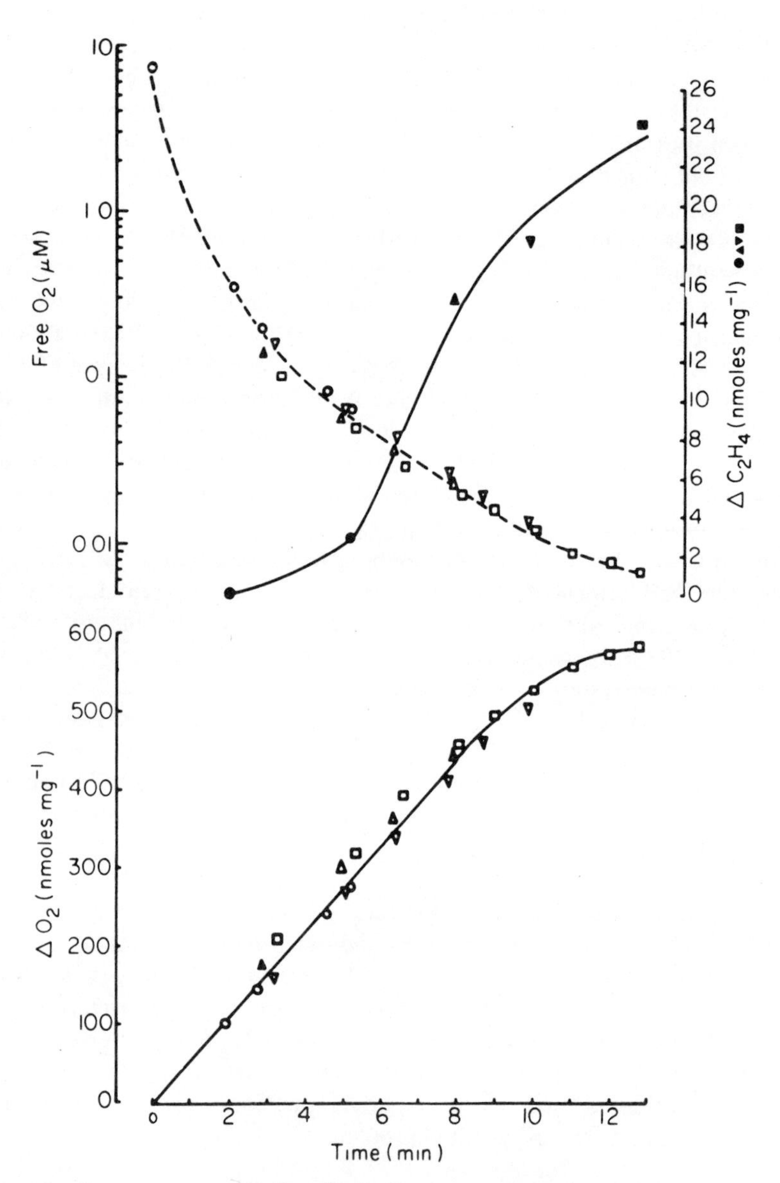

Figure 2. Oxygen consumption, free dissolved oxygen concentration, and nitrogenase activity of *Rhizobium* strain 32H1 from oxygen-limited continuous culture. The reaction mixture contained 100 μM oxyleghemoglobin. Nitrogenase activity is expressed when the Lb is partially desaturated with oxygen. (From Bergersen, 1977.)

the ratio of ATP:ADP (Appleby, Turner, and Macnicol, 1975; Bergersen and Turner, 1975a, 1975b; Appleby et al., 1976). The uncoupling agent carbonyl cyanide *m*-chlorophenylhydrazone (CCCP) dissipates leghemoglobin-dependent ATP generation (Appleby, Turner, and Macnicol, 1975) and dinitrophenol abolishes nitrogenase activity at low pO_2 in the absence of leghemoglobin (Raikhinshtein et al., 1976). These data suggest that ATP is generated by bacterial oxidative phosphorylations or at least that membrane energization is required.

Leghemoglobin-dependent nitrogenase activity, ATP production, and oxygen uptake are inhibited by *N*-phenylimidazole (Appleby, Turner, and Macnicol, 1975; Bergersen and Turner, 1975b) and by carbon monoxide (Bergersen and Turner, 1975b). *N*-Phenylimidazole is a highly specific inhibitor of bacterial cytochrome P-450. Appleby, Turner, and Macnicol (1975) found that the concentration required for half-inhibition of leghemoglobin-dependent nitrogenase activity and ATP accumulation is identical to the dissociation constant for reaction of *N*-phenylimidazole with crude and highly purified high-spin bacteroid cytochrome P-450b. Furthermore, difference spectra of bacteroid supernatants (anaerobic *N*-phenylimidazole *minus* anaerobic) are dominated by the spectrum of high-spin cytochrome P-450. Kretovich and his collaborators (Kretovich et al., 1974; Melik-Sarkisyan et al., 1974; Raikhinshtein et al., 1974) had shown earlier that, in the absence of leghemoglobin, a wide range of inhibitors of cytochrome P-450 inhibited in parallel oxygen uptake and nitrogenase activity of bacteroids produced by effective strains of *Rhizobium*. Oxygen consumption of ineffective bacteroids was unaffected. These facts, taken together, prove that cytochrome P-450 is involved in that respiration that supports nitrogenase activity.

INEFFECTIVE BACTEROID TERMINAL OXIDASE

The low affinity terminal oxidase or terminal oxidase system has little activity below 1 μM free oxygen, is insensitive to inhibition by carbon monoxide or *N*-phenylimidazole, and is totally or largely ineffective in producing ATP in the bacteroids (Appleby, Turner, and Macnicol, 1975; Bergersen and Turner, 1975a, 1975b; Appleby, 1977). Respiration through this system may account for a large fraction of the total nodule respiration and is important in maintaining a low oxygen pressure in continuous chemostat culture of *Rhizobium* (Bergersen et al., 1976).

Appleby (1977) isolated from bacteroids a high molecular weight fraction, possibly derived from the plasma membrane, that vigorously oxidizes cytochrome *c* at the expense of oxygen. This oxidase is not inhibited by *N*-phenylimidazole or carbon monoxide, which suggests that it may be the terminal oxidase of the ineffective, protective respiration. Inhibition by micromolar cyanide or azide and by EDTA suggests iron or copper at the active center. Inhibition by atebrin with partial reversal by flavin mononucleotide raises the possibility of a flavin prosthetic group.

BACTEROID CYTOCHROME P-450

Cytochrome P-450 is released from disrupted bacteroids as a soluble protein. It is abundant in bacteroids from soybeans (Appleby, 1967, 1968, 1969a; Daniel and Appleby, 1972; Appleby and Daniel, 1973), and yellow lupins (Kretovich, Melik-Sarkisyan, and Matus, 1972; Matus, Melik-Sarkisyan, and Kretovich, 1973), but is not easily detected in bacteroids of *Rhizobium leguminosarum* from *Vicia faba* nodules (Kretovich, Romanov, and Korolyov, 1973; Romanov, Korolev, and Kretovich, 1974). The content of cytochrome P-450 in lupin nodules is greatest at the period of active nitrogen fixation (Kretovich, Melik-Sarkisyan, and Matus, 1972; Matus, Melik-Sarkisyan, and Kretovich, 1973). Cytochrome P-450 is absent from lupin nodules produced by an ineffective *Rhizobium* strain (Matus, Melik-Sarkisyan, and Kretovich, 1973). It is also absent from *Rhizobium japonicum* cultured under aerobic conditions (Appleby, 1969b; Appleby and Daniel, 1973). These facts, taken together with the inhibitor studies of Kretovich et al. (1974), of Melik-Sarkisyan et al. (1974), and of Appleby, Turner, and Macnicol (1975), leave no doubt that cytochrome P-450 is essential for the support of nitrogen fixation.

PATTERNS OF OXIDASES IN BACTEROID AND CULTURED RHIZOBIA

Aerobically grown *Rhizobia* have the conventional terminal oxidases, cytochromes aa_3 and o, and, in addition, contain *Rhizobium* hemoglobin (a rhizobial protein distinct from leghemoglobin). These proteins are absent from bacteroids and are replaced by a new set of oxidases, including cytochromes c_{550}, c_{552}, one or more cytochromes b, cytochrome P-428, and cytochromes P-450 (Tuzimura and Watanabe, 1964; Appleby 1969a, 1969b; Daniel and Appleby, 1972; Kretovich, Melik-Sarkisyan, and Matus, 1972; Kretovich, Romanov, and Korolyov, 1973; Matus, Melik-Sarkisyan, and Kretovich, 1973; Romanov et al., 1973; Romanov, Korolev, and Kretovich, 1974; Romanov et al., 1976).

INTRACELLULAR LOCATION OF BACTEROID OXIDASES

Hemoproteins are demonstrated in bacteroid cytomembranes by histochemical staining (Marks and Sprent, 1974). Cytochrome b is membrane bound (Appleby, 1969a, 1977; Daniel and Appleby, 1972; Kretovich, Melik-

Sarkisyan, and Matus, 1972; Kretovich, Romanov, and Korolyov, 1973; Matus, Melik-Sarkisyan, and Kretovich, 1973; Romanov, Korolev, and Kretovich, 1974). Cytochrome c_{550}, which does not react with carbon monoxide, and the carbon monoxide–reactive, air-oxidizable cytochromes c_{552} and c_{554} are variously distributed between membrane bound and soluble fractions. Perhaps they are easily dissociated from membranes to which they are bound in life.

Robertson et al. (1978) found cytochrome c_{550} and cytochrome *b* in isolated bacteroid envelope inner membranes. These authors ascribed an ATPase activity in the inner membrane fraction to 10-nm diameter particles found on the cytoplasmic face of the inner membranes and suggested that these are functionally F_1ATPase particles analogous to those of mitochrondria. NADH oxidase and succinic dehydrogenase activity are located in the inner membrane fraction. These results suggest that the electron transport system of bacteroids is associated with the inner membranes as it is in other gram-negative organisms. Cytochrome P-450 is always found in the soluble fraction.

BACTEROID OXYGEN CARRIER

Recently, Bergersen (1978) studied the oxygen uptake of bacteroid suspensions in the range of oxygen pressure in which the leghemoglobin-dependent, effective terminal oxidase system operates. Leghemoglobin was not present. He found that respiration is oxygen limited to about 200 nM bulk phase oxygen and is oxygen saturated above that level. This component of the respiration was abolished by carbon monoxide or azide, but a much smaller, oxygen-limited respiration persisted in the presence of carbon monoxide. These results may indicate the presence of a carbon monoxide–sensitive carrier that serves to gather oxygen from a dilute surround. In this interpretation, the terminal oxidase itself is considered to be carbon monoxide insensitive; at higher oxygen pressure, oxygen diffusion may bypass the carrier and deliver oxygen directly to the terminal oxidase.

LEGHEMOGLOBIN-ASSOCIATED IRON CHROMOPHORE

The soluble fraction from soybean nodule homogenates contains, in addition to leghemoglobin, 1 g atom of iron per mole of leghemoglobin. The molar ratio, leghemoglobin-associated iron: leghemoglobin, is unity within experimental error. The leghemoglobin-associated iron is not an iron-sulfur protein. The phenomenon is not peculiar to legumes. Wittenberg (1972) found additional iron, equimolar with myoglobin, in muscles and nerves of animals from six phyla. Since leghemoglobin is 1.0–1.5 mM in the plant cell cytoplasm and constitutes perhaps 40% of the soluble cytoplasmic protein,

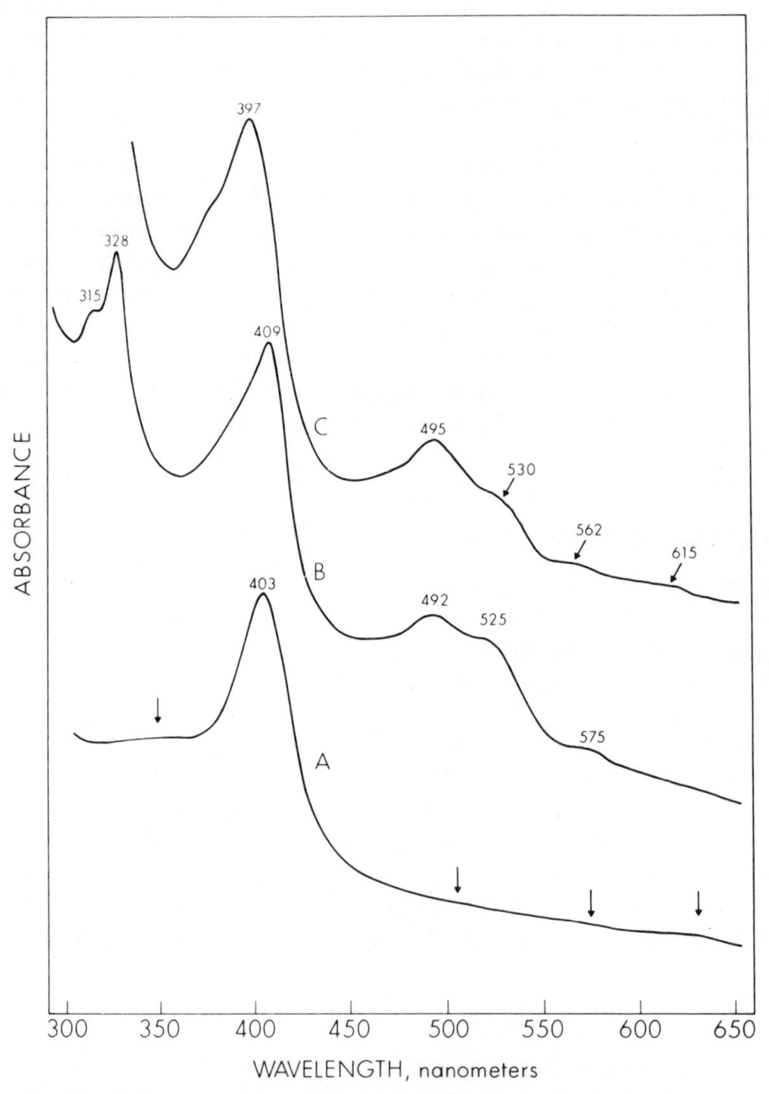

Figure 3. Optical spectra of leghemoglobin-associated iron proteins or polypeptides isolated from soybean root nodules. (A) Partially purified protein; molecular weight 31,000. (B) Polypeptide fraction, molecular weight approximately 5000, isolated on a column of Bio-Gel P-10, a condition that tends to dissociate small molecules from the polypeptide. (C) Polypeptide fraction, molecular weight approximately 5000, isolated on a column of Sephadex G-75.

the leghemoglobin-associated iron must also be a major component of the plant cytoplasm.

A red chromophore containing the leghemoglobin-associated iron may be brought into organic solvents, which shows that ligands contributed by peptides or proteins do not form an integral part of the chromophore. Optical spectra of the free chromophore and of polypeptide-bound fractions (Figure 3) suggest that the chromophore may be an iron tetrapyrrole other than iron protoporphyrin IX.

A red, partially purified iron protein was isolated from one strain of soybean nodules. This component represents the majority of the 403-nm–absorbing material present in the nodule extracts and accounts for all of the leghemoglobin-associated iron. The molecular weight, determined on a calibrated Sephadex column, was 31,000. The ratio Cu:Fe is 1.19; the copper may be extraneous. Neither manganese nor flavin was detected. The leghemoglobin-associated iron chromophore of nodules from other soybean strains accompanies polypeptide fractions with molecular weights of roughly 5000; optical spectra are presented in Figure 3. Pieve, Zhiznevskaya, and Borodenko (1972) reported iron-containing proteins from a variety of legume nodules. A partially purified protein fraction from lupin nodules, molecular weight 66,000–100,000, is reported to contain flavin and manganese in addition to iron and to catalyze the reduction of dichlorophenolindophenol by NADH or NADPH (Borodenko and Zhiznevskaya, 1978). The relation of this protein to that studied by Wittenberg is not clear. No function has been ascribed to the leghemoglobin-associated iron chromophore.

CONCLUSIONS AND SPECULATIONS

Leghemoglobin clearly acts as an oxygen-gathering system that, by virtue of extraordinarily rapid combination with oxygen, can acquire oxygen from an environment of low oxygen pressure at the periphery of the root nodule cell and that, by translational diffusion of oxygenated leghemoglobin, can facilitate diffusion of oxygen to the bacteroid surface. The great abundance of leghemoglobin within the cytoplasm assures a large flux of leghemoglobin-bound oxygen through the cytosol. Simultaneously, the concentration of free oxygen in equilibrium with oxyleghemoglobin is kept vanishingly small because of the extraordinary affinity of leghemoglobin for oxygen.

The location of leghemoglobin within the plant cytoplasm (reviewed by Dilworth, these proceedings) remains controversial. If leghemoglobin is within the peribacteroid membranes, oxygen transfer to the bacteroid is direct and the predicted gradients of oxygen pressure are appropriate for oxygen delivery by facilitated diffusion (Scholander, 1965; Wittenberg, 1970). Operation of a system in which leghemoglobin is located external to the peribacteroid membranes is more difficult to understand; perhaps we do not have all the facts. If Brownian movement of the bacteroids is rapid, their surfaces may be in contact with the inner face of the peribacteroid membranes for a large enough fraction of time to permit oxygen transfer.

Oxygen transfer to the bacteroid surface may be distinct from oxygen transport through the plant cytoplasm. Leghemoglobin serves both functions. Two alternative mechanisms of oxygen transfer may be considered.

Possibly leghemoglobin serves solely to bring oxygen to the bacteroid surface. This would assure a rapidly replenished supply of oxygen at a fixed pressure, but that pressure, set by the oxygen affinity of leghemoglobin, would be very small, less than 10 nM. In this situation, a second bacteroidal oxygen gathering system might be required. Indeed, the recent experiments of Bergersen (1978) may have discovered a bacteroid oxygen carrier serving this function. This carrier may be hemoprotein P-450 and may be located in the bacteroid periplasmic space. If indeed bacteroid hemoprotein P-450 binds oxygen, it could serve to facilitate diffusion of oxygen in the essentially two-dimensional peribacteroid space and might deliver oxygen to terminal oxidases located in the inner bacteroid membrane. The intracellular localization of bacteroid P-450 and the kinetics of its reactions with oxygen become pressing experimental questions.

Oxygen is the ultimate electron acceptor for bacteroid respiration. One must consider the possibility that electron transfer to bound oxygen takes place at the interface between the bacteroid and the plant cytoplasm. This speculation would accommodate two otherwise indigestible experimental facts. Deoxyleghemoglobin, implicated in oxygen transfer to bacteroids, might serve as an intermediate in electron transfer. The leghemoglobin-associated iron chromophore likewise might assume a role in electron transfer reactions at the bacteroid interface.

ACKNOWLEDGMENTS

Much of the work reported here stems from an international collaboration among C. A. Appleby, F. J. Bergersen, and others in Australia and J. Peisach, W. E. Blumberg, B. A. Wittenberg, and myself in the United States. Work in Russia has followed a complementary course. A major pleasure of this symposium has been discussions with some of the architects of that work.

REFERENCES

Akkermans, A. D. L., S. Abdulkadir, and M. J. Trinick. 1978. N_2-fixing root nodules in Ulmaceae: *Parasponia* or (and) *Trema* spp.? Plant Soil 49:711–715.

Appleby, C. A. 1967. A soluble haemoprotein P-450 from nitrogen-fixing *Rhizobium* bacteroids. Biochim. Biophys. Acta 147:399–402.

Appleby, C. A. 1968. Properties of soluble hemoprotein P-450 purified from *Rhizobium japonicum* bacteroids. In: K. Okunuki, M. D. Kamen, and I. Sekuzu (eds.), Structure and Function of Cytochromes, pp. 666–679. University Park Press, Baltimore.

Appleby, C. A. 1969a. Electron transport systems of *Rhizobium japonicum*. I. Hemeprotein P-450, other CO-reactive pigments, cytochromes and oxidases in bacteroids from N_2-fixing root nodules. Biochim. Biophys. Acta 172:71–87.

Appleby, C. A. 1969b. Electron transport systems of *Rhizobium japonicum*. II. Rhizobium hemoglobin, cytochromes and oxidases in free-living (cultured) cells. Biochim. Biophys. Acta 172:88–105.

Appleby, C. A. 1969c. Properties of leghemoglobin *in vivo*, and its isolation as ferrous oxyhemoglobin. Biochim. Biophys. Acta 188:222–229.

Appleby, C. A. 1977. Function of P-450 and other cytochromes in *Rhizobium* respiration. In: H. Degn, D. Lloyd, and G. C. Hill (eds.), F.E.B.S. 11th Meeting, Copenhagen. Vol. 49. Colloquium B6: Functions of Alternative Terminal Oxidases, pp. 11–20. Pergamon Press, Inc., New York.

Appleby, C. A., F. J. Bergersen, P. K. Macnicol, G. L. Turner, B. A. Wittenberg, and J. B. Wittenberg. 1976. Role of leghemoglobin in symbiotic nitrogen fixation. In: W. E. Newton and C. J. Nyman (eds.), Proceedings of the 1st International Symposium on Nitrogen Fixation, pp. 274–292. Washington State University Press, Pullman.

Appleby, C. A., and R. M. Daniel. 1973. *Rhizobium* cytochrome P-450: A family of soluble, separable hemoproteins. In: T. E. King, H. S. Mason, and M. Morrison (eds.), Oxidases and Related Redox Systems, Vol. 2, pp. 515–526. University Park Press, Baltimore.

Appleby, C. A., G. L. Turner, and P. K. Macnicol, 1975. Involvement of oxyleghemoglobin and cytochrome P-450 in an efficient oxidative phosphorylation pathway which supports nitrogen fixation in *Rhizobium*. Biochim. Biophys. Acta 387:461–474.

Bergersen, F. J. 1962. The effects of partial pressure of oxygen upon respiration and nitrogen fixation by soybean root nodules. J. Gen. Microbiol. 29:113–125.

Bergersen, F. J. 1970. The quantitative relationship between nitrogen fixation and the acetylene reduction assay. Aust. J. Biol. Sci. 23:1015–1025.

Bergersen, F. J. 1977. Nitrogenase in chemostat cultures of Rhizobia. In: W. Newton, J. R. Postgate, and C. Rodriguez-Barrueco (eds.), Recent Developments in Nitrogen Fixation, pp. 309–320. Academic Press, London.

Bergersen, F. J. 1978. Leghemoglobin, oxygen supply and nitrogen fixation: Studies with soybean nodules. In: J. Döbereiner, R. H. Burris, and A. Hollaender (eds.), Limitations and Potentials for Biological Nitrogen Fixation in the Tropics, pp. 247–261. Plenum Publishing Corp., New York.

Bergersen, F. J., and D. J. Goodchild. 1973a. Aeration pathways in soybean root nodules. Aust. J. Biol. Sci. 26:229–240.

Bergersen, F. J., and D. J. Goodchild. 1973b. Cellular location and concentration of leghemoglobin in soybean root nodules. Aust. J. Biol. Sci. 26:741–756.

Bergersen, F. J., and G. L. Turner. 1975a. Leghemoglobin and the supply of O_2 to nitrogen-fixing root nodule bacteroids: Studies of an experimental system with no gas phase. J. Gen. Microbiol. 89:31–47.

Bergersen, F. J., and G. L. Turner. 1975b. Leghemoglobin and the supply of oxygen to nitrogen-fixing root nodule bacteroids: Presence of two oxidase systems and ATP production at low free oxygen concentration. J. Gen. Microbiol. 91:345–354.

Bergersen, F. J., G. L. Turner, and C. A. Appleby. 1973. Studies on the physiological role of leghemoglobin in soybean root nodules. Biochim. Biophys. Acta 292:271–282.

Bergersen, F. J., G. L. Turner, A. H. Gibson, and W. F. Dudman. 1976. Nitrogen activity and respiration of cultures of *Rhizobium* spp. with special reference to concentration of dissolved oxygen. Biochim. Biophys. Acta 444:167–174.

Borodenko, L. I., and Y. A. Zhiznevskaya. 1978. The diaphorase activity of a non-hemin, iron-containing protein from lupin nodules. Fiziol. Rastenii 25:455–462.

Burris, R. H., W. E. Magee, and M. K. Bach. 1955. The pN_2 and the pO_2 function for nitrogen fixation by excised soybean nodules. Ann. Acad. Sci. Fenn. Ser. A II Chem. 60:190–199.

Coventry, D. R., M. J. Trinick, and C. A. Appleby. 1976. A search for leghemoglobin-like compound in root nodules from *Trema cannabina* Louv. Biochim. Biophys. Acta 420:105–111.

Criswell, J. G., U. D. Havelka, B. Quebedeaux, and R. W. F. Hardy. 1976. Adaptation of nitrogen fixation by intact soybean nodules to altered Rhizosphere pO_2. Plant Physiol. 58:622–625.

Daniel, R. M., and C. A. Appleby. 1972. Anaerobic nitrate, symbiotic and aerobic growth of *Rhizobium japonicum:* Effects on cytochrome P-450, other haemoproteins, nitrate and nitrite reductases. Biochim. Biophys. Acta 275:347–354.

Davenport, H. E. 1960. Hemoglobin in the root nodules of *Casuarina cunninghamiana*. Nature 186:653–654.

Evans, H. J., and S. A. Russell. 1971. Physiological chemistry of symbiotic nitrogen fixation by legumes. In: J. R. Postgate (ed.), The Chemistry and Biochemistry Nitrogen Fixation, pp. 191–244. Plenum Press, London.

Hardy, R. W. F., J. G. Criswell, and U. D. Havelka. 1977. Invesigations of possible limitations of nitrogen fixation by legumes. In: W. Newton, J. R. Postgate, and C. Rodriguez-Barrueco (eds.), Recent Developments in Nitrogen Fixation, pp. 451–467. Academic Press, London.

Kretovich, V. L., S. S. Melik-Sarkisyan, and V. K. Matus. 1972. Cytochrome composition of yellow lupin nodules. Biokhimiya 37:711–719.

Kretovich, W. L., S. S. Melik-Sarkisyan, M. V. Raikchinstein, and A. I. Archakov. 1974. The binding of microsomal hydroxylation substrates to cytochrome $P\text{-}450_{Rh}$ and its effect on the nitrogen fixation by lupin bacteroids. FEBS Letters 44:305–308.

Kretovich, V. L., V. I. Romanov, and A. V. Korolyov. 1973. *Rhizobium leguminosarum* cytochromes (*Vicia faba*). Plant Soil 39:619–634.

Marks, I., and J. I. Sprent. 1974. The localization of enzymes in fixed sections of soybean root nodules by electron microscopy. J. Cell Sci. 16:623–637.

Matus, V. K., S. S. Melik-Sarkisyan, and V. L. Kretovich. 1973. Cytochromes and respiration rate of bacteroids from nodules of lupin inoculated with effective and ineffective strains of *Rhizobium lupini.* Mikrobiologiya 42:112–118.

Melik-Sarkisyan, S. S., M. V. Raikhinshtein, A. I. Archakov, and V. L. Kretovich. 1974. Influence of substrates and inhibitors of microsomal hydroxylation on the nitrogen fixing activity of lupin bacteroids. Dokl. Akad. Nauk SSSR 216:1410–1412.

Melik-Sarkisyan, S. S., M. V. Raikhinshtein, L. P. Vladzievskaya, N. F. Bashirova, and V. L. Kretovich. 1976. Effect of legoglobin on respiration and nitrogen-fixing activity of lupin bacteroids during growth of the plants. Fiziol. Rastenii 23:274–278.

Pankhurst, C. E., and J. I. Sprent. 1975. Surface features of soybean root nodules. Protoplasma 85:85–98.

Pieve, Y. V., G. Y. Zhiznevskaya, and L. I. Borodenko. 1972. Cytoplasmic iron proteins from nodules of legumes. Fiziol. Rastenii 19:1053–1059.

Raikhinshtein, M. V., S. S. Melik-Sarkisyan, A. I. Archakov, and V. L. Kretovich. 1974. Spectral properties of cytochrome P-450 from the bacteroids of lupin nodules. Dokl. Akad. Nauk SSSR 216:1185–1187.

Raikhinshtein, M. V., S. S. Melik-Sarkisyan, G. G. Zaigraeva, and V. K. Kretovich. 1976. Inhibitory analysis of the respiration of bacteroids from yellow lupin nodules. Mikrobiologiya 45:190–195.

Riveros-Moreno, V., and J. B. Wittenberg, 1972. The self-diffusion coefficients of myoglobin and hemoglobin in concentrated solutions. J. Biol. Chem. 247:895–901.

Robertson, J. G., M. P. Warburton, P. Lyttleton, A. M. Fordyce, and S. Bullivant. 1978. Membranes in lupin root nodules. II. Preparation and properties of peribacteroid membranes and bacteroid envelope inner membranes from developing lupin nodules. J. Cell Sci. 30:151–174.

Romanov, V. I., N. G. Fedulova, A. V. Korolev, and V. L. Kretovich. 1976. Effect of nitrate and nitrite on respiration and the cytochrome system of *Rhizobium lupini*. Mikrobiologiya 45:85–91.

Romanov, V. I., A. V. Korolev, and V. K. Kretovich. 1974. The cytochrome composition of bacteroids of *Rhizobium leguminosarum* (*Vicia faba*). Biokhimiya 39:719–724.

Romanov, V. I., V. K. Matus, A. V. Korolev, and V. L. Kretovich. 1973. Influence of oxygen conditions on the respiration and cytochrome composition in *Rhizobium leguminosarum*. Mikrobiologiya 42:976–982.

Scholander, P. F. 1965. Tension gradients accompanying accelerated oxygen transport in a membrane. Science 149:876–877.

Smith, J. D. 1949. Haemoglobin and the oxygen uptake of leguminous root nodules. Biochem. J. 44:591–598.

Sprent, J. I. 1972. The effects of water stress on nitrogen-fixing root nodules. II. Effects on the fine structure of detached soybeans. New Phytol. 71:443–450.

Stokes, A. N. 1975. Facilitated diffusion: The elasticity of oxygen supply. J. Theor. Biol. 52:285–297.

Tjepkema, J. D., and C. S. Yocum. 1973. Respiration and oxygen transport in soybean nodules. Planta 115:59–72.

Tjepkema, J. D., and C. S. Yocum. 1974. Measurement of oxygen partial pressure within soybean nodules by oxygen microelectrodes. Planta 119:351–360.

Trinick, M. J. 1973. Symbiosis between *Rhizobium* and the non-legume, *Trema aspera*. Nature 244:459–460.

Trinick, M. J., and J. Galbraith. 1976. Structure of root nodules formed by *Rhizobium* on the non-legume *Trema cannabina*. Arch. Microbiol. 108:159–166.

Tuzimura, K., and I. Watanabe. 1964. Electron transport systems of *Rhizobium* grown in nodules and in laboratory medium. Plant Cell Physiol. 5:157–170.

Wittenberg, J. B. 1970. Myoglobin-facilitated oxygen diffusion: Role of myoglobin in oxygen entry into muscle. Physiol. Rev. 50:559–636.

Wittenberg, J. B. 1972. An ubiquitous association between myoglobin and equimolar quantities of iron protein(s). Fed. Proc. 31:923.

Wittenberg, J. B. 1976. Facilitation of oxygen diffusion by intracellular leghemoglobin and myoglobin. In: F. F. Jöbsis (ed.), Oxygen and Physiological Function, pp. 228–246. Professional Information Library, Dallas, Texas.

Wittenberg, J. B., F. J. Bergersen, C. A. Appleby, and G. L. Turner. 1974. Facilitated oxygen diffusion. The role of leghemoglobin in nitrogen fixation by bacteroids isolated from soybean root nodules. J. Biol. Chem. 249:4057–4066.

Nitrogen Fixation, Volume II
Edited by W. E. Newton and W. H. Orme-Johnson

Hydrogen Metabolism in the Legume-*Rhizobium* Symbiosis

H. J. Evans, D. W. Emerich, T. Ruiz-Argüeso, R. J. Maier, and S. L. Albrecht

The relationship of hydrogenase to the N_2 fixation process in legumes has been reviewed recently (Evans et al., 1977; Schubert and Evans, 1977; Evans, Ruiz-Argüeso, and Russell, 1978). This paper briefly summarizes some major points discussed previously and presents an overview of the present status of research in this field.

Professor P. W. Wilson and associates at the University of Wisconsin (Wilson and Burris, 1947) discovered that a relationship existed between hydrogenase and the N_2-fixing process. Hoch, Schneider, and Burris (1960) concluded that N_2 fixation and H_2 evolution from soybean nodules were catalyzed by closely related enzyme systems, but the early observations could not be completely interpreted until the basic properties of the N_2-fixing system were elucidated. Characterization of the nitrogenase system from several sources (Bulen and LeComte, 1966; Winter and Burris, 1968; Burns and Hardy, 1975) revealed that all were capable of catalyzing reductant- and ATP-dependent H_2 evolution. Dixon (1967, 1968, 1972) conducted a series of experiments with pea root nodules from which he concluded that at least two separate hydrogenase systems were present. One of these (the nitrogenase system) was involved in H_2 evolution, whereas the other catalyzed O_2-dependent H_2 uptake. The hydrogenase system responsible for H_2 uptake was associated with the bacteroids of pea nodules and it exhibited properties similar to those of the particulate hydrogenase complex from

This research was supported by NSF Grant 77-08784, USDA Grant 701-15-30, and the Oregon Agricultural Experiment Station, from which this is paper No. 4874. D. E. and R. M. acknowledge the receipt of postdoctoral fellowships from the Rockefeller Foundation.

Azotobacter vinelandii (Hyndman, Burris, and Wilson, 1953). Hydrogen oxidation via the hydrogenase from pea bacteroids or *A. vinelandii* supported ATP synthesis (Hyndman, Burris, and Wilson, 1953; Dixon, 1968). Dixon (1972) listed the following advantages of the unidirectional hydrogenase system: 1) H_2 oxidation may utilize excess O_2 and thereby contribute toward the maintenance of O_2-labile nitrogenase; 2) H_2 removal may decrease the possibility of H_2 accumulation to concentrations that could be inhibitory to nitrogenase; and 3) evolution of H_2 from the nitrogenase system is ATP dependent and therefore represents energy wastage, but H_2 oxidation via the hydrogenase complex leads to ATP synthesis and therefore conserves some energy that otherwise would be lost.

In recent discussions (Evans et al., 1977; Schubert and Evans, 1977; Evans, Ruiz-Argüeso, and Russell, 1978) of the nitrogenase-hydrogenase relationship several points have been emphasized. Cell-free nitrogenase from all known sources produces H_2 concomitant with N_2 reduction. This H_2 production during N_2 fixation usually accounts for between 25% and 35% of the electron flow through the nitrogenase system. When C_2H_2 is reduced by nitrogenase, H_2 evolution is greatly inhibited and, in many cases, not measurable. In the absence of acetylene, not only N_2 but also other acceptors for nitrogenase allow reductant- and ATP-dependent formation of H_2 from protons. From kinetics analyses, Rivera-Ortiz and Burris (1975) have concluded that even an infinite N_2 concentration would not prevent H_2 evolution. Some of the factors that influence electron allocation to different nitrogenase substrates include temperature, the ratio of nitrogenase components, the ATP-to-ADP ratio, and the magnitude of the electron flux through the nitrogenase system (Burns and Hardy, 1975; Zumft and Mortenson, 1975; Winter and Burris, 1976). Because H_2 evolution during N_2 reduction was not completely suppressed, H_2 evolution during N_2 reduction may be an integral part of the nitrogenase reaction.

According to recent reviews (Zumft and Mortenson, 1975; Winter and Burris, 1976; Burris, 1977) 4 to 5 moles of ATP are utilized per pair of electrons transported during the nitrogenase reaction. The ratio of ATP utilized per electron pair transported is not appreciably influenced by the different nitrogenase acceptors. From this information and an assumed minimum consumption of 4 moles of ATP per pair of electrons transported, the stoichiometry of the nitrogenase reaction may be written as follows:

$$N_2 + 8e^- + 8H^+ + 16ATP \rightarrow 2NH_3 + 16ADP + H_2 + 16P_i$$

If we assume that each pair of low potential electrons that are utilized for the reduction of either N_2 or H^+ is equivalent to 3 molecules of ATP in an aerobic organism, then a total of 28 moles of ATP are theoretically consumed in the reduction of 1 mole of N_2 (Table 1). Seven (or 25%) of

Table 1. Utilization of ATP or equivalent energy during nitrogenase reaction[a]

Portion of reaction	ATP needed (moles)
$N_2 \rightarrow 2NH_3$	12
$6e^-$ for N_2 reduction (equivalent to)	9
$2H^+ \rightarrow H_2$	4
$2e^-$ for $2H^+$ reduction (equivalent to)	3
Total reaction	28
Loss due to H_2 evolution	7

[a] It is assumed that 4 ATP are needed for each electron pair transferred through the nitrogenase reaction and that each pair of electrons used as reductant is equivalent to 3 ATP in an aerobic organism.

these are utilized in the H_2 evolution process. The energy expenditure in H_2 evolution by the nitrogenase system obviously is substantial.

H_2 LOSS FROM NODULATED SYMBIONTS

We have devised a method to estimate, under in vivo conditions, the proportion of the electron flux through the nitrogenase system that is used for N_2 reduction (Evans et al., 1977; Schubert and Evans, 1977). The total electron flux through nitrogenase is measured as the rate of C_2H_2 reduction because H_2 evolution under C_2H_2 is minimal (Schubert and Evans, 1976). Nitrogenase-dependent H_2 evolution also is measured, and from these data relative efficiency values are calculated. Relative efficiency is simply an estimate of the proportion of the electron flow used for N_2 reduction (Evans et al., 1977; Schubert and Evans, 1977; Evans, Ruiz-Argüeso, and Russell, 1978). The total electron flux through nitrogenase may also be measured as the rate of H_2 evolution in an O_2-argon gas mixture, provided that the organism does not possess a hydrogenase to recycle H_2. When the hydrogenase status of nodule samples is unknown, the rate of C_2H_2 reduction is a more reliable measure of the nitrogenase electron flux than the rate of H_2 evolution under O_2 and argon.

In 1975, it was stated that there was no evidence for ATP-dependent H_2 evolution in vivo (Burns and Hardy, 1975). In a survey of 77 samples of several different legumes inoculated with native or commercial inocula, H_2 losses from nodules accounted for an average of 44% of the electron flow through nitrogenase (Schubert and Evans, 1976). This is equivalent to a mean relative efficiency of 0.56. In contrast, the mean relative efficiency of 29 samples of nodulated nonlegumes (i.e., alder, purshia, ceanothus, myrica) was 0.96. Cowpeas inoculated with strain 32H1 lost no H_2 from nodules and had a relative efficiency of 0.99. Also, no measurable quantity of H_2 was lost from the nodules of soybeans inoculated with *Rhizobium japonicum* 3I1b 110 (Schubert and Evans, 1977; Schubert et al., 1977).

After completion of the initial survey, a series of different legumes were cultured where bacteriologically controlled conditions were maintained by use of the Leonard jar technique (Evans et al., 1977; Schubert et al., 1977; Carter et al., 1978; Ruiz-Argüeso, Hanus, and Evans, 1978). Over 90 *Rhizobium* strains were examined with several legumes and a mean relative efficiency of 0.73 was found for all strains in the experiment (Table 2). The cowpea rhizobia were unique among the species tested. Ten of thirteen strains produced nodules on cowpeas with relative efficiencies greater than 0.90. In contrast, only seven *R. japonicum* strains of 32 surveyed in replicate cultures produced nodules with relative efficiencies greater than 0.90, of which most were near 1.0 (Carter et al., 1978). The mean relative efficiency of all 32 *R. japonicum* strains examined was 0.71. Samples of commercial inoculum of *R. japonicum* tested on Anoka soybeans had a mean relative efficiency of 0.74. Among the series of strains of *Rhizobium trifolii* and *Rhizobium meliloti* tested, three *R. meliloti* strains formed nodules with relative efficiency values between 0.80 and 0.83 and all others lost H_2 at considerable rates. The mean relative efficiency value of all strains of *R. meliloti* and *R. trifolii* was 0.71 (Table 2). Of the 15 strains of *Rhizobium leguminosarum* examined, only one produced nodules with a relative efficiency above 0.80 and the mean of all strains was 0.66 (Ruiz-Argüeso, Hanus, and Evans, 1978). It is apparent, therefore, that most nodulated legumes used in agriculture are losing an appreciable portion of the energy provided to the nitrogenase system as evolved H_2. This is an important factor in attempts to improve N_2 fixation by legumes because the supply of energy to the N_2-fixing apparatus has been considered a major factor limiting N_2 fixation in legumes (Hardy and Havelka, 1975).

OCCURRENCE OF HYDROGENASE IN NODULES

It was reported (Schubert et al., 1977; Schubert, Jennings, and Evans, 1978) that soybeans inoculated with 3I1b 110 and cowpeas inoculated with strain 176A28 produced nodules that took up H_2 under O_2/H_2 atmospheres. Furthermore, nodules from red alder (*Alnus rubra*) and several other nodulated nonlegumes that failed to evolve H_2 during N_2 fixation exhibited O_2-dependent H_2 uptake (Schubert and Evans, 1977). This was also true of soybean nodules formed by seven different strains of *R. japonicum* (Table 2; Carter et al., 1978) and is illustrated (Figure 1) for soybeans inoculated with one of the non-H_2 evolving strains (USDA 122). Figure 1 also shows O_2-dependent H_2 uptake by bacteroid suspensions prepared from nodules formed by strains USDA 122 DES. In our experience, nodules that actively fix N_2

Table 2. Identification of *Rhizobium* strains that form nodules that efficiently recycle the hydrogen produced during nitrogen fixation

Legume	*Rhizobium* species[a]	Strains examined (number)	Strains with efficiencies of[b]	
			0.80–0.90	>0.90
Alfalfa, cv. Vernal	*R. meliloti*	19	SU51, 102F65, ATCC10312	none
Alfalfa, cv. Dupuit	*R. meliloti*	5	none	none
White clover, cv. New Zeland	*R. trifolii*	11	none	none
Subterranean clover, cv. Mt. Barker	*R. trifolii*	6	none	none
Pea, cv. Austrian Winter	*R. leguminosarum*	15	128C53	none
Cowpea, cv. Whippoorwill	cowpea rhizobia	13	12703	8A12, 17A6A22, 21A1, 176827, 2589, 22H1, 3223, 47A1, 61B11, 130C5
Soybean, cv. Anoka	*R. japonicum*	8	R54a	3I1b 143, 3I1b 6, 3I1b 142, 3I1b 110
Soybean, cv. Portage	*R. japonicum*	24	none	USDA 122, USDA 136, 3I1b 110, WA5099-1-1

[a] *Rhizobium* strains were obtained from Drs. Deane Weber of USDA; Joe Marlow, The Celpril Corp.; Joe Burton, The Nitragin Co.; George Ham, University of Minnesota; and Winston Brill, University of Wisconsin.

[b] Relative efficiencies were determined as described by Schubert and Evans (1976) using the rate of C_2H_2 reduction as a measure of the total electron flux through nitrogenase. The values reported are the fractions of the rate of electron flux that are used in N_2 reduction. Plants were grown in Leonard jar assemblies, except Mt. Barker subterranean clover, which was grown in 70-ml vials with agar containing nitrogen-free nutrient solution. Results are means of at least four replicate cultures inoculated with each strain. Experiments with alfalfa, white clover, subterranean clover, and peas were conducted by Ruiz-Argüeso, Hanus, and Evans (1978), cowpeas by Schubert, Jennings, and Evans (1978), and soybeans by Carter et al. (1978).

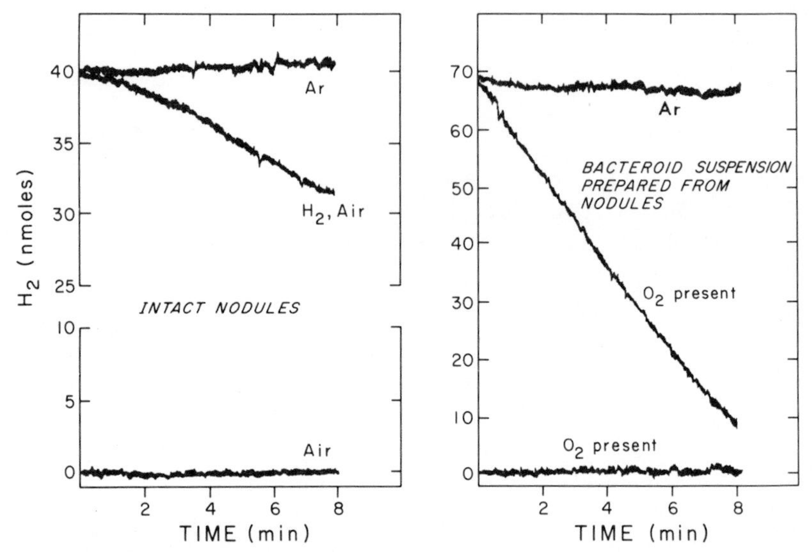

Figure 1. Hydrogenase activity of *R. japonicum* in nodules and as bacteroids. Nodules from Wilkin cv. of soybeans inoculated with *R. japonicum* USDA 122 (DES). The assay vessel contained 170 mg of fresh nodules per 2.8 ml volume; 40 nmol of H_2 and other gases as indicated were added. The H_2 uptake was measured amperometrically (Schubert and Evans, 1976; Evans et al., 1978) and the rate was 1.5 nmol per min. Bacteroids were prepared from nodules of the same pot culture of soybeans indicated above and were suspended at 34 μg of bacterial protein per ml of Dixon's (1972) buffer under argon or air and about 70 nmol H_2 injected into the vessel. The O_2-dependent H_2 uptake rate was 7.25 μmol per min. Bacteroids showed considerably greater hydrogenase activity than nodules because all nitrogenase was deliberately destroyed during their extraction, and consequently recycling of nitrogenase-dependent hydrogen no longer competed with uptake of exogenous hydrogen. (Courtesy of J. Hanus.)

without H_2 loss and bacteroid suspensions from these nodules also show a capacity to take up H_2.

Some nodules that evolve H_2 contain hydrogenase activity in bacteroids but at a level that is insufficient to recycle all of the H_2 produced by nitrogenase. This was demonstrated using 15 *R. leguminosarum* strains to inoculate Austrian Winter peas (Ruiz-Argüeso, Hanus, and Evans, 1978). By the exposure of pea nodules to a high O_2 atmosphere to suppress H_2 evolution through inactivation of the nitrogenase, nodules formed by several *R. leguminosarum* strains with efficiencies from 0.76 to 0.80 took up H_2 when it was supplied externally, whereas those nodules with relative efficiencies from 0.45 to 0.68 showed no H_2 uptake. Using the O_2 inactivation technique, Evans, Ruiz-Argüeso, and Russell (1978) demonstrated that nodules of red clover (*Trifolium pratense*) from the field catalyzed O_2-dependent H_2 uptake and therefore possessed hydrogenase activity that was insufficient to recycle all the H_2 produced by nitrogenase.

All of the 19 strains of *R. meliloti* that have been tested (Table 2) produced nodules that evolved appreciable H_2. Strains SU51, 102F65, and

ATCC 10312 on Alfalfa formed nodules that had the highest relative efficiencies, ranging between 0.80 and 0.83. Since C_2H_2 reduction rates were higher than rates of H_2 evolution under argon for nodules formed by these strains, Maier (unpublished data) in our laboratory has examined a series of *R. meliloti* strains for H_2 uptake under free-living conditions using methods found optimal for *R. japonicum* (Maier et al., 1978). Strain 102F65 showed a capacity for H_2 uptake and therefore could produce hydrogenase. Preliminary results indicate that some other strains may produce hydrogenase. Clearly, the level of hydrogenase activity in legume nodules is influenced greatly by the strain of *Rhizobium* used.

EFFECT OF PLANT HOST ON H_2 LOSS

Experiments by Dixon (1972) with *R. leguminosarum* strain 311 as an inoculum for *Pisum sativum* indicated that the legume host played a role in the extent of H_2 loss from nodules. Nodules formed on *P. sativum* by strain 311 exhibited hydrogenase activity but nodules produced by the same strain on *Vicia bengalensis* had no measurable activity (Dixon, 1972). In our laboratory, a limited number of alfalfa and clover cultivars have been compared on a series of several strains of rhizobia, but no consistent effect of cultivar on H_2 loss was observed (Table 2). Carter et al. (1978) have systematically examined the effect of soybean cultivars on H_2 loss and relative efficiencies of nodules. Five of seven cultivars inoculated with *R. japonicum* 3I1b 123 formed nodules, all of which lost H_2 at considerable rates; relative efficiency values ranged from 0.69 to 0.79 (Table 3). The same cultivars inoculated with 3I1b 143 formed nodules that lost no H_2 and had relative efficiency values near 1.0. Since these cultivars varied appreciably in genetic background, it was concluded that *R. japonicum* strain rather than plant cultivar plays a major role in controlling H_2 loss from soybean nodules. Undoubtedly, the environment within the nodule has a crucial role in the expression of hydrogenase activity in the bacteroids. This argument is supported by the conditions necessary for expression of hydrogenase in free-living rhizobia (see below).

ENVIRONMENTAL EFFECTS ON H_2 LOSS

Environmental conditions also influence the extent of H_2 evolution from nodules. Jennings (unpublished data) has studied the effect of temperature on H_2 evolution from nodules formed by hydrogenase-positive *R. japonicum* strain 3I1b 143 and observed no H_2 loss at 10°–25°C, but measurable loss of H_2 occurred at temperatures above 30°C. Bethlenfalvay and Phillips (1977a) reported a rise in relative efficiency values of pea nodules from 0.4 during early stages of growth to 0.7 after plants reached flowering. Ruiz-Argüeso, Hanus, and Evans (1978) have demonstrated that strain 128C53

Table 3. Effect of soybean cultivars and *R. japonicum* strains on hydrogen evolution from root nodules

Cultivar	H_2 evolution (μmol·hr^{-1}·g^{-1})[a] 3I1b 123	H_2 evolution (μmol·hr^{-1}·g^{-1})[a] 3I1b 143	Relative efficiency[b] 3I1b 123	Relative efficiency[b] 3I1b 143
Portage	4.1	0.0	0.71	1.0
Anoka	5.5	0.0	0.69	1.0
Amsoy 71	3.6	0.0	0.79	1.0
Peking	—	0.0	—	1.0
Bonus	4.1	0.0	0.74	1.0
Kent	—	0.0	—	1.0
Clark 63	4.2	0.0	0.77	1.0
Least significant difference (0.05)	1.3		0.15	

[a] Data are means of five replicate cultures in Leonard jar assemblies each inoculated with the strains as indicated. Control cultures without inoculum were not nodulated. H_2 evolution rates reported as 0.0 were below the minimum detectable rate of 0.1 μmol·hr^{-1}·g of fresh nodule.

[b] Relative efficiency is defined as the fraction of the electron flow through the nitrogenase system (rate of C_2H_2 reduction) that is used for N_2 reduction (Schubert and Evans, 1976). Data from Carter et al. (1978).

used in these experiments (Bethlenfalvay and Phillips, 1977a) possesses hydrogenase activity. Bethlenfalvay and Phillips' results may be explained if the hydrogenase system is inadequate to recycle all H_2 produced during the early part of the growth cycle, but recycles an increasing proportion of the H_2 evolved from nitrogenase during the latter part of the growth period when nitrogenase activity begins to decline. Bethlenfalvay and Phillips (1977b) also described an increase in the rate of H_2 evolution by inoculated peas as the intensity of the light was increased from 200 to 800 μEinstein·m^{-2}·sec^{-1}. In order to interpret these results, measurements of both nitrogenase and hydrogenase activities in nodules are needed. Differential inactivation of nitrogenase and hydrogenase in nodules of peas and red clover has been demonstrated (Evans, Ruiz-Argüeso, and Russell, 1978). The extent of H_2 evolution from nodules obviously depends on several factors, including the capacity of the organism to synthesize nitrogenase and hydrogenase and the effect of the environment and other factors on the stability and activity of the two enzymes during the growth period of the legume.

HYDROGENASE IN BACTEROIDS

Some of the basic properties of the hydrogenase reaction catalyzed by soybean nodule bacteroids have been established by McCrae, Hanus, and Evans (1978). All the activity within the nodule is located in the bacteroids. Also, hydrogenase activity in macerated bacteroids is limited to the particulate fraction and no evidence for a reversible hydrogenase has been observed. In addition to O_2, methylene blue, ferricyanide, and phenazine methosulfate functioned as electron acceptors for the hydrogenase complex in *R. japonicum* bacteroids (McCrae, Hanus, and Evans, 1978) and in particulate preparations of hydrogenase (Emerich et al., 1978). The ratio of H_2 uptake to H_2-dependent O_2 uptake was approximately 2, a value that is consistent with previous data for *R. leguminosarum* (Dixon, 1968). The estimated K_m values for H_2 and O_2 were 2.8 and 1.3 μM, respectively (McCrae, Hanus, and Evans, 1978). From the effect of carbon substrates and inhibitors of the electron transport chain on hydrogenase activity in intact bacteroids from *R. japonicum* USDA 122 DES, Emerich et al. (1978) concluded that the O_2-dependent H_2 uptake pathway involved cytochromes and interacts with the respiratory electron transport system (see also "Benefits of the Hydrogenase System").

HYDROGENASE IN FREE-LIVING RHIZOBIA

Until recently, no one had successfully demonstrated expression of hydrogenase activity by rhizobia grown on laboratory media. Schubert and Evans (1977) stated that McCrae (unpublished data) had observed H_2-dependent methylene blue reduction by *R. japonicum* 3I1b 110 and cowpea *Rhizobium* 32H1. The medium for growth of rhizobia in these tests contained yeast extract, mineral salts, mannitol, and 1% H_2 supplied in the gas phase. Maier et al. (1978) have recently defined the conditions for reproducible expression of hydrogenase activity by *R. japonicum*. Lim (1978) has also reported some conditions necessary for expression of hydrogenase activity by free-living *R. japonicum*. The optimum conditions for expression of hydrogenase in *R. japonicum* (Maier et al., 1978) include a low level of carbon substrates, an appropriate concentration of glutamate, and an O_2 concentration near 1%. *Rhizobium japonicum* grown in a medium where N_2-fixing activity was not detected required a preincubation period with H_2 for expression of hydrogenase activity. After optimum conditions were established for hydrogenase expression by free-living *R. japonicum*, cells were harvested and activity was easily demonstrated during a 3-min assay by use of the amperometric technique (Figure 2). H_2 utilization was rapid in a suspension supplied with O_2 but stopped abruptly when the O_2 supply was exhausted. Obviously, the bacteria carry out the oxyhydrogen reaction. Properties and regulation of the hydrogenase system in free-living rhizobia are now under investigation.

Development of methods whereby hydrogenase activity may be consistently demonstrated in cultures of free-living rhizobia has contributed greatly toward the isolation of *R. japonicum* mutants lacking hydrogenase activity (Maier, 1978; Maier et al., 1978). Strains of *R. japonicum* that have

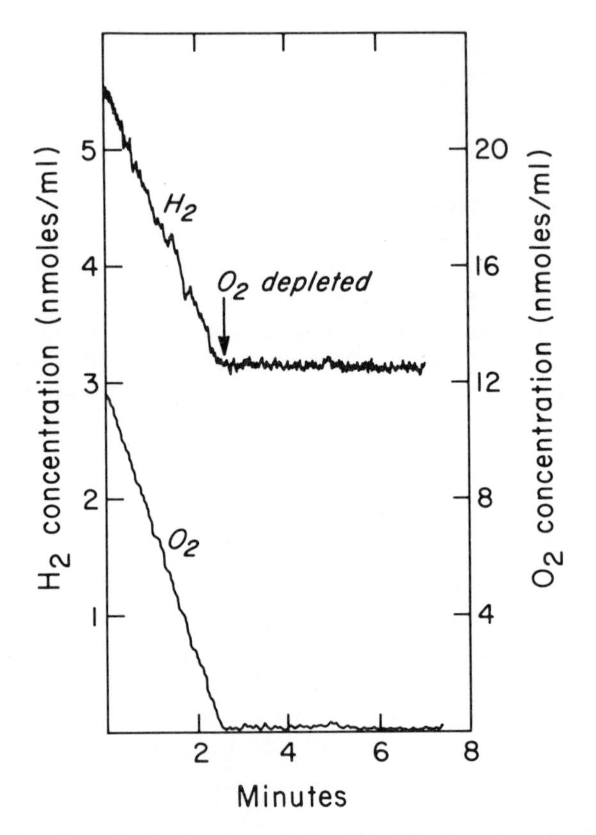

Figure 2. Oxygen-dependent hydrogen uptake by *Rhizobium japonicum* from free-living cultures. *R. japonicum* strain 651 (an antibiotic-resistant mutant of strain 122 DES) was cultured on a medium for expression of hydrogenase as described by Maier et al. (1978). Cells were grown for 2 days, removed from slants, suspended in 0.05 M phosphate buffer, exposed to 1% H_2, 1% O_2, and 0.5% CO_2 for 15 hr, then assayed for H_2 and O_2 uptake as described by Maier et al. (1978). The assay mixture of 4.8 ml contained 3.2×10^8 cells per ml. (This experiment was conducted by F. J. Hanus of our laboratory.)

the capacity to express hydrogenase activity reduce triphenyl tetrazolium chloride more rapidly on an agar medium than those that fail to express hydrogenase activity either as bacteroids or as free-living cultures (Maier et al., 1978). Selection of isolates by Maier (Maier, 1978; Maier et al., 1978) on the basis of their limited capacities to reduce triphenyl tetrazolium chloride has been used in a procedure for selecting mutants that fail to oxidize H_2. The symbiotic properties of the mutants appear no different from those of the parent organism with the exception that nodules formed by the mutants evolved H_2 and their bacteroids had no hydrogenase activity. Mutants lacking hydrogenase are essential in the evaluation of benefits of the hydrogenase system to the N_2-fixing process in legumes. Also, they are needed in the elucidation of the biochemical properties of the hydrogenase complex.

RELATIONSHIP OF H_2 LOSS TO HYDROGENASE ACTIVITY

That the H_2 loss from nodules is associated with an insufficiency of hydrogenase activity in nodule bacteroids is indicated by experiments with soybeans (Carter et al., 1978; McCrae, Hanus, and Evans, 1978) and peas (Ruiz-Argüeso, Hanus, and Evans, 1978). An example of this relationship is presented in Table 4, where four hydrogenase-positive strains of *R. japonicum* are compared with four randomly selected hydrogenase-negative strains. Nodules produced on Portage soybeans by the hydrogenase-negative strains lost H_2 in air at substantial rates; bacteroids from nodules formed by these strains produced no detectable hydrogenase activities, and free-living cultures of the hydrogenase-negative strains either produced no hydrogenase activity or very low levels of activity. In contrast, the hydrogenase-positive strains produced nodules that lost little or no H_2, and both bacteroids and cells from free-living cultures of these strains exhibited easily measurable hydrogenase activities. This and other information provide strong evidence that the rhizobia possess the genetic information for hydrogenase synthesis, that hydrogenase activity in nodules is located in the bacteroids, and that H_2 recycling via the bacteroid hydrogenase prevents or minimizes H_2 loss from nodules. The experiments of Maier (1978) in which an antibiotic-marked

Table 4. Relationship of hydrogen loss from soybean nodules to hydrogenase activities in bacteroids and free-living cultures of *R. japonicum*[a]

Strains[b]	H_2 loss from nodules (μmol/hr/g fresh wt)	Hydrogenase activity of bacteroids (μmol/hr/mg of protein)	Hydrogenase activity of free-living cultures (ηmol/hr/mg of protein)
USDA 16 −	6.9	<0.002	2.6
USDA 117 −	7.0	<0.002	0.0
USDA 120 −	7.2	<0.002	0.0
USDA 135 −	4.5	<0.002	—
3I1b 110 +	0.2	8.770	46
USDA 122 +	0.0	3.730	39
3I1b 6 +	0.1	9.680	12
3I1b 143 +	0.0	6.134	45

[a] Rates of H_2 loss from nodules from Carter et al. (1978), hydrogenase activities of nodule bacteroids from McCrae, Hanus, and Evans (1978), and hydrogenase activities of cultures of free-living *R. japonicum* by Maier et al. (1978).

[b] Hydrogenase-negative and hydrogenase-positive strains are identified by the symbols − and +, respectively.

hydrogenase-negative mutant of *R. japonicum* 122 DES has been utilized are consistent with this conclusion.

BENEFITS OF THE HYDROGENASE SYSTEM

There is considerable evidence that metabolism of H_2 via the hydrogenase in blue-green algal heterocysts may supply reducing power and ATP for support of the nitrogenase reaction (Bothe, Tennigkeit, and Eisbrenner, 1977; Bothe et al., 1977; Tel-Or, Luijk, and Packer, 1977; Peterson and Burris, 1978). Also, the oxyhydrogen reaction in blue-green algae provides a means of protecting nitrogenase from O_2 damage (Bothe, Tennigkeit, and Eisbrenner, 1977; Bothe et al., 1977). H_2 utilization via the hydrogenase system in *Azotobacter chroococcum* also provides respiratory protection for nitrogenase, especially when carbon substrates are limiting and can produce the energy to support the nitrogenase reaction in *Azotobacter* (Walker and Yates, 1978). At this conference, Emerich et al. (1978) have presented data showing that H_2 supports C_2H_2 reduction by nitrogenase in the hydrogenase-containing bacteroids formed by USDA 122 DES. H_2 oxidation in suspensions of these bacteroids definitely protected the nitrogenase from O_2 inactivation and resulted in a 40% increase in the steady state level of ATP within the cells (Emerich et al., 1978). When the oxidation of endogenous substrates within bacteroids was suppressed by use of iodoacetate, H_2 uptake was not appreciably inhibited, but the steady state level of ATP was increased fourfold or more from the addition of H_2. In parallel experiments in which hydrogenase-negative USDA 117 bacteroids from soybean nodules were utilized, no effects of adding H_2 were observed. McCrae, Hanus, and Evans (1978) have shown that the addition of 70 μM H_2 to bacteroid suspensions resulted in an average of 60% decrease in the rate of CO_2 evolution from the metabolism of endogenous substrates. The oxidation of H_2 therefore resulted in the conservation of endogenous substrates. Previously, it had been shown that a supply of H_2 to *R. leguminosarum* bacteroids had a sparing effect on the oxidation of added succinate (Dixon, 1972).

Figure 3 shows a very low level of nitrogenase activity in USDA 122 DES bacteroids that were not provided with H_2 or other exogenous substrates. Without H_2 the optimum O_2 level for nitrogenase activity was 0.002 atm, and greater concentrations inhibited strongly. In contrast, when H_2 was added (10% in the gas phase), the C_2H_2 reduction rate increased strikingly as the O_2 concentration increased, reaching an optimum level at 0.01 atm. Further increases in pO_2 caused decreases in activity. Thus, the presence of hydrogenase enabled H_2 to support nitrogenase activity in bacteroids and provided a means of respiratory protection. Other evidence by Emerich et al. (1978) indicates that maintenance of ATP supply is a major benefit from H_2 oxidation. The utilization of H_2 via the hydrogenase

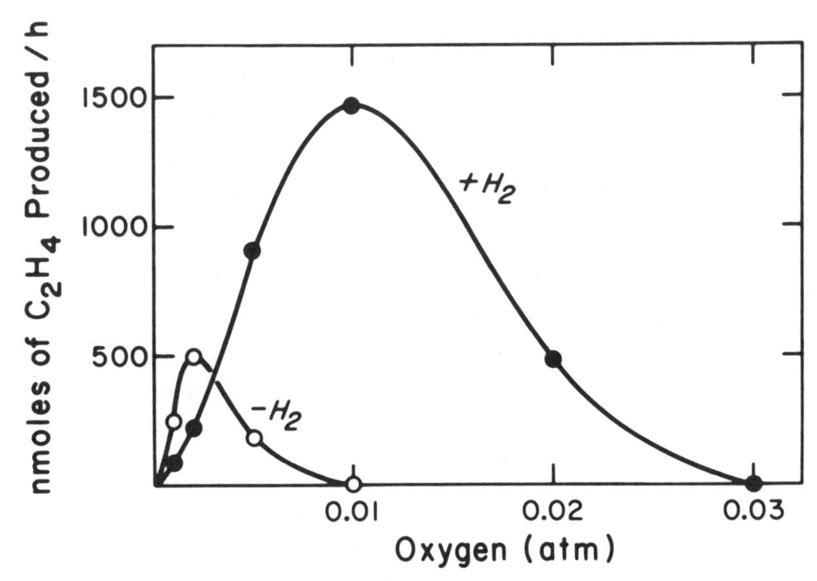

Figure 3. Effect of hydrogen on the nitrogenase activity of soybean nodule bacteroids from *R. japonicum* strain USDA 122 DES. The graph represents the rates of C_2H_2 reduction by bacteroids provided with different O_2 concentrations in the presence (●) and in the absence (○) of 10% H_2. Assays were conducted in 21-ml vials containing 2.4 ml of HEPES-Mg-phosphate buffer (50 mM HEPES, 1 mM $MgCl_2$, 1 mM K_2HPO_4; pH 7.5). The atmosphere consisted of 10% C_2H_2, O_2 and H_2 as indicated, and sufficient argon to obtain 1 atm. Reactions were started by injection of 0.1-ml bacteroids (7.9 mg dry wt). Vials were incubated on a shaker at 23°C. Gas samples were removed for C_2H_4 determination by gas chromatography. Rates of C_2H_4 reduction at O_2 concentrations below the optimum were linear for 1 hour (Emerich et al., 1978).

system in soybean bacteroids therefore provides a means for respiratory protection of nitrogenase, produces energy for support of nitrogenase activity, and conserves carbon substrates. The interrelationships in a legume nodule of carbohydrate utilization, generation of ATP and reductant for support of nitrogenase activity, evolution of H_2 as a by-product of the nitrogenase reaction, and recycling of this H_2 through the oxyhydrogen reaction via a particulate hydrogenase are illustrated in Figure 4. Nodules without the unidirectional hydrogenase system lose H_2 to the atmosphere and cannot benefit from all the potential advantages of the H_2-recycling process.

The crucial test of whether H_2-recycling capability increases the N_2-fixing process in legumes must be determined in comparisons of hydrogenase-positive and -negative rhizobia as inocula in replicated growth experiments. A summary of three greenhouse experiments and one field trial in which the effect of hydrogenase-positive and -negative rhizobial strains on yield and total nitrogen contents of legumes was measured is presented in Table 5. In greenhouse experiments, cowpeas and soybeans inoculated with hydrogenase-positive strains showed statistically significant increases in yield of dry matter and total nitrogen fixed in all cases. In the field, the mean nitrogen content of soybean seed produced by plants inoculated with hydrogenase-positive strains was 13.7% higher than that of beans produced by plants inoculated with the hydrogenase-positive strains. This difference, which was obtained in a comparison of 20 replicate field plots inoculated with a group of hydrogenase-positive strains versus the same number of plots inoculated with hydrogenase-negative strains, is statistically significant at the 5% confidence level. A significant positive correlation was observed for the relationship between nitrogen content of plants from pot cultures and relative efficiency values for nodules inoculated with a series of different *R. japonicum* strains (correlation coefficient = 0.92). An analogous field experiment also showed a significant positive correlation (correlation coefficient = 0.81) (see Table 5; Albrecht et al., 1978). Considering the results of the four experiments, a consistent positive effect of inoculation with hydrogenase-positive strains of rhizobia on the N_2-fixing process has been observed. The mutant strains of *R. japonicum* that lack the hydrogenase phenotype are now available in our laboratory and these are being compared with the parent strain for more accurate assessment of the benefits of the hydrogenase complex to the N_2-fixing process.

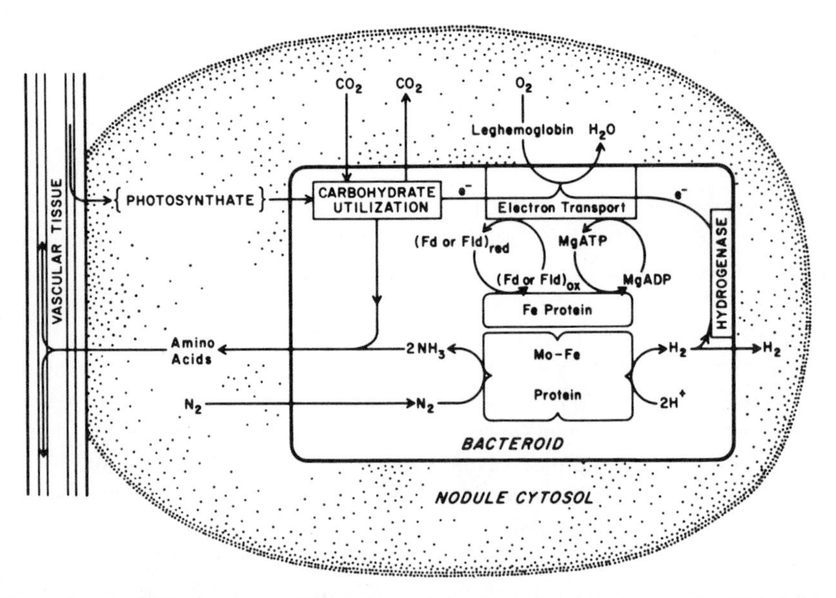

Figure 4. A schematic representation of the relationships of carbohydrate utilization, electron transport, nitrogenase activity, and hydrogen recycling in a legume root nodule. Nodules without the unidirectional hydrogenase system lose energy through H_2 evolution.

Table 5. Effect of hydrogenase-positive *Rhizobium* strains on yield and nitrogen content of legumes

Type of experiment	Inocula compared	Increase from H_2-positive inocula: Yield (%)	Increase from H_2-positive inocula: Nitrogen content (%)
Soybeans, greenhouse[a]	USDA 31 (H_2ase negative) vs. 3I1b 110 (H_2ase positive)	24[c]	31[c]
Soybeans, greenhouse[b]	5 H_2ase positive vs. 5 H_2ase negative[b]	13[c]	23[c]
Soybeans, field[b]	4 H_2ase positive vs. 4 H_2ase negative[b]	5 (beans)	13[c] (beans)
Cowpea, greenhouse[a]	176A27 (H_2ase negative) vs. 176A28 (H_2ase positive)	12[c]	15[d]

[a] Details of these experiments are described by Schubert, Jennings, and Evans (1978).

[b] The field experiment consisted of 20 replicate plots (five each of four hydrogenase-positive and four hydrogenase-negative strains). The hydrogenase-negative strains were: USDA 16, USDA 117, USDA 135, and USDA 20. These were compared with the following hydrogenase-positive strains: 3I1b 110, USDA 122, 3I1b 143, and 3I1b 6. The strain comparisons made in the greenhouse (with six replicate cultures of each strain) were the same as those used in the field, except that USDA 136 was included in the group of hydrogenase-positive strains and USDA 3 in the hydrogenase-negative strains (Albrecht et al. 1978).

[c] Statistically significant at the 5% level.

[d] Statistically significant at the 1% probability level. All data are yields and nitrogen contents of entire plant shoots unless otherwise indicated.

CONCLUSIONS

1. Cell-free preparations of nitrogenase from different sources evolve H_2 concomitant with N_2 reduction, and the magnitude of this loss usually amounts to 25% to 35% of the electron flow through the nitrogenase system.
2. A survey of 77 samples of nodules from legumes inoculated with commercial or native rhizobia lost H_2 equal to 44% of the electron flow through nitrogenase. With over 90 *Rhizobium* strains used for inoculation of a series of legumes grown under controlled conditions, a mean of 27% of the electron flow through nitrogenase was lost as H_2. Since the energy supply to nodules is a major factor limiting N_2 fixation by legumes, ATP- and reductant-dependent H_2 produced by nitrogenase represents inefficient energy utilization by nodulated legumes.
3. Seven strains of *R. japonicum*, 10 strains of cowpea rhizobia, and the endophytes from three nonlegumes form nodules that lose little or no H_2. All of these nodules exhibit a capacity for O_2-dependent H_2 uptake.

4. Some strains of *R. leguminosarum*, *R. meliloti*, and *R. trifolii* form nodules that lose H_2 but possess a capacity to recycle some of the H_2 that is produced by the nitrogenase system.
5. The bacteroids from nodules that take up exogenous H_2 exhibit a capability to carry out the unidirectional oxyhydrogen reaction. The capacity to recycle the H_2 produced from nitrogenase in bacteroids may conserve carbon substrates, protect nitrogenase from O_2 damage, and produce energy during H_2 oxidation via the electron transport chain for use in support of the nitrogenase reaction.
6. Environmental and other factors influence H_2 production from nitrogenase and H_2 recycling via hydrogenase. A major factor controlling H_2 evolution from nodules, however, is the capacity of the *Rhizobium* strain to synthesize the hydrogenase system.
7. Conditions have been developed whereby free-living cultures of *R. japonicum* consistently express hydrogenase activity. A low level of carbon substrates, an optimum concentration of glutamate, a low O_2 concentration, and a source of H_2 are the major factors necessary for hydrogenase expression in laboratory media.
8. Strains of *R. japonicum* that produce nodules that fail to lose appreciable H_2 exhibit hydrogenase activity in bacteroids and in cells grown under free-living conditions. In contrast, *Rhizobium* strains that produce nodules that lose H_2 at rapid rates (relative efficiencies less than 0.75) show little or no hydrogenase activity as bacteroids from nodules or as cell suspensions from free-living cultures.
9. In two greenhouse experiments with soybeans, one with cowpeas, and a field experiment with soybeans, individual comparisons of hydrogenase-positive and -negative strains or comparisons of groups of randomly selected hydrogenase-positive and -negative strains have been made. Inoculation with the hydrogenase-positive strains has produced consistent and statistically significant increases in yield and in total nitrogen content of plants in greenhouse experiments and in total nitrogen content of soybean seed in the field experiment. In both field and greenhouse experiments where groups of hydrogenase-positive and -negative strains were compared, significant positive correlations ($r = 0.92$) have been demonstrated between nitrogen content of plants and relative efficiency values in greenhouse experiments and between nitrogen content of harvested beans and relative efficiency values ($r = 0.81$) in the field experiment.
10. Mutant strains of *R. japonicum* that lack the hydrogenase phenotype have been developed and are being used in the assessment of the benefits of the hydrogenase system for N_2 fixation. Also, they are being used in the biochemical characterization of the hydrogenase system.

11. Selection of rhizobial strains for inocula has been based on desirable characteristics that include effectiveness, competitiveness, and ability to survive under adverse conditions. In addition to these qualities, we recommend that rhizobial strains chosen for inocula have a capacity to form nodules that recycle H_2.

ACKNOWLEDGMENTS

The authors express appreciation to Mr. Joe Hanus, Mr. Sterling Russell, Ms. Nancy Jennings, Mr. Kevin Carter, and Mr. Frank Simpson for contributions to the research and to Mrs. Flora Ivers for typing the manuscript. T.R.A. wishes to thank the Program of Cultural Cooperation between the United States and Spain for support.

REFERENCES

Albrecht, S. L., S. A. Russell, J. Hanus, D. W. Emerich, and H. J. Evans. 1978. In: Proceedings of the Steenbock-Kettering International Symposium on Biological Nitrogen Fixation, June 12–16, 1978, Madison, Wisconsin. Abstract A-81, p. 41.

Bethlenfalvay, G. J., and D. Phillips. 1977a. Plant Physiol. 60:868–871.

Bethlenfalvay, G. J., and D. A. Phillips. 1977b. Plant Physiol. 60:419–421.

Bothe, H., J. Tennigkeit, and G. Eisbrenner. 1977. Arch. Microbiol. 114:43–49.

Bothe, H., J. Tennigkeit, G. Eisbrenner, and M. G. Yates. 1977. Planta 133:237–242.

Bulen, W. A., and J. R. LeComte. 1966. Proc. Natl. Acad. Sci. USA 56:979–986.

Burns, R. C., and R. W. F. Hardy. 1975. Nitrogen Fixation in Bacteria and Higher Plants. Springer-Verlag, New York.

Burris, R. H. 1977. In: A. San Pietro and A. Mitsui (eds.), Biological Solar Energy Conversions. Academic Press, Inc. New York.

Carter, K. R., N. T. Jennings, J. Hanus, and H. J. Evans. 1978. Can. J. Microbiol. 24:307–311.

Dixon, R. O. D. 1967. Ann. Bot. 31:179–188.

Dixon, R. O. D. 1968. Arch. Microbiol. 62:272–283.

Dixon, R. O. D. 1972. Arch. Microbiol. 85:193–201.

Emerich, D. W., T. Ruiz-Argüeso, TeMay Ching, and H. J. Evans. 1978. In: Proceedings of the Steenbock-Kettering International Symposium on Biological Nitrogen Fixation, June 12–16, 1978, Madison, Wisconsin. Abstract B-97, p. 49.

Evans, H. J., T. Ruiz-Argüeso, N. T. Jennings, and J. Hanus. 1977. In: A. Hollaender (ed.), Genetic Engineering for Nitrogen Fixation, pp. 333–335. Plenum Publishing Corp., New York.

Evans, H. J., T. Ruiz-Argüeso, and S. A. Russell. 1978. In: J. Dobereiner, R. H. Burris, and A. Hollaender (eds.), Limitations and Potentials for Biological Nitrogen Fixation in the Tropics, pp. 209–222. Plenum Publishing Corp., New York.

Hardy, R. W. F., and U. D. Havelka. 1975. Science 188:633–643.

Hoch, G. E., K. C. Schneider, and R. H. Burris. 1960. Biochim. Biophys. Acta 37:273–279.

Hyndman, L. A., R. H. Burris, and P. W. Wilson. 1953. J. Bacteriol. 65:522–531.

Lim, S. T. 1978. In: Abstracts of the Annual Meeting of the American Society for Microbiology, May 14–19, Las Vegas, Nevada.
Maier, R. J. 1978. In: Proceedings of the Steenbock-Kettering International Symposium on Biological Nitrogen Fixation, June 12–16, 1978, Madison, Wisconsin. Abstract C-57, p. 38.
Maier, R. J., N. E. R. Campbell, F. J. Hanus, F. B. Simpson, S. A. Russell, and H. J. Evans. 1978. Proc. Natl. Acad. Sci. USA 75:3258–3262.
McCrae, R. E., J. Hanus, and H. J. Evans. 1978. Biochem. Biophys. Res. Commun. 80:384–390.
Peterson, R. B., and R. H. Burris. 1978. Arch. Microbiol. 116:125–132.
Rivera-Ortiz, J. M., and R. H. Burris. 1975. J. Bacteriol. 123:537–545.
Ruiz-Argüeso, T., J. Hanus, and H. J. Evans. 1978. Arch. Microbiol. 116:113–118.
Schubert, K. R., J. A. Engelke, S. A. Russell, and H. J. Evans. 1977. Plant Physiol. 60:651–654.
Schubert, K. R., and H. J. Evans. 1976. Proc. Natl. Acad. Sci. USA 73:1207–1211.
Schubert, K. R., and H. J. Evans. 1977. In: W. E. Newton, J. R. Postgate, and C. Rodriguez-Barrueco (eds.), Recent Developments in Nitrogen Fixation, pp. 469–487. Academic Press, London.
Schubert, K. R., N. T. Jennings, and H. J. Evans. 1978. Plant Physiol. 61:398–401.
Tel-Or, E., L. W. Luijk, and L. Packer. 1977. FEBS Lett. 78:49–52.
Walker, C., and M. G. Yates. 1978. Biochimie 60:225–231.
Wilson, P. W., and R. H. Burris. 1947. Bacteriol. Rev. 11:41–73.
Winter, H. C., and R. H. Burris. 1968. J. Biol. Chem. 243:940–944.
Winter, H. C., and R. H. Burris. 1976. Annu. Rev. Biochem. 45:409–426.
Zumft, W. G., and L. E. Mortenson. 1975. Biochim. Biophys. Acta 416:1–52.

Nitrogen Fixation, Volume II
Edited by W. E. Newton and W. H. Orme-Johnson

Control of Morphogenesis and Differentiation of Pea Root Nodules

W. Newcomb

There have been few recent reports on the morphogenesis and differentiation of leguminous root nodules. This is somewhat surprising because we have relatively meager knowledge of the morphogenetic factors and controls involved in their development. Root nodules are unique plant organs not only because they are the site of symbiotic nitrogen fixation but also because nodule development involves the physical and chemical interactions of two organisms—a soil bacterium and a vascular plant, a dicotyledonous angiosperm. Nodules are complex organs structurally, metabolically, and developmentally and are the product of an intimate association of two very different organisms. The leguminous plant whose nodular development has been most intensively studied is the garden pea, *Pisum sativum* L. This paper sets out to present a coherent account of the current knowledge of the developmental biology of pea nodules and is not intended to be a comprehensive survey of the literature of leguminous nodules, which has been superbly reviewed recently by Dart (1975, 1977). There is considerable interest in using the exciting and still developing techniques of somatic cell genetics to convert some non–nitrogen-fixing crop plants into plants capable of meeting some of their nitrogen requirements via symbiotic nitrogen fixation. This approach may well involve the development of nodules on new

The observations of the author reported in this paper were made in the Laboratories of Professor Torrey at Harvard Forest, Dr. Otto Stein at the Amherst campus of the University of Massachusetts, and Dr. R. L. Peterson at the University of Guelph, and were made possible by the financial support of the Maria Moors Cabot Foundation for Botanical Research at Harvard University as well as grants from the U.S. National Science Foundation to J. G. T. and the National Research Council of Canada to R. L. P. Additional financial support was in the form of a Queen's University Research Award.

species, and thus an understanding of the developmental biology of nodule morphogenesis could prove to be valuable in attaining that goal.

NODULE ONTOGENY

Initial morphological changes involve deformation of growing root hairs by substances excreted by rhizobia in the rhizosphere and the subsequent invasion of these hairs by rhizobia. The rhizobia enter the deformed hairs at the base of a fold, possibly through a pore (Newcomb, Sippell, and Peterson, 1979), although there is no ultrastructural evidence of this for the pea, as there is for *Trifolium* (Napoli and Hubbell, 1975). Within the root hair cell, the rhizobia are encapsulated within an infection thread, a tubelike structure consisting of an outer plant cell wall and an inner mucopolysaccharide matrix (Dart, 1975) in which the rhizobia are embedded. In other species, formation of the infection thread may be due to a reorientation of cell wall growth at the site of microbial entry (Nutman, 1956; Napoli and Hubbell, 1975). Unfortunately, no fine structural study showing development of growing infection threads in infected root hairs has been published. Admittedly such a study would be technically difficult but, coupled with cytochemical and enzyme histochemical procedures, it could resolve some of the mysteries of this stage. For instance, is the so-called pore really an aperture or merely a chemically altered region of the cell wall? Can the pore form only at certain sites on the root hair cell wall? If the site of penetration is chemically altered, which organism, the plant or the bacterium, produces the necessary enzymes? What organelles are involved with the synthesis and deposition of the infection thread cell wall? Many accounts for other species indicate that the root hair cell nucleus is situated near the growing tip of the infection thread and the two move together toward the base of the cell (Dart, 1975, 1977). What organelles (if any) are associated with the nuclear movement? What happens to nuclear movement in root hairs having branched infection threads or more than one infection thread?

Almost all leguminous root nodules develop from an infected root hair and in many, but not all, nodules rhizobia are distributed to various cortical cells by the invasion of infection threads and host cell mitoses (Dart, 1975, 1977). Pea cells may divide after being invaded by an infection thread but not after rhizobia are liberated from the thread into the pea cytoplasm. Therefore, in pea nodules only cells invaded by an infection thread become infected with rhizobia. What substance(s) are responsible for the attraction of the infection thread to certain cortical cells? Are the same biochemical changes involved in the invasion of root hair and cortical cells? So far, the answers to these questions have eluded us.

In 5-day-old pea nodules, the rhizobia are restricted to the infection threads within the deformed root hair (Figure 1). Already, considerable cytological changes have occurred within the root cortex, the most notable being a marked size increase in the outer cortical cells, much mitotic activity in the inner and middle cortical cells, and an increase in nucleolar size and number. Seven- and eight-day-old pea nodules contain more cells and are just beginning to become macroscopic as they emerge from the root tissue (Newcomb, Sippell, and Peterson, 1979). The infection threads have now grown between and through the cortical cells and their derivatives penetrated the innermost cortical cells, into which rhizobia are released from the infection thread. Mitotic activity continues in the middle cortex in a group of cells known collectively as the nodule meristem. Infection threads continue to invade cells derived from the nodule meristem, and rhizobia are released into the host cytoplasm (see Figure 3). These cells then undergo a phase of symbiotic growth and differentiation during which the rhizobia increase in number while the host cells increase in size, numbers of organelles, and starch granules in amyloplasts. The rhizobia enlarge and become pleomorphic at the approximate time that the amount of starch in

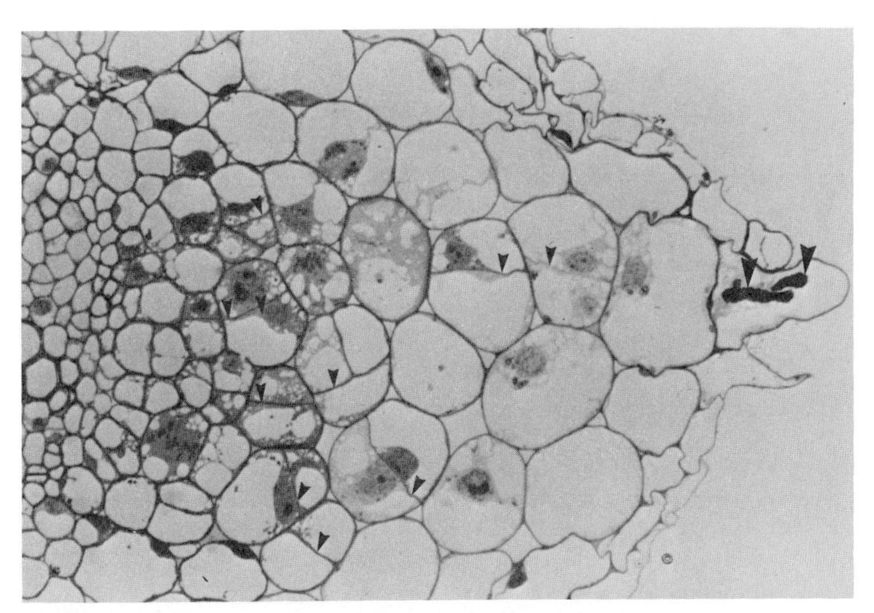

Figure 1. Light micrograph of a 5-day-old pea nodule illustrating the infection thread (large arrows) restricted to the infected root hair cell, the enlarged outer and middle cortical cells, new cell walls (small arrows) in cortical cells that have recently undergone mitosis, several mitotic figures, many prominent nucleoli, and intercellular spaces. (×440)

the amyloplasts diminishes; the enlarged rhizobia cells are known as bacteroids (see Figure 4) and are the stage that reduces molecular dinitrogen. The differentiated infected cells have a functional life of about 7 to 10 days prior to senescing. The enlargement of the infected cortical cells plus the continued invasion of cells derived from the nodule meristem push the nodule meristem outward, producing an emergent cylindrically shaped root nodule. Such a nodule consists of a central zone of infected cells and a distal region of nodule meristem cells, both of which are surrounded by a covering of large vacuolate nodule cortex cells (Figure 2).

Fine structural studies have provided more detailed descriptions of the various stages of pea nodule development, but the physiological and biochemical changes associated with the structural changes remain unclear. The infection thread appears to start as an invagination of the host primary cell wall and sometimes may traverse a cell or branch into adjacent cells (Figure 3). The source of the wall components for this growth is unclear, as are the biochemical changes (if any) occurring within the wall, but it appears evident that the host cell wall is not plastic enough to wholly account for intracellular infection thread growth. Undoubtedly, enzymatically mediated biochemical changes of the wall occur, but so far these have not been demonstrated in pea nodules. Since these changes are likely to involve very localized reaction sites, they may be difficult to assay and document (Hunter and Elkan, 1975). It is conceivable that plant growth substances, especially auxins, could mediate the wall changes associated with the formation of the infection thread within the deformed root hair and the subsequent invasion of the cortex. The release of the rhizobia only occurs from the unwalled regions of the infected thread, but what is not clear is why these regions are lacking a wall: lytic vesicles may be involved with rhizobial release (Truchet and Coulomb, 1973). Is the wall lacking because the infection thread grows faster than the deposition of the cell wall components or has the wall been removed (enzymatically)? In soybean nodules, it has been suggested that the rhizobium release is accompanied by wall degradation and compartmentalization of the wall components in membranous vesicles (Bassett, Goodman, and Novacky, 1977). Biochemical studies coupled with histochemical techniques suggest that the rhizobia produce pectinase and the plant cytoplasm synthesizes cellulase to degrade the host cell wall in a controlled cooperative manner (Verma, Zogbi, and Bal, 1978).

The release of the rhizobia into the pea cytoplasm brings about a number of events. Pea cortical cells will undergo cell division when they are invaded by an infection thread but not after rhizobia have escaped from the infection thread (Newcomb, 1976). This is true even if the number of "free" bacteria is relatively low and thus they are obviously not capable of physically restricting mitosis. In addition to the apparent inhibition of mitosis, possibly by a supraoptimal cytokinin concentration similar to what

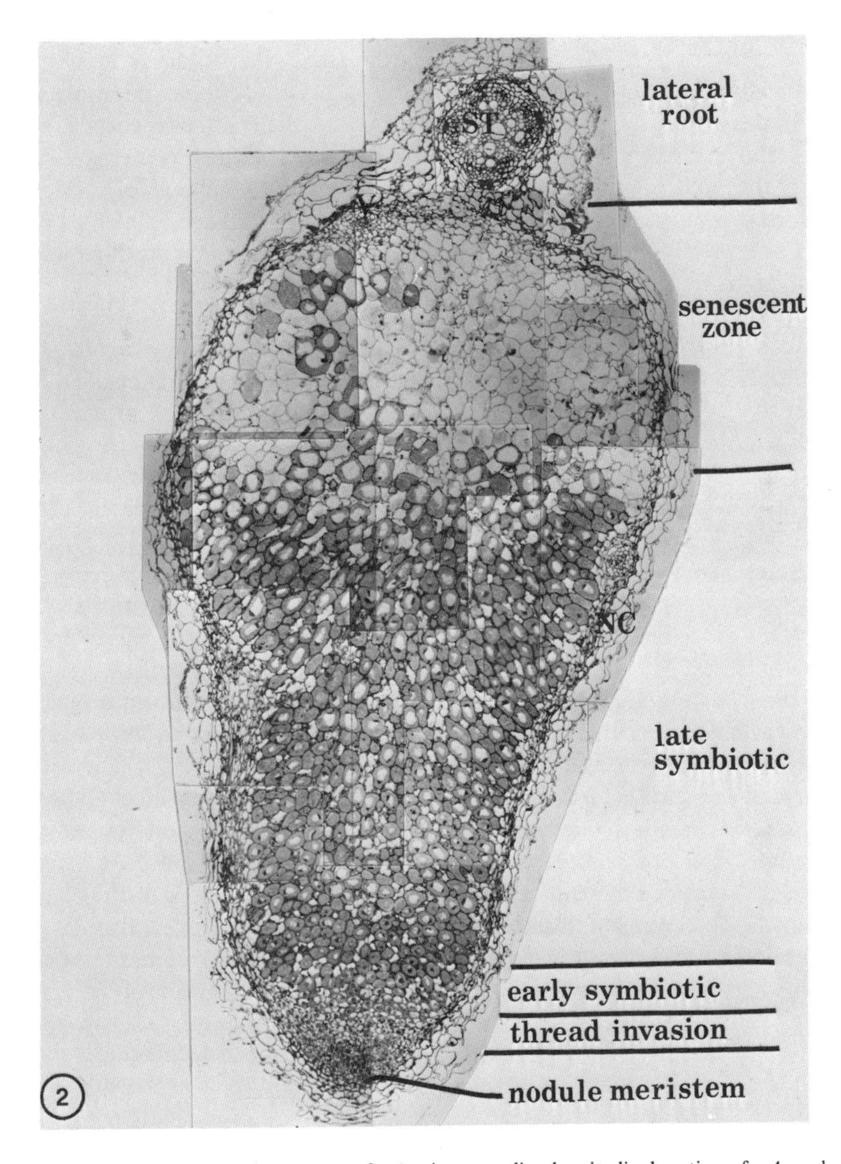

Figure 2. Light micrograph montage of a 1-micron median longitudinal section of a 4-week-old pea nodule illustrating the nodule meristem, the region of infection thread invasion, the infected cells of the early and late symbiotic growth phases, and the senescent region. The nodule is attached to a lateral root, cut transversely revealing a tetrarch pattern in the stele, and surrounded by several layers of nodule cortex (NC) cells. (×60) (Reprinted with permission of the *Canadian Journal of Botany*.)

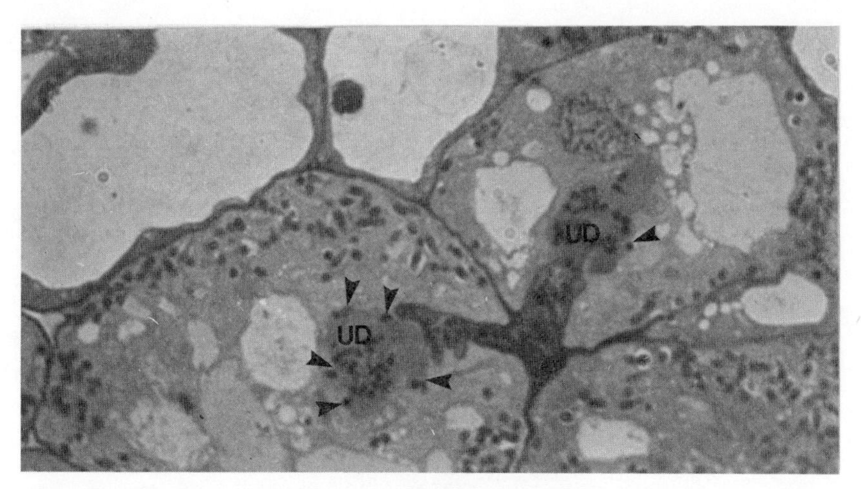

Figure 3. Light micrograph illustrating a branched infection thread invading two adjacent cells. Unwalled droplets (UD) of thread matrix are located at the ends of the thread and contain numerous rhizobia, some of which are in the process of escaping (small arrows) into the host cytoplasm. Many "free" rhizobia are already present in the pea cytoplasm. (×1800)

occurs in soybean callus cultures (Feldman, 1975), the biosynthetic activities of the pea cell are stimulated, resulting in more cytoplasmic organelles and inclusions, particularly rough endoplasmic reticulum, free ribosomes, mitochondria, and the membrane envelopes surrounding the growing rhizobia (Figure 4) (Newcomb, 1976). In an ineffective pea nodule caused by a mutation in the *Rhizobium leguminosarum* genome, membrane biosynthesis is apparently not sufficiently stimulated. The newly released rhizobia are surrounded by a membrane envelope but, after rhizobia growth and division plus biochemical turnover of membrane envelope, the rhizobia lack membrane envelopes (Figure 5); thus, it appeared that membrane biosynthesis occurred at a lower rate than bacteria growth and membrane degradation (Newcomb, Syōno, and Torrey, 1977).

INVOLVEMENT OF POLYPLOID CELLS IN NODULE INITIATION

Approximately 40 years ago, Wipf and Cooper (1940) put forth their now famous hypothesis that there was a relationship between the normal occurrence of polyploid cells and the ontogeny of pea nodules. Because of their effects on subsequent research in this field, their evidence and conclusions must be critically examined. Occasional polyploid cells were observed in the inner cortex of uninoculated roots in the region where root hairs are found. Wipf and Cooper also observed dividing nuclei, some of which were polyploid, in the inner cortex just beyond the growing tip of the infection thread. Also present in the inner cortex were interphase nuclei of various sizes. These workers suggested that the larger nuclei were polyploid and the smaller ones diploid. They reported infection threads passing through cortical cells, sometimes without stimulating mitosis, and on the basis of nuclear size and staining suggested that the infection thread had a pathological effect on diploid cells. On the basis of these observations, they suggested that when the infection thread approached a group of tetraploid cells, both the tetraploid and adjacent diploid cells are stimulated to divide and that this stimulation of tetraploid cells was evoked by a secretion or hormonal factor from the bacteria. Eventually, as they had observed earlier (Wipf and Cooper, 1938; Wipf, 1939), the infection thread continues growing toward and into the tetraploid meristematic cells, which become infected with rhizobia and subsequently develop into nitrogen-fixing bacteroid-containing cells.

A number of workers verified and added further details on the role of polyploid cells in the development of pea root nodules (Bond, 1948;

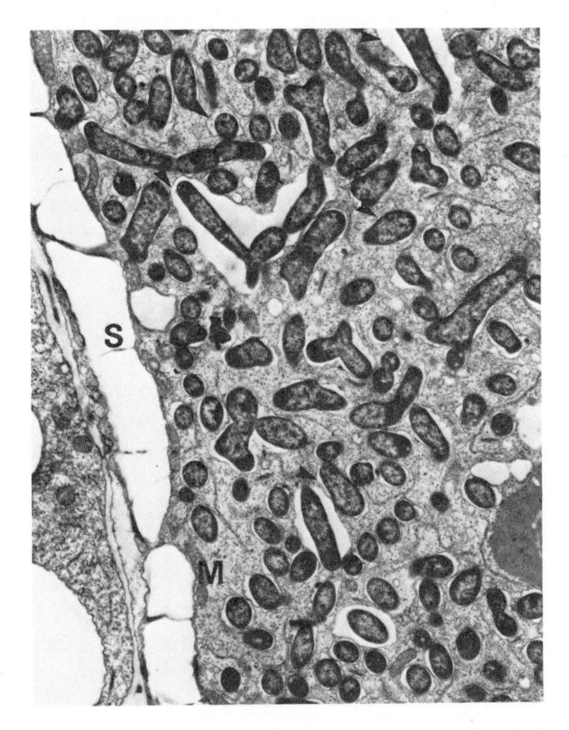

Figure 4. Transmission electron micrograph showing portions of an infected cell and the adjacent uninfected cell (UI). The infected cell contains many rhizobia differentiated into bacteroids that are surrounded by membrane envelopes (arrows), amyloplasts with flattened starch granules (S), and mitochondria (M). (×4260) (Reprinted with permission of the *Canadian Journal of Botany*.)

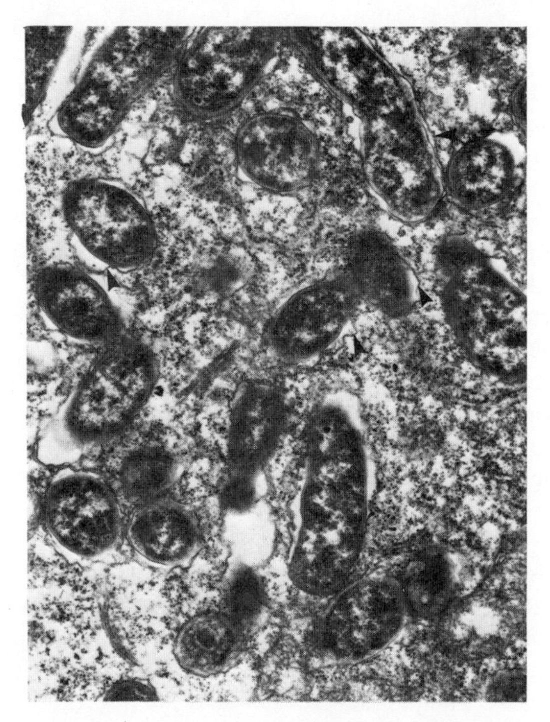

Figure 5. Transmission electron micrograph illustrating a portion of an infected cell in an ineffective pea nodule infected with a mutant strain (#1019) of *R. leguminosarum*. Shown are rhizobia that are not completely surrounded by a membrane envelope (arrows). (×13720) (Reprinted with permission of the *Canadian Journal of Botany*.)

Oinuma, 1948; Fujita and Mitsuishi, 1953; Torrey and Barrios, 1969). Bond (1948) reported that every infection thread she observed produced some effect on cortical cells; often it was the occurrence of tetraploid division figures near the tip of the growing thread. Torrey and Barrios (1969) utilized a squash technique and set out to determine when the first polyploid mitoses occurred during nodule initiation. They were unable to observe polyploid mitoses earlier than 6 days after inoculation; in some cases, these nuclei were endomitotic, mainly tetraploid with paired diplochromosomes and were regularly associated with an infection thread. Often, the thread terminated in cells undergoing a polyploid mitosis; infection threads were not associated with diploid nuclei. Unfortunately, but understandably, they had difficulty in identifying the earliest stages of nodules and thus did not report any information on the mitoses that occur when the infection thread is restricted to the infected root hair. Although they concluded that the infection thread specifically and selectively stimulated preexisting endomitotic tetraploid cortical cells to divide, the fact remains that many cortical cells are stimulated to divide prior to the entry of an infection thread into the dividing cell (Figure 1; Wipf and Cooper, 1940; Bond, 1948; Libbenga and Harkes, 1973; Newcomb, Sippell, and Peterson, 1978). Torrey and Barrios (1969) did, however, point out the lack of information regarding the point at which DNA synthesis in cortical cells is stimulated by the penetration of the infection thread. Mitchell (1965) used microspectrometric measurements to determine the relative DNA contents in the nuclei of an emergent developed pea nodule. The meristematic cells had 4 c and 8 c levels of DNA, the bacteroid-containing cells had 8 c and 16 c levels, and the nodule cortex 2 c and 4 c levels. These data also substantiated the polyploid cell hypothesis of Wipf and Cooper (1940) that tetraploid cells proliferate in the nodule meristem, and eventually become infected with rhizobia. Evidently, the latter cells undergo further endoreduplication, accounting for their higher values of DNA.

Thus, there is no doubt that preexisting polyploid cells occur in the root cortex, endomitotic nuclei with diplochromosomes are associated with the nodule meristem, and subsequently only polyploid cells become infected. What is unclear is whether infection of the root cells by the rhizobia, in addition to the stimulation of preexisting endomitotic cells, stimulates diploid cortical cells to become endomitotic and to subsequently divide or nodules are only initiated in regions of the root having preexisting endomitotic cortical cells. The pathological effect of the infection thread on diploid cells observed by Wipf and Cooper (1940) was not substantiated by either Libbenga and Harkes (1973) or Newcomb, Sippell, and Peterson (1979). The latter workers did not observe any infection threads that did not stimulate mitoses in the cortex, whereas the two other groups did. One must question the reliability of using nuclear size as an indication of the ploidy level of nuclei, because at best nuclear volume is only an approximate measure of DNA content. In pea cortical explants treated with auxins and cytokinins, the nuclei swelled prior to mitosis and thus an increase in nuclear volume may be the result rather than the cause of the induction or noninduction of cell division (Libbenga and Bogers, 1974). In ungerminated pea seeds, diploid meristematic cells in the plumule contain about 15% more Feulgen-stainable material than the diploid epicotyl cortical cells (van Oostveldt and van Parijs, 1976); this difference is believed to be associated with an under-replication of some highly repetitive DNA sequences in cells preparing for endoreduplication.

INVOLVEMENT OF GROWTH SUBSTANCES IN NODULE DEVELOPMENT

The possibility that growth substances secreted by rhizobia stimulate mitotic activity and subsequently mitosis has been suggested (Wipf and

Cooper, 1940; Libbenga and Torrey, 1973; Libbenga et al., 1973, and others). It is known that *R. leguminosarum* produces both IAA and cytokinins in vitro (see below). Evidence derived from in vitro studies of pea root tissues strongly suggests that auxins and cytokinins are involved in the induction of endoreduplication (Torrey, 1961) and are required for the mitotic activity of these polyploid cells (Matthysse and Torrey, 1967). Similarly, in pea root tissues cultured in the presence of auxins and cytokinins, the DNA values of the cortical nuclei increased from 2 c and 4 c levels at day 0 to 8 c and 16 c levels at days 3 and 7 (Libbenga and Torrey, 1973). From these data, these workers suggested that cytokinin in the presence of auxin induces two rounds of DNA synthesis prior to the first mitoses; the first round of DNA synthesis is associated with endoreduplication and the second is connected with mitosis. The same workers demonstrated that polyploid mitoses occur in higher frequencies away from the root tip, whereas the converse is true for diploid mitoses (Libbenga and Torrey, 1973). Interestingly, pea root nodules do not form near the root tip. Thus, the tissue culture system of inducing endoreduplication may be similar to the stimulus provided by the rhizobium in vivo. Presumably, the rhizobia secrete auxins, cytokinins, and possibly other compounds that stimulate the DNA synthesis associated with endoreduplication and mitotic activity.

Evidence for the involvement of bacterial-derived growth substances in the early stages of pea nodule morphogenesis is indirect. Cytokinins occur in high concentrations in pea root nodules and are associated with the nodule meristem (Syōno, Newcomb, and Torrey, 1976). Cytokinin levels are highest in young nodules and for the most part are localized in the nodule meristem. Five-week-old nodules, which lack a nodule meristem and thus have a mitotic index of zero, have no detectable cytokinins. The extractable cytokinins of pea root nodules include zeatin and its riboside and ribotide in addition to small amounts of isopentyladenine (2iP) and its derivatives (Syōno and Torrey, 1976). Uninfected pea root meristems contain zeatin compounds (Short and Torrey, 1972), whereas *R. leguminosarum* in vitro produces 2iP, isopentenyladenosine (IPA), and two unknown cytokinins (Giannatasis and Coppola, 1969; Phillips and Torrey, 1972). Similarly, within the nodule the bacteria may produce 2iP and IPA, which may serve as precursors for zeatin compounds (Miura and Miller, 1969) that are probably synthesized by the plant cytoplasm. Whether or not 2iP is functional in nodule development, or if it has to be converted to zeatin prior to gaining activity are not known (Syōno, Newcomb, and Torrey, 1976). An ineffective pea nodule caused by a mutant strain of *R. leguminosarum* contains an unidentified cytokinin in addition to those present in effective pea nodules and provides further evidence for the involvement of both partners in cytokinin biosynthesis (Newcomb, Syōno, and Torrey, 1977). Mitotic activity and cytokinin levels peaked and declined simultaneously in the same ineffective pea nodule and earlier than in effective pea nodules (Newcomb, Syōno, and Torrey, 1977); the smaller size of these ineffective nodules may be due to a premature shutdown of cytokinin biosynthesis and the consequent production from the nodule meristem of derivative cells that mostly develop into infected cells. Pea nodules also contain high levels of auxins (Pate, 1958), and *R. leguminosarum* synthesizes indoleacetic acid in vitro (Hartmann and Glombitza, 1967), but which partner synthesizes the auxins in vivo is unknown.

In addition to the exogenous cytokinins and auxins supplied by infecting rhizobia, endogenous growth substances play an important regulatory role in nodule morphogenesis. A transverse gradient of endogenous cell division factors occurring in the cortical tissue has been proposed (Libbenga et al., 1973). A crude extract of the central cylinder or cortex of the root explant resulted in the stimulation of mitoses throughout the cortex. Normally, without exogenous growth substances three prominent meristematic areas arise in cultured cortical explants opposite the xylem poles of the excised central cylinder. Libbenga et al. (1973) hypothesized that the transverse gradient of unknown cell division factors interacts with the rhizobial growth substances in the initiation of root nodules. Inhibitors are also involved in the regulation of nodule development. Exogenous abscisic acid (ABA) inhibits nodulation in intact pea seedlings without altering shoot or root growth and will also inhibit mitosis in pea root explants and cytokinin-induced cell divisions in soybean callus (Phillips, 1971a). An inhibitor present in the cotyledons of pea seedlings also inhibits nodulation because removal of a cotyledon releases this inhibition, which may be due to endogenous ABA (Phillips, 1971b).

RELATION BETWEEN TRANSFER CELLS AND NODULE DEVELOPMENT

Transport of substances containing fixed nitrogen out of the nodule to other growing regions of the plants is important for plant growth and seed development. Much excitement was generated by the discovery of specialized cells, termed transfer cells, in the pericycle of many leguminous plants, including pea (Pate, Gunning, and Briarty, 1969). Transfer cells have wall ingrowths or projections over which is situated the plasma membrane, whose area is increased many-fold; this increase is believed to facilitate the short-distance transport of solutes (Pate and Gunning, 1972), although direct evidence for this is lacking. Transfer cells occur in vascular plants at many sites at which high rates of solute transport are believed to occur (Gunning and Pate, 1969). In *Trifolium* nodules, the wall ingrowths developed at approximately the same time as the onset of nitrogen fixation (Pate, Gunning, and Briarty, 1969). On the basis of this observation and similar findings in other plants, it has been suggested that the solute to be transported induces the formation of the wall ingrowths (Gunning, 1977). Recently, it was observed that the xylem parenchyma cells in the root tissue adjacent to effective and two kinds of ineffective pea nodules were transfer cells (Newcomb, Syōno, and Torrey, 1977; Newcomb, Sippell, and Peterson, 1979). In addition, the pericycle cells occurring in ineffective pea nodules are transfer cells. Since no substances containing fixed nitrogen would be expected to be transported from ineffective nodules, it seemed unlikely that such compounds were the inducer substances, and thus it would be desirable to relate the ontogeny of the wall ingrowths to nodule development, i.e., to the onset of nitrogen fixation. Furthermore, in effective pea nodules, the wall ingrowths in both the pericycle and xylem parenchyma transfer cells begin developing prior to the onset of nitrogen fixation (Newcomb and Peterson, 1979). However, the nitrogen requirements of young growing nodules prior to the commencement of nitrogen fixation are met by amino acids and organic acids transported from the shoot along with photosynthetate (J. S. Pate, unpublished data); some of these nitrogenous substances are recycled out of the nodule and thus *might* be involved in the induction of the wall ingrowths. Xylem parenchyma transfer cells also occur near soybean nodules and show a similar ontogeny to those of pea (Newcomb and Peterson, 1979). At the present time, it is not possible to realistically assess the role of transfer cells in the export of nitrogenous substances, as attractive as this hypothesis may be, because of the lack of direct quantitative data. However, in pea nodules, it is an anatomical necessity that substances such as amino acids leaving the nodule and subsequently transported in the xylem must pass through the pericycle transfer cells. The critical question is not whether or not the passage of the solutes occurs across the area of wall ingrowths but whether or not the increased surface of the plasmalemma in this area actually does facilitate solute movement.

GENETIC CONTROL OF NODULE DEVELOPMENT

Pea nodule development involves the cooperative interaction of both the *R. leguminosarum* and pea genomes, but until recently the influence of the host genome has received little attention. Zobel (1975) demonstrated, by means of mutagenic treatments, that the pea genome exerts considerable control over nodulation. In 96 M_2 seedlings examined, 75 differed in their nodulation whereas only 5 had modified shoot systems. Among the variants he observed were normal nodules that lacked leghemoglobin, nodules containing leghemoglobin that were unable to reduce acetylene, varying nodule size and number, and lack of nodulation. Two thousand cultivars in the World Pea Collection were screened for variants that included 10 cultivars that did not fix nitrogen (Holl, 1975; Holl and LaRue, 1975). In examining *Pisum sativum* cv. Afghanistan, which resists nodulation, they crossed it with a normal cultivar Trapper and demonstrated two recessive genes in Afghanistan, one controlling nodulation and the other controlling nitrogen fixation. Grafting experiments indicated that the non-nodulating factor was produced in the root and was associated with proliferation of the cortical cells, not with the infection of root hairs (Degenhardt, LaRue, and Paul, 1976). An F_2 nodulating nonfixing segregate (PRL-72HI-2) was isolated; these nodules contained a functional nitrogenase that was only demonstrable when nodule slices were incubated in succinate or pyruvate (Holl and LaRue, 1975). Initial fine structural studies suggested an altered carbohydrate deposition (Holl and LaRue, 1975). However, recent studies (Newcomb and Holl, unpublished data) indicate a normal cytology and show that buildup in starch granules is associated with bacteroid differentiation (Figures 6 and 7), as occurs in effective pea nodules (Newcomb, 1976), and is not related to the altered increased depositions found in a *Rhizobium* mutant-induced pea nodule (Newcomb, Syōno, and Torrey, 1977). A con-

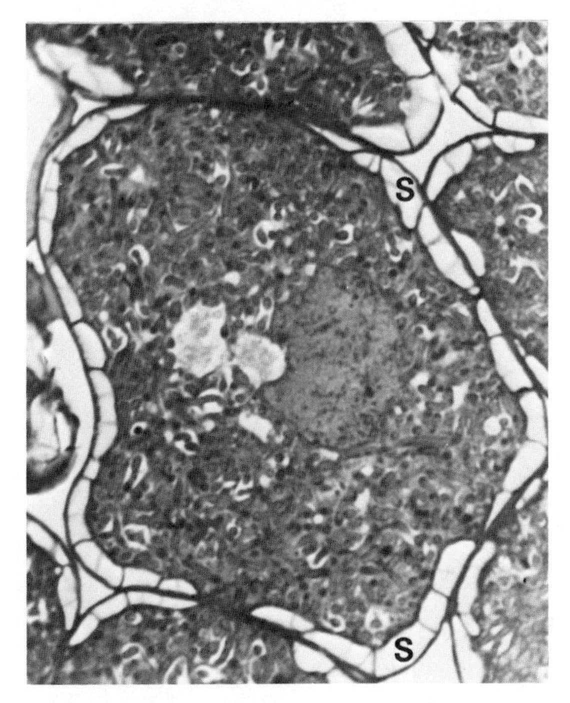

Figure 6. Light micrograph of an infected cell from a differentiating ineffective pea nodule (PRL-72HI-2) caused by a host mutation. Many developing bacteroids and amyloplasts with flattened starch granules are present. (×1320)

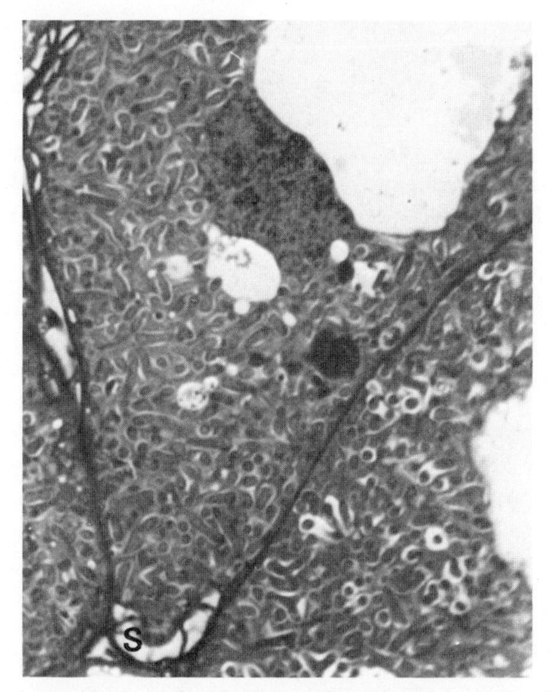

Figure 7. Light micrograph of an infected cell from the same nodules as that shown in Figure 6 in a more advanced stage of differentiation. Bacteroids are well developed and little starch is present. (×1320)

servative estimate of the host genes affecting symbiotic nitrogen fixation is 10 (Holl and LaRue, 1975). Obviously, mutants in either the host or the microbial genome, affecting various stages in the infection process, nodule development, and the metabolic aspects of nitrogen fixation, could be valuable research tools in future investigations of the developmental biology of leguminous root nodules. It is hoped that this review will revive interest and activity in this important area that might be exploited in the future crop improvements involving symbiotic nitrogen fixation.

ACKNOWLEDGMENTS

The author wishes to express his gratitude to Professor J. G. Torrey for first interesting him in this problem and for his continued interest and enthusiasm in it.

REFERENCES

Bassett, B., R. N. Goodman, and A. Novacky. 1977. Ultrastructure of soybean nodules. I. Release of rhizobia from the infection thread. Can. J. Microbiol. 23:573–582.

Bond, L. 1948. Origin and developmental morphology of root nodules of *Pisum sativum*. Bot. Gaz. 109:411–434.

Dart, P. J. 1975. Legume root nodule initiation and development. In: J. G. Torrey and D. T. Clarkson (eds.), The Development and Function of Roots, pp. 467–506. Academic Press, London.

Dart, P. 1977. Infection and development of leguminous nodules. In: R. W. F. Hardy and W. S. Silver (eds.), A Treatise on Dinitrogen Fixation, Section III, Biology, pp. 367–472. John Wiley & Sons, Inc., New York.

Degenhardt, T. L., T. A. LaRue, and E. A. Paul. 1976. Investigation of a non-nodulating cultivar of *Pisum sativum*. Can. J. Bot. 54:1633–1636.

Feldman, L. J. 1975. Cytokinins and quiescent center activity in roots of *Zea*. In: J. G. Torrey and D. T. Clarkson (eds.), The Development and Function of Roots, pp. 55–72. Academic Press, London.

Fujita, T., and S. Mitsuishi. 1953. Cytological studies on the root nodules of peas. Proc. Crop Sci. Japan 22:97–98.

Giannatasis, M., and S. Coppola. 1969. Isolamento di citochimine dal *Rhizobium leguminosarum* Frank. G. Bot. Ital. 103:11–17.

Gunning, B. E. S. 1977. Transfer cells and their roles in the transport of solutes in plants. Sci. Prog. Oxf. 64:539–568.

Gunning, B. E. S., and J. S. Pate. 1969. Transfer cells: Plant cells with wall ingrowths, specialized in relation to short distance transport of solutes—their occurrence, structure, and development. Protoplasma 68:107–133.

Hartmann, T., and K. W. Glombitza. 1967. Der Tryptophanabbau bei *Rhizobium leguminosarum*. Arch. Mikrobiol. 56:1–8.

Holl, F. B. 1975. Host plant control of the inheritance of dinitrogen fixation in the *Pisum-Rhizobium* symbiosis. Euphytic 24:767–770.

Holl, F. B., and T. A. LaRue. 1975. Genetics of legume plant hosts. In: W. E. Newton and C. J. Nyman (eds.), Proceedings of the 1st International Symposium on Nitrogen Fixation, Volume II, pp. 391–399. Washington State University Press, Pullman.

Hunter, J., and G. H. Elkan. 1975. Role of pectic and cellulytic enzymes in the invasion of the soybean by *Rhizobium japonicum*. Can. J. Microbiol. 21:1254–1258.

Libbenga, K. R., and R. J. Bogers. 1974. In: A. Quispel (ed.), The Biology of Nitrogen Fixation, pp. 430–472. North-Holland Publishing Company, Amsterdam.

Libbenga, K. R., and P. A. A. Harkes. 1973. Initial proliferation of cortical cells in the formation of root nodules in *Pisum sativum* L. Planta 114:17–28.

Libbenga, K. R., F. van Iren, R. J. Bogers, and M. F. Schraag-Lamers. 1973. The role of hormones and gradients in the initiation of cortex proliferation and nodule formation in *Pisum sativum* L. Planta 114:29–39.

Libbenga, K. R., and J. G. Torrey. 1973. Hormone-induced endoreduplication prior to mitosis in cultured pea nodule cortex cells. Am. J. Bot. 60:293–299.

Matthysse, A. G., and J. G. Torrey. 1967. Nutritional requirements for polyploid mitoses in cultured pea root segments. Physiol. Plant. 20:661–672.

Miura, G. A., and C. O. Miller. 1969. 6-(γ,γ-Dimethylallylamino) purine as a precursor of zeatin. Plant Physiol. 44:372–376.

Napoli, C. A., and D. H. Hubbell. 1975. Ultrastructure of *Rhizobium*-induced infection threads in clover root hairs. Appl. Micribiol. 20:1003–1009.

Newcomb, W. 1976. A correlated light and electron microscopic study of symbiotic growth and differentiation in *Pisum sativum* root nodules. Can. J. Bot. 54:2163–2186.

Newcomb, W., and R. L. Peterson. 1979. The occurrence and ontogeny of transfer cells associated with lateral roots and root nodules in certain Leguminosae. Can. J. Bot. In press.

Newcomb, W., D. Sippell, and R. L. Peterson. 1979. The early morphogenesis of *Glycine max* and *Pisum sativum* root nodules. Can. J. Bot. In press.

Newcomb, W., K. Syōno, and J. G. Torrey. 1977. Development of an ineffective pea root nodule: Morphogenesis, fine structure and cytokinin biosynthesis. Can. J. Bot. 55:1891–1907.

Nutman, P. S. 1956. The influence of the legume in root-nodule symbiosis. A comparative study of host determinents and functions. Biol. Rev. 31:109–151.

Oinuma, T. 1948. Cytological and morphological studies on root nodules of garden pea, *Pisum sativum* L. Seitbusu 3:155–161.

Pate, J. S. 1958. Studies of the growth substances of legume nodules using paper chromatography. Aust. J. Biol. Sci. 11:516–528.

Pate, J. S., and B. E. S. Gunning. 1972. Transfer cells. In: L. Machlis (ed.), Annual Review of Plant Physiology, Volume 23, pp. 173–196. Annual Reviews, Inc., Palo Alto, California.

Pate, J. S., B. E. S. Gunning, and L. G. Briarty. 1969. Ultrastructure and functioning of the transport system of the leguminous root nodule. Planta 85:11–34.

Phillips, D. A. 1971a. Abscisic acid inhibition of root nodule initiation in *Pisum sativum*. Planta 100:181–190.

Phillips, D. A. 1971b. A cotyledonary inhibitor of root nodulation in *Pisum sativum*. Physiol. Plant. 25:482–487.

Phillips, D. A., and J. G. Torrey. 1972. Studies on cytokinin production by *Rhizobium*. Plant Physiol. 49:11–15.

Short, K. C., and J. G. Torrey. 1972. Cytokinins in seedling roots of pea. Plant Physiol. 49:155–160.

Syōno, K., W. Newcomb, and J. G. Torrey. 1976. Cytokinin production in relation to the development of pea root nodules. Can. J. Bot. 54:2155–2162.

Syōno, K., and J. G. Torrey. 1976. Identification of cytokinins in root nodules of the garden pea, *Pisum sativum* L. Plant Physiol. 57:602–606.

Torrey, J. G. 1961. Kinetin as trigger for mitosis in mature endomitotic plant cells. Exp. Cell Res. 23:281–299.

Torrey, J. G., and S. Barrios. 1969. Cytological studies on rhizobial nodule initiation in *Pisum*. Caryologia 22:47–61.

Truchet, G., and P. Coulomb. 1973. Mise en évidence et évolution du systeme phytolysosomal dans les cellules des différentes zones de nodules radiculaires de Pois (*Pisum sativum* L.). Notion d'hétérophagie. J. Ultrastruct. Res. 43:36–57.

Van Oostveldt, P., and R. van Parijs. 1976. Underreplication of repetitive DNA in polyploid cells of *Pisum sativum*. Exp. Cell Res. 98:210–221.

Verma, D. P. S., V. Zogbi, and A. K. Bal. 1978. The cooperative action of plant and *Rhizobium* to dissolve the host cell wall during development of root nodule symbiosis. Plant. Sci. Lett. 13:137–142.

Wipf, L. 1939. Chromosome numbers in root nodules and root tips of certain Leguminosae. Bot. Gaz. 101:51–67.

Wipf, L., and D. C. Cooper. 1938. Chromosome numbers in nodules and roots of red clover, common vetch and garden pea. Proc. Natl. Acad. Sci. USA 24:87–91.

Wipf, L., and D. C. Cooper. 1940. Somatic doubling of chromosomes and nodular infection in certain leguminosae. Am. J. Bot. 27:821–824.

Zobel, R. W. 1975. The genetics of root development. In: J. G. Torrey and D. T. Clarkson (eds.), The Development and Function of Roots, pp. 261–275. Academic Press, London.

Nitrogen Fixation, Volume II
Edited by W. E. Newton and W. H. Orme-Johnson

Factors Controlling the Legume-*Rhizobium* Symbiosis

J. M. Vincent

Whether or not the outcome of an encounter between a legume and a rhizobium will result in an effective dinitrogen-fixing symbiosis depends on the intrinsic properties of the partners, their mutual genetically controlled compatibility, and the influence exercised by the environment on each separately and on their interaction. The suitability of a legume to a natural or agriculturally modified environment will be a major determinant of its nitrogen-fixing potential. If the supply of combined nitrogen in the soil is limiting, the N_2-fixing capacity of rhizobial-formed root nodules will be equally important. Environmental factors can operate at many points of the symbiosis, ranging from determining survival of the rhizobia outside the plant and the growth of the host plant to more subtle effects on the formation and functioning of the root nodule. Several recent accounts are available dealing with the symbionts and their interrelationship (Quispel, 1974; Newton and Nyman, 1976; Nutman, 1976; Hardy and Silver, 1977; Newton, Postgate, and Rodriguez-Barrueco, 1977; Döbereiner et al., 1978), including an account of a workshop concerned with tropical legumes (Vincent, Whitney, and Bose, 1977). The present paper concentrates on genetic aspects of the legume-rhizobium symbiosis and, more briefly, on physiological and environmental considerations.

THE NATURE OF THE SYMBIOSIS

The process that begins on the root surface and culminates in the establishment of an effective N_2-fixing nodule is a multistage sequence of interdependent steps. Attempts to define its genetic, physiological, or environmental control by such broad expressions as invasion or effectiveness are therefore bound to fail and confuse. Unfortunately, most reports of the symbiosis are still in such general, nonanalytical terms. There is need for direct evidence as to the affected stage of the symbiosis; generally, once infection has occurred such evidence requires careful observation with both light and electron microscopes. An older scheme that details stages of failure in the clover–*Rhizobium trifolii* symbiosis (Bergersen, 1957b) provides a start to the development of a more informative phenotype sequence. By isolating currently recognizable stages, a framework is produced within which genotypic and environmental effects can be systematically located and conveniently coded (Table 1). The scheme is based on infections via root hairs. For legumes with different portals of entry and method of invasion within the developing nodule (Dart, 1974,

Table 1. Analysis of symbiotic sequence

Stage	Abridged description	Phenotypic code
I. Preinfection		
1. Multiplication on root surface ("rhizoplane")	Root colonization	Roc
2. Attachment to root surface	Root adhesion	Roa
3. Branching of root hairs	Hair branching	Hab
4. "Marked" curling of root hairs	Hair curling	Hac
II. Infection and Nodule Formation		
5. Formation of infection thread	Infection	Inf
6. Development of polyploid (disomatic) meristem; nodule development and differentiation	Nodule initiation	Noi
7. "Intracellular" release of rhizobia from infection thread	Bacterial release	Bar
8. "Intracellular" multiplication of rhizobia and development of full bacteroid form	Bacteroid development	Bad
III. Nodule Function		
9. Reduction of N_2 to NH_4^+ (nitrogenase)	Nitrogen fixation	Nif
10. Complementary biochemical and physiological functions	Complementary functions	Cof
11. Persistence of nodule function	Nodule persistence	Nop

1977) it will need modification. Also, some of the presently designated stages will have to be further divided as more information comes to hand.

The detailed stages have been grouped to conform with the three major steps that are commonly specified: preinfection, infection and nodule development, and nodule function.

Preinfection

Rhizobial colonization of the root surface (Roc) will generally be a necessary first step for both a minority population and an applied inoculum. The rhizobia probably need not build up to the multilayered level found in liquid-cultured *Medicago* (Dart and Mercer, 1964), since the numbers recovered from roots of other species are considerably less (representative data cited by Broughton, 1978). The colonizing rhizobia come into close proximity to the possible infection sites in a polar fashion, with the bacterial cell at right angles to the plane of attachment (Roa) and possibly linked through a lectin molecule (Hamblin and Kent, 1973; Bohlool and Schmidt, 1974, 1976). The working hypothesis of Dazzo and Hubbell (1975) suggests that lectin provides a linkage that holds the rhizobium end-on to the root. There is, however, some confusion as to the nature of the interaction with the rhizobial surface. Enzymatic degradation of a rhizobial surface lipopolysaccharide by lectin (Albersheim and Wolpert, 1976) is itself some evidence of combination. The interaction might also expose a combining group not previously accessible nor readily demonstrated serologically. Also, polysaccharide preparations can carry entrapped smaller polysaccharides that may be the true basis for specific lectin combination (Planqué and Kijne, 1977).

Situated close to the root and with sufficient numbers, the rhizobia cause striking plant responses that are either a necessary prelude to infection or, at least, the signal that invasion is possible. Where root hairs are involved, this reaction to rhizobia takes two distinct forms; root hair branching (Hab) and curling. When the latter is pronounced ("marked" curling, Hac in Table 1), the invasion system is likely to operate.

Infection and Nodule Formation

When invasion is via root hairs, some, although generally a minority, of the markedly curled hairs develop an infection thread (Inf) that contains entrapped rhizobia bound by a plant-synthesized membrane. The process begins at an unthickened part of the hair with the development of a bright refractile spot, and requires proximity of the root hair nucleus for its initiation and persistence (Sahlman and Fåhraeus, 1963; Higashi, 1969; Napoli and Hubbell, 1975). In this situation, sufficient localized enzyme and/or hormone production "loosens" the root hair wall and stimulates growth to initiate an invaginating infection thread. The two separately caused features of root hair deformation, branching and marked curling, seem obligatory for invasion (Yao and Vincent, 1969, 1976); branching could reflect lateral wall "loosening" and marked curling could reflect a localized asymmetric growth response to the presence of viable specific rhizobia.

The earlier proposal that specific induction of pectolytic enzymes had a role in the initiation of infection (Fåhraeus and Ljunggren, 1959) has not been regularly confirmed, possibly because of technical shortcomings, including the possible toxic effects of seedlings (Fahraeus and Sahlman, 1976/77). However, direct production of a pectolytic enzyme has been demonstrated recently with fast-growing rhizobia (Hubbell, Morales, and Umali-Garcia, 1978). At no stage of the normal infection are the rhizobia naked within the root hair cytoplasm. They are embedded in a large amount of matrix where they divide continually as the infection thread advances through several layers of the host cortex, probably by repetition of the first invaginating "entry." A variable, but often very large, proportion of the infection threads abort. The question of multiple invasion of the root hair is difficult to resolve. Restricted occupation of an individual nodule could reflect restricted opportunity for multiple entry, particularly in a natural particulate environment such as soil, where individual invasions could arise from one or more members of a well separated microcolony formed by the progeny of the one strain. There are many reports of mixed nodule occupancy when conditions favor continuous mixed growth over the root, but these could be caused by double entry at one point of infection, invasion of more than one neighboring hair, or a second invasion of the same hair. The associated occurrence of rhizobialike, but not invasive, bacteria in nodules formed by independently invasive *Rhizobium japonicum* on soybean (Van Rensburg and Strijdom, 1971, 1972a, 1972b) and the assisted entry of a *Rhizobium* with an otherwise different host requirement (Johnston and Beringer, 1975, 1976a, 1976b) give some support to the simultaneous invasion proposition.

The fact that the bulk of the specialized nodule tissue is polypoid has been explained as the product of an encounter between the penetrating infection thread and a preexistent disomatic cell. Evidence against this and in favor of rhizobial-directed development of polyploidy has been convincingly presented and is now generally accepted (Libbenga and Bogers, 1974; Dart, 1977). A cotyledon factor also assists nodulation.

The establishment of one or several centers of disomatic meristem, generally within the cortex but sometimes in the axils of lateral roots, is the key step (Noi) in the commencement of nodule growth. Further normal development requires thorough infection thread ramification throughout the new tissue, "release" of bacteria into the host cells (although still surrounded by a plant-formed isolating membrane), and their ordered multiplication and conversion to the N_2-fixing bacteroid condition. Bacterial release (Bar) depends on relatively thin, unwalled regions of the surrounding plant-produced infection thread. Some convincing electron micrographs (Newcomb, 1976) show thin "ballooning" portions of the otherwise thickened infection thread. The full development of the bacteroid (Bad) follows a period of extensive multiplication such that the young enlarged disomatic cells of the developing nodule have their cytoplasmic region tightly packed with bacteria. The bacteria are often much larger than the vegetative form and often grossly swollen and branched. The packeting of bacteroids is single or multiple. Biochemical changes in the maturing bacteroid are as profound as the morphological ones, and produce the enzyme systems needed for the plant-utilizable N_2 fixation. The symbionts cooperate to produce leghemoglobin, which protects against excessive O_2 concentration and acts as a steady supplier of this element. The bacteroid is less robust than the corresponding cultured cell, being low in lipopolysaccharide and directly susceptible to lysozyme (Van Brussel, Planqué, and Quispel, 1977), which could permit a freer metabolic exchange, possibly an important requirement for efficient functioning. Indeed, the similarity between bacteroids and calcium-deficient cultured cells might be explained if the plant's demands in the host cells were to create a shortage of divalent cation. Conversely, lost susceptibility to D-amino acids and antimetabolites might be associated with too robust an envelope in the resistant, commonly ineffective bacteroid mutant (Yu and Jordan, 1971; MacKenzie and Jordan, 1972). Additional membranes have been reported in fully formed bacteroids of *Rhizobium meliloti*, some due to invagination of the cell's own envelope and others developing within the bacterial cell (Hornez et al., 1974).

Although the matured bacteroid differs in physiological detail from the normally cultured rhizobial cell, it is not short of DNA (Bisseling et al., 1977; Paau, Lee, and Cowles, 1977), and the long-repeated claim of nonviability will also have to be reviewed (Greshoff et al., 1977; Sutton, Jepsen, and Shaw, 1977; Tsien, Cain, and Schmidt, 1977). Rather, earlier failures to secure colony growth from bacteroids can be attributed to osmotic fragility, due to a desirably leaky membrane system, and to the use of unsuitable media for their recovery.

Nodule Function

The final determination of N_2-fixation efficiency depends on the development of the functional nitrogenase complex (Nif), its linkage with the electron and energy supply systems, with some sharing of responsibility between the symbionts (Cof), and the length of time that this finely balanced partnership is maintained (Nop). The several simultaneous demonstrations of N_2 fixation in culture by various slow-growing rhizobia (Kurz and LaRue, 1975; McComb, Elliott, and Dilworth, 1975; Pagan et al., 1975) have confirmed the indications from *nif* transfer experiments from

R. trifolii to *Klebsiella* (Dunican and Tierney, 1974) that at least some rhizobia carry the full genetic *nif* complement. However, the key word for the legume and its rhizobia is *symbiosis* and it is symbiosis that continues to demand the close analytical attention of the geneticist, physiologist, and biochemist both for the immediate benefit of the farmer and for the long term good of mankind.

SPECIFICITY BETWEEN SYMBIONTS

Success in N_2-fixing symbiosis is largely a matter of genetic compatibility between the symbionts. There are legumes that seem not to nodulate (Caesalpinioideae, notably *Cassia*; Allen and Allen, 1976). In the laboratory, there are rhizobia that lack an invadable host. They are also likely to exist outside the laboratory; perhaps the non-nodulating cohabitor of soybean nodules is a "deprived" rhizobium? Rhizobia nodulate legumes almost exclusively. The well authenticated typically slow-growing *Rhizobium*, which is able to nodulate a nonlegume *Parasponia* (*Trema*) as well as several legumes (Trinick, 1973, 1976), is so exceptional as to highlight the specificity of the legume-rhizobium relationship. The genetic barriers that generally prevent rhizobial nodulation of nonlegumes must indeed be considerable. However, the considerable taxonomic differences between "fast"- and "slow"-growing rhizobia could indicate two separate evolutionary developments of the symbiosis.

It is possible to think of the association between these symbionts in terms of specifically matching genotypes. For most legumes there are rhizobia that are able to produce N_2-fixing nodules, and for most naturally occurring rhizobia we can expect to find a legume species with which each will effectively associate. There are hosts that are particularly selective in their acceptance of an effectively nodulating partner (e.g., *Lotononis bainesii*, *Leucaena latisiliqua* [formerly *L. leucocephala*]). Others, such as *Macroptilium atropurpureum*, are notably promiscuous. Specificity becomes even more apparent when effectiveness of N_2 fixation, and not merely nodule-forming ability, is considered. Nodulation that occurs outside the common cross-inoculation groups is seldom effective in N_2 fixation. Even within a cross-inoculation group, it is unusual for the one rhizobial strain to be fully effective with more than relatively few of the species and not even all of the cultivars it nodulates. Well documented cases of specificity among cultivars are found with *Medicago sativa*, Northam First Early and Woogenellup varieties of subterranean clover (Gibson, 1962, 1964, 1968), and with a range of soybeans. The relative nodulating success of competing strains may also be markedly affected by the host species or variety (Vincent and Waters, 1953; Marques Pinto, Yao, and Vincent, 1974).

Unfortunately, most information about specificity is described in the broadest categories, such as invasive or noninvasive (alternatively virulent or avirulent) or effective or ineffective. I shall attempt to take the analysis further using the phenotypic framework already outlined. This framework is tentative and is for others to test, reduce, or expand, and the examples are not claimed to be exhaustive. Difficulties in reinterpreting published accounts can leave areas of disagreement, but the catalytic purpose of the present article will have been well served if others are stimulated to reevaluate their own data and undertake fresh experiments directed to a more penetrating analysis.

INFLUENCE OF THE HOST

Table 2 records stages at which the host genome affects the symbiosis. The same framework is used as in Table 1.

Preinfection

The rhizobia are generally stimulated by root secretions and more by legumes than by nonlegumes. There are, however, few published data and most work was done some years ago (notably Rovira, 1961). Similarly, the evidence of specificity between legumes and rhizobia at the stage of root colonization (Roc) is old and slight. *R. meliloti* seems to have a preference for *Medicago* over *Trifolium* but *R. trifolii* does perfectly well on a heterologous host (*Medicago*) (Dart, 1974). Perhaps *Rhizobium leguminosarum* is relatively favored by homoserine secreted from the region of lateral root emergence in *Pisum sativum* (Van Egeraat, 1975), but the relationship is by no means clear. Nor does relative colonization of the root provide any explanation of host-dependent nodulating success of competing strains (Vincent and Waters, 1953; Marques Pinto, Yao, and Vincent, 1974; Labandera and Vincent, 1975; Franco and Vincent, 1976).

The lectin hypothesis of facilitated root adhesion (Roa) has to be supported by evidence that the host root to be nodulated produces a lectin-like protein (not necessarily a hemagglutinin) able to bind rhizobia. Most reported cases with soybean seed lectin (Bohlool and Schmidt, 1974) conformed with this expectation, although some reactions were weak, and for two nodulating rhizobia were negative. Studies with clover roots introduced the concept of specifically ambivalent lectin that links sites on the roots to sites on infective rhizobia (Dazzo and Hubbell, 1975). Specific nodulating rhizobia adsorbed strongly on the roots and root hairs of its clover host, but not on alfalfa (*M. sativa*). This reaction was blocked by 2-deoxyglucose, a hapten associated with the active binding sites of the bacterium, but not that between alfalfa root extract and its nodulating *R. meliloti*. 2-Deoxyglucose therefore appears specific to the clover–*R. trifolii* interaction (Bohlool and

Table 2. Influence of host genome on symbiosis

Phenotypic Stage	Examples of host effects	References
I. Preinfection		
1. Roc	Host stimulation or inhibition; low specificity	
2. Roa	Possible lectin specificity	Dazzo, Napoli, and Hubbell (1976)
3. Hab	Homologous legumes (++)	Nutman (1959)
	Heterologous legumes (+)	Haack (1964)
4. Hac	Homologous and "near" homologous	Yao and Vincent (1969)
II. Infection and Nodule Formation		
5. Inf	Homologous (+++); heterologous (±)	
	Number of infection threads and mode of entry	Dart (1977)
	Resistant lines [rr (ρ) *T. pratense*, non-nodulating soybean]	Nutman (1949) Williams and Lynch (1954)
6. Noi	Degree of polyploidy	Dart (1977)
	Location of meristem	
	Proportion of successful infection threads	Nutman (1959)
7. Bar		
8. Bad	Bacteroids per envelope; leghemoglobin i_1–*ie* lines *T. pratense*	Dart (1977) Bergersen and Nutman (1957)
	Ineffectivity with *T. ambiguum*	Bergersen (1957a)
III. Nodule Function		
9. Nif	Apt environment and photosynthate	
	Relative "efficiency" (NH_3 comp. H_2)	Schubert and Evans (1977)
10. Cof	Photosynthate	
	Enzyme systems	
	Nature of —NH_2 compounds	
	Onset of reproduction	Pate (1977)
11. Nop	Growth habit, onset of maturation	

Schmidt, 1976; Dazzo, Napoli, and Hubbell, 1976; Dazzo and Brill, 1977). Correspondingly, soybean lectin binding to cells of *R. japonicum* is reversible by *N*-acetylgalactosamine, its specific hapten inhibitor (Bhuvaneswari, Pueppke, and Bauer, 1977).

Failures to achieve the general relationship required for unqualified support of the hypothesis (Bohlool and Schmidt, 1974; Law and Strijdom, 1977; Rougé and Labroue, 1977) need some consideration. Binding of rhizobia by a lectin from a host not nodulated by that rhizobium would not invalidate the hypothesis because restriction of nodule formation could well occur at some other stage in the overall process. Furthermore, the lack of lectin binding for a nodulating partnership may be due to the use of the wrong member of a group of heterogeneous lectins produced by the one plant. A comprehensive review by Broughton (1978) considers these aspects and others, including the polymeric form of lectins that may determine if they are phytohemagglutinins or not. This variation could explain cases where host susceptibility to nodulation is not accompanied by phytohemagglutination. Isolectins provide the possibility of different physiological roles for lectins within the one plant. There is no reason to assume that root lectins would be identical with the predominant seed lectin. Although there is still plenty of room to maneuver within the lectin hypothesis, some doubt remains as to its complete validity (e.g., Chen and Phillips, 1976). More root lectins and corresponding rhizobial receptors need to be isolated and purified.

Root hair branching (Hab) is moderately specific; marked curling (Hac) is much more host specific. Apart from heterologous associations and a noncharacteristic response of clover to one of the strains of *Rhizobium lupini*, the only marked departure from nodulating specificity is between clover and *R. leguminosarum*. This breach of the host specificity barrier also occurs in the converse situation and reflects the well established taxonomic similarity of *R. trifolii* and *R. leguminosarum* (Haack, 1964; Yao and Vincent, 1969).

Infection and Nodule Formation

The susceptibility of a particular legume to infection by a given group of rhizobia is the common basis for cross-inoculation groups, rhizobial species, or other groupings, e.g., "cowpea rhizobia" and "Lotus rhizobia." Imperfections in such categorization are well documented and need not be dwelt on here.

The site of initial infection (Inf), whether root-hair, epidermal, or wound, is host determined, as is the frequency of infection thread formation (Dart, 1977). Resistant ("non-nodulating") lines have been obtained with red clover (Nutman, 1949), for which infection threads are not developed; presumably, the non-nodulating soybean (Williams and Lynch, 1954) also similarly resists the first step in invasion. Nodule initiation (Noi) is achieved generally by only a small proportion of the infections. The host seems able to control the degree of polyploidy developed in response to the advancing infection thread and location of meristem (Dart, 1977), and influences both the number of bacteroids per membrane envelope ("peribacteroid membrane") and the protein moiety of leghemoglobin. The *i* and *ie* lines of *Trifolium pratense* fail to mature normal bacteroids (Bergersen and

Nutman, 1957). In the case of *i* and some rhizobial strains, the released bacteria remain rodlike. Strain *ie/ie* plants showed renewed division of the host cells immediately following release of the bacteria, which generally remained in the rod form. In this case, division of both invaded and nearby noninvaded host cells resulted in a tumorlike condition. The *ie* response was relatively independent of the nature of the rhizobial strain. An ineffective response of *Trifolium ambiguum* to a rhizobial strain effective with clovers of different geographical origin was also due to failure at this stage (Bergersen, 1957a).

Nodule Function

The efficiency and persistence of the nodule must be markedly affected by the plant's capacity to maximize and maintain its growth under a given set of environmental conditions. There is therefore a need to include symbiotically maintained host lines in breeding and introduction programs. Although some mutations cause loss of symbiotic capacity with any host, there are indications, as there often are with changed plant genotype, that the extent of residual property may be host dependent (Pankhurst, 1977). It would seem inevitable that host species and varieties would differ in the suitability of the "physiological environment" they supply for the functioning nitrogenase complex (Nif). Degrees of nutritional demand for molybdenum and intrinsic photosynthetic capacity would seem to come into this category. The influence of host on the energy loss as H_2 (Schubert and Evans, 1977) could perhaps be located here, although it could be part of the complementary function (Cof) insofar as it depends on hydrogenase production for reutilization of the H_2 (Dixon, 1972, cited by Holl and LaRue, 1976).

Several complementary functions (Cof) are host affected. Many of the substrates, enzymes, and conditions needed to take the fixed nitrogen (NH_4^+) through to nitrogen metabolites can be considered subject to plant control (e.g., Rokosh and Kurz, cited by Holl and LaRue, 1976). This could reflect the efficiency of its photosynthetic system, the nature and exchange of nitrogen compounds between shoot and root, and plant-provided enzymes. The globin (but not the heme) portion of leghemoglobin appears to be under plant genetic control (Dilworth and Coventry, 1977; Nadler and Avissar, 1977).

The physiological characteristics of the host can be expected to be a major factor determining nodule persistence (Nop) and the useful N_2-fixing life of the nodules it carries, as shown by pea cultivars that maintain fixation during pod fill (Holl and LaRue, 1976).

General Significance of Manipulating the Host Genome

Improvements in the host could come from those genetic changes that make it more agronomically suitable to particular needs (pasture, forage, grain), to particular soil environments (fertility, toxic effects such as unfavorable pH, manganese, and aluminum, and available nitrogen), and to the plant's physical environment (light, temperature, and water supply) and that make it resistant to pests and disease. Also important are genetic modifications both to permit greater tolerance of combined nitrogen, perhaps by augmenting the photosynthate supply to the nodules, and to cause an earlier start and delayed closing down of the N_2-fixation system, particularly during seed fill. Holl and LaRue (1976) screened pea cultivars for the influence of nitrate on fixation. Unfortunately, those that initially appeared promising failed to demonstrate similar "nitrate resistance" in the field. However, encouraged by the continued fixation into pod filling in faba bean (*Vicia faba*), they found several pea cultivars that also possessed this property. Field and breeding trials were to be undertaken.

The immediate problem confronting the plant breeder and the rhizobiologist is to make sure that superior host lines retain sufficient symbiotic compatibility with rhizobia that are already widely distributed or commonly used as inoculant to secure superior N_2 fixation. Another possibility is to develop cultivars that are relatively immune to an ineffective rhizobial population so that they could be successfully introduced with an appropriate compatible inoculum. Experience suggests that such a procedure might be successful initially, but the longer-term survival of introduced effective rhizobia is more doubtful, and this merits investigation.

GENETIC BASES AND MECHANISMS IN RHIZOBIUM

The first basic genetic information about rhizobia came from the use of easily recognized laboratory-selected marker characteristics. Symbiotic capacity is too complex and too dependent on interaction with the host plant to be helpful, although it is the most significant feature of this bacterium. This paper first deals briefly with marker characteristics, then with the symbiotic capacity.

Studies with Nonsymbiotic Markers

Mutations, natural or by mutagenic agents, are as readily encountered with rhizobia as with other bacteria. They may involve changes in colony size and form, in the production and possibly the nature of extracellular gum, and, less frequently, in the antigenic surface (Humphrey and Vincent, 1975; Wilson, Humphrey, and Vincent, 1975). Resistant mutants can be selected using bacteriophage, antibiotics, rather high concentrations of amino acids (particularly D forms), and antimetabolites such as amino acid analogs. Auxotrophs lacking the capacity to synthesize particular vitamins, amino acids, and purine or pyrimidine bases have been regularly obtained. Selection against chlorate gives mutants with reduced nitrate reductase activity (Sik et al., 1976). Nitrosoguanidine (NTG) has been used to obtain closely linked mutations in both *R. meliloti* and *R. trifolii*, and then to construct low resolution replication maps to a streptomycin resistance marker. These maps can increase the likelihood that closely located, co-transducible markers are selected before transduction tests are undertaken (Dénarié and Truchet, 1976). The first and many subsequent reports of genetic exchange between rhizobia were concerned with DNA transformation, particularly using streptomycin resistance as a marker. More recently, conjugation-facilitating plasmids, such as RP_4 from *Escherichia coli* and R1-19drd from *Pseudomonas aeruginosa*, have been transferred to rhizobia by this method (Table 1 in Dénarié and Truchet, 1976; Table 12.3 in Schwinghamer, 1977). Generalized transduction has been demonstrated with *R. meliloti* with streptomycin resistance and auxotrophic markers for lysine and leucine (Kowalski and co-workers, cited by Schwinghamer, 1977). Specialized transduction of cysteine independence was also obtained (Sik and Orosz, 1971).

Exchange of chromosome markers between authenticated rhizobial cultures has been facilitated by the incorporation of P group R factor plasmids (Beringer, 1976). Complementation of auxotrophs and transfer of antibiotic resistance have been used as markers and mapping of the circular chromosome of *R. leguminosarum* has been initiated (Beringer and Hopwood, 1976; Beringer and Johnston, 1977; Beringer, Hoggan, and Johnston, 1978). A circular chromosome has also been demonstrated for *R. meliloti* (Meade and Signer, 1977) and a circular genetic map of 20 genes, none of which involved symbiotic characters, has been constructed for the same species (Kondorosi et al., 1977a). Interspecific exchange of chromosomal genes has also been reported among *R. leguminosarum*, *R. trifolii*, and *Rhizobium phaseoli* (Johnston and Beringer, 1977).

Using R 68.45 as a conjugative plasmid, the *R. meliloti* → *R. leguminosarum* cross gave progeny that harbored R primes with sections of the *R. meliloti* chromosome inserted. The chromosomal marker and others linked to it could thus be transferred between strains at high frequency, which facilitated understanding of the gene organization and function, complementation, and dominance (Johnston, Setchell, and Beringer, 1978). Transfer of RP_4 has been reported between strains of *R. leguminosarum* jointly occupying pea nodules (Johnston and Beringer, 1975, 1976a, 1976b) but not between this species and *R. trifolii* in the same circumstances. Chromosomal replication in *R. trifolii* appears to be bidirectional (Zurkowski and Lorkiewicz, 1977).

Extrachromosomal (plasmid) DNA has been demonstrated in some rhizobial strains by "curing" and, more reliably, by physical methods. One, carrying antibiotic resistance, has been reported in *R. japonicum* (Cole and Elkan, 1973), but seems not to have been exploited in facilitating genetic transfer. Polygalacturonase activity has been found to be linked with a plasmid of 60×10^6 daltons in *R. meliloti* (Olivares, Montoya, and

Palomares, 1977). R1-19drd plasmid was conjugatively transferred from *E. coli* to *R. trifolii* and stably maintained. Another endogenous plasmid of unknown function was found in several other strains of this species (Kowalczuk and Lorkiewicz, 1977).

The material and techniques for investigating the genetics of *Rhizobium* are now available. Its application to the multistep phenomenon of symbiotic N_2 fixation is of much greater practical importance but is of course also much more difficult. Below, the same stepwise analysis of the symbiosis is used to locate the stage at which various mutations interfere with the process.

MUTATIONS AFFECTING RHIZOBIAL PARTICIPATION IN THE SYMBIOTIC SEQUENCE

Mutational loss of symbiotic capacity (avirulence or reduced N_2-fixing capacity of formed nodules) is not uncommonly encountered in rhizobia. It may occur naturally or after mutagenesis and may be associated with other changed characteristics such as colony form (large "gummy" → small "nongummy"), resistance to a bacteriophage, excess metabolite, antimetabolite, or antibiotic, or additional nutritional requirements (auxotrophy). A causal relationship between a loss of symbiotic capacity and any pleiotropic effects requires consistent simultaneous reversion to both normal cultural and symbiotic performance (by back-mutation or genetic manipulation). However, this may be due to different phenotypic expression of a common gene change, or to a generalized or specific role in the symbiotic step.

Auxotrophy with *R. meliloti* toward adenine, uracil, and leucine is regularly associated with lost symbiotic capacity (Dénarié, Truchet, and Bergeron, 1976). Leu revertants or transductants simultaneously reverted to effectiveness, although some remained ineffective or yielded revertants (20%) in which the nodules contained hemoglobin and bacteroids but failed to reduce acetylene (Kowalski and Dénarié, 1972, and Kowalski, 1974, cited by Dénarié and Truchet, 1976).

If resistance to bacteriophage, high amino acid, and antimetabolite concentrations and to antibiotics involves a change in the bacterial cell envelope (as often occurs) the associated lost symbiotic capacity might be due to an unfavorable effect on either invasion itself or the development of a functioning bacteroid-efficient nodule. Lost infectiveness in small nongummy variants of *R. leguminosarum* and restoration of virulence in gummy revertants suggested to Sanders, Carlson, and Albersheim (1978) that reduced production of extracellular carbohydrate was related to lost nodulating ability. However, this view has to be modified because small colony forms of *R. trifolii* caused typical root-hair response and remained fully infective even though their nodules commonly lacked any detectable N_2-fixing capacity (Vincent, 1954; Yao and Vincent, 1976).

An apparent relationship between nitrate reductase and nitrogenase suggested that selection for the former might predict activity of the latter (Sik et al., 1976). Associated loss of N_2-fixing capacity was indeed found for mutants selected in the in vitro nitrate reductase test. However, the data look less convincing when C_2H_2 reduction versus nitrate reductase activity are plotted. The five revertants in nitrate reductase activity show a similar wide scatter in N_2-fixing capacity. More recently, any structural relationship between nitrate reductase and *nif* mutants has been ruled out (Sik and Barabas, 1976). Pagan et al. (1977) also noted a poor correlation in culture between nitrate reductase mutation and loss of N_2-fixing activity.

Temperature-sensitive, conditional, ineffective mutants of *R. leguminosarum* and *R. trifolii* that affect postinfection stages of the symbiosis at two distinct steps (Beringer, Johnston, and Wells, 1977; Rolfe, unpublished data) hold some promise for experimental manipulation.

Examples of Located Effects

Table 3 lists examples of authentic rhizobial mutants where the effect of the mutation on the symbiosis has been located. Some involve loss of a preinfection capacity, whereas most are involved with infection and nodule development and have had their deficiencies revealed by microscopy. Many are failures to initiate a nodule meristem (Noi); others inhibit release from the infection thread (Bar) or result in abnormal development of the bacteroid after release (Bad). Examples of Nif and Cof lesions are less common, possibly because they are harder to demonstrate. Several mutations are reflected in early nodule senescence (Nop).

Preinfection

Rhizobium trifolii NA34 (also used by Dazzo and Hubbell as strain 2) produces noninvasive mutants. Two mutant forms have been regularly encountered: a large, noninvasive, gummy colony form that is indistinguishable from the effectively nodulating parent (culture 2L on the American scene); and a small, nongummy form that is fully invasive, but ineffective. These mutants have been used, together with other paired substrains, to determine relative adsorption to clover root hairs (Roa). The results agree well with the lectin binding hypothesis and with the demonstration of an "infective" rhizobial antigen by means of exhaustively immunized antiserum (Dazzo and Hubbell, 1975; Dazzo, Napoli, and Hubbell, 1976).

Australia-held NA34 cultures and other cases of lost invasiveness have also been used to determine respective root hair responses. With the SU304 series of NA34 (Yao and Vincent, 1976), the avirulent mutants lost the capacity to cause either branching (Hab) or marked curling (Hac) while still

Table 3. Influence of rhizobial genome

Phenotypic stage	Examples of rhizobial effects[a]	References
I. Preinfection		
1. Roc	Rt 0435, avir	Macgregor and Alexander (1972)
2. Roa	Rt NA 34, avir etc.	Dazzo and Hubbell (1975) Dazzo, Napoli, and Hubbell, (1976)
3. Hab	Rt NA34, avir etc.	Yao and Vincent (1969, 1976)
4. Hac	Rt NA34, avir etc.	
II. Infection and Nodule Formation		
5. Inf	Rt NA34, avir etc.	
6. Noi	Rt T1ts-54	Rolfe (pers. comm.)
	Rt T1-D-ala^{r}2 etc.	Pankhurst (1974)
7. Bar	Rt 220/1	Bergersen (1957b)
	Rt vio^r – 1 etc.	Pankhurst (1974)
	Rt Spc^r – 6	Rolfe (pers. comm.)
	Rt T1/D-his^r – 15(ribo$^-$)	Pankhurst, Schwinghamer, and Bergersen (1972)
	Rl 603/7106	Beringer, Johnston, and Wells (1977)
	Rm leu$^-$	Truchet and Dénarié (1973b)
8. Bad	Rt T42/D-his^r – 7	Pankhurst (1974)
	Rt T1/DL-5-fluorotry^r – 3	
	Rt T1/D-cyc^r – 1	
	Rm ura$^-$	Truchet and Dénarié (1973a)
	Rm 21vio^r	Mackenzie and Jordan (1974)
III. Nodule Function		
9. Nif	Rj sm5	Maier and Brill (1976)
10. Cof	Rj 110-L1 and 76S	Upchurch and Elkan (1978)
	Rsp 32H1, GS$^-$	Kondorosi et al. (1977b)
	Rm GS$^-$	Ludwig and Signer (1977)
11. Nop	Rt T1Neo^r	Rolfe (pers. comm.)
	Rt T1Rif^r	
	Rt f12	Bergersen and Nutman (1957)
	Rl 603/7115	Beringer, Johnson, and Wells (1977)
	Rm ade$^-$	Truchet and Dénarié (1973a)

[a] Rj = *Rhizobium japonicum*; Rl = *R. leguminosarum*; Rm = *R. meliloti*; Rt = *R. trifolii*; Rsp = *Rhizobium* sp. *Resistances* (^{r}): D-ala, D-alanine; D-cyc, D-cycloserine; DL-fluorotry, DL-fluorotryptophan; D-his, D-histidine; Neo, neomycin; Rif, rifamycin; Spc, spectinomycin; Vio, viomycin. *Auxotrophs* ($^-$): ade, adenine; leu, leucine; ribo, riboflavin; ura, uracil. Other features: avir, avirulent; GS, glutamine synthetase.

serologically indistinguishable. The small non- (or low-) gummy mutants of *R. trifolii* were fully invasive and adsorbed on the root. They cause more marked root hair deformation than the gummy parent. These observations differ from those with *R. leguminosarum*, where a close connection exists between loss of virulence and lost gum production (Sanders, Carlson, and Albersheim, 1978).

Infection and Nodule Formation

Resistance-selected mutants of *R. trifolii*, some temperature-conditional mutants (Beringer, Johnston, and Wells, 1977; Rolfe, unpublished data), and some mutants of other species have formed the basis for allocation of these mutations to the various stages following infection. Failure at Noi (Pankhurst, 1974) was reflected in the absence of a nodule meristem and the development of tumorized tissue. Many mutants were restricted in their release from the infection thread (Bar) whereas others, although released into the infected cell, failed to develop further or completely (Bad), e.g., multiple bacteria per envelope (in a host normally producing singles) or conversion to abnormally long thin bacteroids (Pankhurst, 1974).

Nodule Function

Ineffectiveness found among leu^+-revertant leu^- of strain B11 appeared to be due to failure at the Nif step (Dénarié and Truchet, 1976). A direct mutational lesion affecting component II of nitrogenase (*nif*) has been reported for *R. japonicum* (Maier and Brill, 1976).

The plasmid-mediated transfer of the Nif character from an appropriately constructed strain of *R. trifolii* to *nif*-deficient *Klebsiella* (Dunican and Tierney, 1974), *Escherichia* (Skotnicki and Rolfe, 1978), and *Azotobacter* (Bishop et al., 1977, and references therein) seemed to directly involve this gene cluster. In the case of *Azotobacter*, independent transfer of a *R. trifolii* lectin-binding antigen also occurred. Not surprisingly, the converted *Azotobacter* was still not able to nodulate clover. In the *Escherichia* work, the plasmid-modified rhizobium (*R. trifolii* T1K) carried, as well as the expected symbiotic features, other properties characteristic of *E. coli*, from which the plasmid had been originally derived. The same strain also had the *Agrobacterium* capacity to cause tumors when applied by wounding tomato stems. *R. trifolii* T1K is an interesting, if difficult to place, bacterium because it is different in so many ways from the T1 from which it was derived.

Interaction between Nif and Cof is indicated in other somewhat contradictory results. A comparison of two ineffective mutants of *R. japonicum* with the effective parent showed a correlation between lost N_2 fixation and increased capacity to assimilate NH_4^+ in pure culture and as bacteroids (Upchurch and Elkan, 1978). In *R. meliloti*, a glutamate synthase (GOGAT⁻) mutant yielded effective nodules while a glutamine synthetase (GS⁻) mutant was ineffective (Kondorosi et al., 1977b). A similar correlated deficiency occurs with the slow grower 32H1 (Ludwig and Signer, 1977).

Several mutant-completed symbioses fail because of early senescence (Nop). This observation is reminiscent of the naturally occurring, ineffective Coryn strain of *R. trifolii* (Chen and Thornton, 1940) in that the symbiosis appears to have been normal up to this point but fails to continue long enough to achieve much N_2 fixation.

Possible Role of Plasmid (Extrachromosomal) Genes

The idea that extrachromosomal genes may be involved in the symbiotic capability of some rhizobia provides a possible explanation of the frequency with which ability to nodulate and fix N_2 can be lost in rhizobial cultures. Earlier evidence from severe "curing" treatments (Higashi, 1967) and other contradictory evidence (Pariiskaya, 1973, and Zurkowski, Hoffman, and Lorkiewicz, 1973, cited by Dénarié and Truchet, 1976) were problematical. Recently, large plasmids were found in four fast-growing and one of two slow-growing strains. However, infectivity was found in the plasmid-negative slow strain and was lacking in two of the four that contained plasmids (Nuti et al., 1977). A plasmid was not directly involved in the transfer of *nif* to *Azotobacter* (Bishop et al., 1977). Extrachromosomal DNA in *R. meliloti* was linked with polygalacturonase (PGA) induction but all cured strains, despite their low PGA induction ability, retained full symbiotic capacity (Olivares, Montoya, and Palomares, 1977). This observation argues against a role for PGA induction in invasion.

Prospects for Rhizobial Genetics

Techniques are now available for conjugation and chromosome mapping, using the numerous mutants having reasonably well defined symbiotic lesions. Genetic engineering may provide rhizobial strains with better practical performance, but the task of maximizing their benefit goes well beyond the confines of laboratory, greenhouse, or phytotron and points to many unsolved problems, ranging from farmer psychology and manufacturer cupidity to ecological and environmental limitations.

Certain precautions need to be observed. First, complete assurance as to the authenticity of the cultures being used is vital. Are they rhizobia? In the case of claimed mutants, are they in fact genetically derived from the wild parent strain? Much study of "rhizobial" genetics may have been wasted by failure to observe these basic precautions, or by working with a mixed culture; for example, the work on "*R. lupini*" (Heumann et al., 1976), the studies by Balassa (Balassa, 1963), and earlier attempts to identify symbiotic capacity with extrachromosomal DNA. An error in the "plasmid controlled" infectiveness of a soybean culture was fortunately avoided by the awareness and competence of the investigators (Van Rensburg and Strijdom, 1971, 1972a, 1972b). Second, the phenotypic stage of the symbiosis primarily affected by the mutation or genetic manipulation should be closely analyzed. Without such an analysis as that presented herein, more confusion could result. The availability of free-living fixation with some rhizobia permits the study of Nif and Cof stages in an otherwise symbiotically incompetent strain (e.g., see references above and Maier and Brill, 1976).

PHYSIOLOGICAL AND ENVIRONMENTAL EFFECTS

The environment, as much as the host or rhizobial genetic constitution, will determine the success or failure of the symbiosis. Some effects can be recognized by their influence on the symbionts separately (e.g., rhizobia survival and general plant growth vigor). Others determine the establishment or functioning of the symbiosis. Recognition of both the needs of the nodulated legume and the interaction of the symbiotic state and its environment can provide considerable benefit in both the short and long term (see Nutman, 1976, Part IV). An earlier attempt to analyze environmental impact on the major stages of the symbiosis (Vincent, 1965) can be extended within the same framework used for genetic purposes (Table 4).

Preinfection

A good many biotic (inhibitory or stimulatory) factors affect the opportunity for root colonization (Roc) and adsorption (Roa). The first of several stages susceptible to combined nitrogen are those associated with root hair deformation (Hab and Hac), and here a compensating photosynthate (or sugar) supply has a marked effect (Thornton, 1936). Root hair deformation is also pH sensitive (Munns, 1968).

Infection and Nodule Initiation

The first detectable infection step (Inf) is markedly affected by combined nitrogen. However, a very low level of NO_3^- (or NO_2^-), different from the level of NH_4^+ or urea, produces more total infections. Again, glucose reverses this stimulatory effect of nitrate, but only at moderately high concentration. Elimination of the light prevents the nitrate response (Darbyshire, 1966). NO_3^- inhibition of Inf occurred just as frequently with nitrate reductase–deficient mutants; in such cases, any NO_2^- effect would have to be due to the plant reductase (Gibson, Scowcroft, and Pagan, 1977). Temperature also affects infection thread formation: if too low, it causes delay (Roughley, 1970); if too high, it causes abnormal threads in temperature-sensitive strains of *R. trifolii* (Pankhurst and Gibson, 1973). Further

Table 4. Influence of environment[a]

Phenotypic stage	Examples of effects (direct and interacting)	References
I. Preinfection		
1. Roc	Host secretions (stimulating/ inhibitory)	
	Associated microorganisms	
	pH	
	Divalent cations (?)	
	Toxic chemicals	
2. Roa	Ca^{2+}	
3. Hab	↑ NO_3^-	Thornton (1936)
	b Sugar/photosynthate	Munns (1968)
4. Hac	↓ pH	
II. Infection and Nodule Formation		
5. Inf	pH	Munns (1968)
	NO_3^-: sugar/photosynthate	Darbyshire (1966)
	Temperature	Pankhurst and Gibson (1973)
6. Noi	Ca^{2+}	Lowther and Loneragan (1970)
	Temperature-conditional mutant	(Table 3)
7. Bar	Temperature (including conditional mutants)	(Table 3)
	Riboflavin "cure" (with ribo⁻ mutant)	Pankhurst, Schwinghamer, and Bergersen (1972)
8. Bad	Temperature optima	Roughley (1970)
		Roughley, Dart, and Day (1976)
		Pankhurst and Gibson (1973)
III. Nodule Function		
9. Nif	NO_2^-	Gibson, Scowcroft, and Pagan (1977)
	Molybdenum	Bergersen (1977)
10. Cof	NO_2^-	Rigaud and Puppo (1977)
	Photosynthates, CO_2	Hardy et. al. (1976)
	O_2, H_2O	Sprent (1976)
	Sulfur, cobalt, phosphorus, potassium, etc.	
11. Nop	Photosynthates, CO_2	Hardy et. al. (1976)
	Boron	
	NO_3^- etc.	
	Temperature	Pankhurst and Gibson (1973)

[a] General references: Vincent (1965), Pate (1977).

[b] Interacting.

development of the infection process toward nodule initiation (Noi) is calcium dependent (Lowther and Loneragan, 1970). The failure of a temperature-sensitive mutant at this stage has already been noted (Table 3). Too high a temperature also prevents bacterial release (Bar) in other conditional mutants (Table 3) and low temperature causes delay at this step also (Roughley, 1970; Roughley, Dart, and Day, 1976). The supply of riboflavin to the clover host releases from the infection thread of the auxotroph mutant that lacks this synthesizing ability (Table 3). Temperature, too high or too low, also adversely affects bacteroid development (Bad).

Nodule Function

The harmful effects of molybdenum deficiency result directly in nitrogenase deficiency (Nif). The efficiency of N_2 fixation changes during development in pea so that demonstrations of photosynthetic contributions from leaves to root nodules will require careful timing (Bethlenfalvay and Phillips, 1977). Complementary functions (Cof) are subject to many environmental effects, such as the plant's ability to supply photosynthate, oxygen availability to the bacteroid, water for the nodule (Sprent, 1976) and a range of nutrient elements that supply the needs of the nodule system. Some of these factors (photosynthesis, CO_2, combined nitrogen) also determine nodule persistence (Nop). Boron plays a long established role in the vascular system that provides two-way transportation between nodule and shoot. Too high a temperature is unfavorable for the more sensitive strains of *R. trifolii* and for the temperature-conditional mutants that fail at this step (Table 3).

CONCLUSIONS

The period since the first of these symposia has been one of heightened interest and considerable progress in our understanding of the legume-*Rhizobium* symbiosis, particularly in the rapidly advancing field of rhizobial genetics. Some interesting clues as to the basis of nodulating specificity have emerged. The immensely important area of environmental and physiological influences continues as probably the best immediate hope of bridging the gap between laboratory and practice and between the scientist, the agronomist, and the farmer. Many old problems remain, e.g., how to maximize symbiotic contribution with strains already available to the plant and how to introduce, establish, and maintain superior strains. "Super" strains will not get far in practice unless the ecological problems they will encounter in fiercely competitive situations in the soil and on the root are understood and solved. Moreover, even this stage will not be reached until inoculant producers can guarantee a quality product at the point of use. How unfortunate it would be if the hopes aroused and claims made for likely benefit from these symbioses should collapse for want of sufficient attention to more practical issues.

ACKNOWLEDGMENTS

The author appreciates the generosity of the symposium sponsors and the advice and practical help provided by Dr. B. Rolfe and his associates of the Research School of Biological Sciences, Australian National University.

REFERENCES

Albersheim, P., and J. S. Wolpert. 1976. The lectins of legumes are enzymes which degrade the lipopolysaccharides of their symbiont rhizobia. Plant Physiol. 57(Suppl.):416.

Allen, E. K., and O. N. Allen. 1976. The nodulation profile of the genus *Cassia*. In P. S. Nutman (ed.), Symbiotic Nitrogen Fixation in Plants. International Biological Programme 7, pp. 113–121. Cambridge University Press, London.

Balassa, G. 1963. Genetic transformation of *Rhizobium:* A review of the work of R. Balassa. Bacteriol. Rev. 27:228–241.

Bergersen, F. J. 1957a. The occurrence of a previously unobserved polysaccharide in immature infected cells of root nodules of *Trifolium ambiguum* M. Bieb. and other members of the Trifolieae. Aust. J. Biol. Sci. 10:17–24.

Bergersen, F. J. 1957b. The structure of ineffective root nodules of legumes: An unusual new type of ineffectiveness, and an appraisal of present knowledge. Aust. J. Biol. Sci. 10:233–242.

Bergersen, F. J. 1977. Physiological chemistry of dinitrogen fixation by legumes. In: R. W. F. Hardy and W. S. Silver (eds.), A Treatise on Dinitrogen Fixation, Section III, Biology, pp. 519–555. John Wiley & Sons, Inc., New York.

Bergersen, F. J., and P. S. Nutman. 1957. Symbiotic effectiveness in nodulated red clover. IV. The influence of the host factors i and ie upon nodule structure and cytology. Heredity II:175–184.

Beringer, J. E. 1976. The demonstration of conjugation in *Rhizobium leguminosarum*. In: P. S. Nutman (ed.), Symbiotic Nitrogen Fixation in Plants. International Biological Programme 7, pp. 91–97. Cambridge University Press, London.

Beringer, J. E., S. A. Hoggan, and A. W. B. Johnston. 1978. Linkage mapping in *Rhizobium leguminosarum* by means of R plasmid-mediated recombination. J. Gen. Microbiol. 104:201–207.

Beringer, J. E., and D. A. Hopwood. 1976. Chromosomal recombination and mapping in *Rhizobium leguminosarum*. Nature 264:291–293.

Beringer, J. E., and A. W. B. Johnston. 1977. Recent advances in Rhizobium genetics. Soc. Gen. Microbiol. Proc. 4:145–146.

Beringer, J. E., A. W. B. Johnston, and B. Wells. 1977. The isolation of conditional ineffective mutants of *Rhizobium leguminosarum*. J. Gen. Microbiol. 98:339–343.

Bethlenfalvay, G. J., and D. A. Phillips. 1977. Ontogenic interactions between photosynthesis and symbiotic nitrogen fixation. Plant Physiol. 60:419–421.

Bhuvaneswari, T. V., S. G. Pueppke, and W. D. Bauer. 1977. Differential binding of soybean lectin to *Rhizobium*. Plant Physiol. 60:486–491.

Bishop, P. E., F. B. Dazzo, E. R. Appelbaum, R. J. Maier, and W. J. Brill. 1977. Intergeneric transfer of genes involved in the *Rhizobium*-legume symbiosis. Science 198:938–940.

Bisseling, T., R. C. Van Den Bos, A. Van Kommen, M. Van Der Ploeg, P. Van Duijn, and A. Houwers. 1977. Cytofluorometrical determination of the DNA contents of bacteroids and corresponding broth-cultured *Rhizobium* bacteria. J. Gen. Microbiol. 101:79–84.

Bohlool, B. B., and E. L. Schmidt. 1974. Lectins: a possible basis for specificity in the *Rhizobium*-legume root nodule symbiosis. Science 185:269–271.

Bohlool, B. B., and E. L. Schmidt. 1976. Immunofluorescent polar tips of *Rhizobium japonicum*: Possible site of attachment or lectin binding. J. Bacteriol. 125:1184–1194.

Broughton, W. J. 1978. A review: Control of specificity in legume-*Rhizobium* associations. J. Appl. Bacteriol. 45:165–194.

Chen, A. P. T., and D. A. Phillips. 1976. Attachment of Rhizobium to legume roots as the basis for specific interactions. Physiol. Plant. 38:83–88.

Chen, H. K., and H. G. Thornton. 1940. The structure of ineffective nodules and its effect on nitrogen fixation. Proc. R. Soc. Lond. [Biol.] 129:208–229.

Chen, P.-C., and D. A. Phillips. 1977. Induction of root nodule senescence by combined nitrogen in *Pisum sativum* L. Plant Physiol. 59:440–442.

Cole, M. A., and G. H. Elkan. 1973. Antibiotic resistance transfer in *Rhizobium japonicum*. Antimicrob. Agents Chemother. 4:248–253.

Darbyshire, J. F. 1966. Studies on the physiology of nodule formation. IX. The influence of combined nitrogen, glucose, light intensity and day length on root hair infection of clover. Ann. Bot. 30:623–638.

Dart, P. J. 1974. Development of root-nodule symbiosis: The infection process. In: A. Quispel (ed.), The Biology of Nitrogen Fixation. North-Holland Research Monographs, Frontiers of Biology, Vol. 33, pp. 381–429. North-Holland Publishing Company, Amsterdam.

Dart, P. 1977. Infection and development of leguminous nodules. In: R. W. F. Hardy and W. S. Silver (eds.), A Treatise on Dinitrogen Fixation, Section III, Biology, pp. 367–472. John Wiley & Sons, Inc., New York.

Dart, P. J., and F. V. Mercer. 1964. The legume rhizosphere. Arch. Mikrobiol. 47:344–378.

Dazzo, F. B., and W. J. Brill. 1977. The recepter site on clover and alfalfa roots for *Rhizobium*. Appl. Environ. Microbiol. 33:132–136.

Dazzo, F. B., and D. H. Hubbell. 1975. Antigenic differences between infective and noninfective strains of *Rhizobium trifolii*. Appl. Microbiol. 30:172–177.

Dazzo, F. B., C. A. Napoli, and D. H. Hubbell. 1976. Adsorption of bacteria to roots as related to host specificity in the *Rhizobium*-clover symbiosis. Appl. Environ. Microbiol. 32:166–171.

Dénarié, J., and G. Truchet. 1976. Genetics of *Rhizobium:* A short survey. In: W. E. Newton and C. J. Nyman (eds.), Proceedings of the First International Symposium on Nitrogen Fixation, pp. 343–357. Washington State University Press, Pullman.

Dénarié, J., G. Truchet, and B. Bergeron. 1976. Effects of some mutations on symbiotic properties of *Rhizobium*. In: P. S. Nutman (ed.), Symbiotic Nitrogen Fixation in Plants. International Biological Programme 7, pp. 47–61. Cambridge University Press, London.

Dilworth, M. J., and D. R. Coventry. 1977. Stability of leghaemoglobin in yellow lupin nodules. In: W. E. Newton, J. R. Postgate, and C. Y. Rodriguez-Barrueco (eds.), Recent Developments in Nitrogen Fixation, pp. 431–442. Academic Press, London.

Döbereiner, J., R. H. Burris., A. Hollaender, A. A. Franco, C. A. Neyra, and D. B. Scott (eds.). 1978. Limitations and Potentials for Biological Nitrogen Fixation in the Tropics. Basic Life Sciences, Vol. 10. Plenum Publishing Corp., New York.

Dunican, L. K., and A. B. Tierney. 1974. Genetic transfer of nitrogen fixation from *Rhizobium trifolii* to *Klebsiella aerogenes*. Biochem. Biophys. Res. Commun. 57:67–72.

Fåhraeus, G., and H. Ljunggren. 1959. The possible significance of pectic enzymes in root hair infection by nodule bacteria. Physiol. Plant. 12:145–154.

Fåhraeus, G., and K. Sahlman. 1976/77. The infection of root hairs of leguminous plants by nodule bacteria: A review. Ann. Acad. Reg. Sci. Upsaliensis 20:103–131.

Franco, A. A., and J. M. Vincent. 1976. Competition amongst rhizobial strains for the colonization and nodulation of two tropical legumes. Plant Soil. 45:27–48.

Gibson, A. H. 1962. Genetic variation in the effectiveness of nodulation of lucerne varieties. Aust. J. Agric. Res. 13:388–399.

Gibson, A. H. 1964. Genetic control of strain specific ineffective nodulation in *Trifolium subterraneum* L. Aust. J. Agric. Res. 15:37–49.

Gibson, A. H. 1968. Nodulation failure in *Trifolium subterraneum* L. cv Woogenellup (syn. Marrar). Aust. J. Agric. Res. 19:907–918.

Gibson, A. H., W. R. Scowcroft, and J. D. Pagan. 1977. Nitrogen fixation in plants: An expanding horizon? In: W. E. Newton, J. R. Postgate, and C. Rodriguez-Barrueco (eds.), Recent Developments in Nitrogen Fixation, pp. 387–417. Academic Press, London.

Greshoff, P. M., M. L. Skotnicki, J. F. Eadie, and B. G. Rolfe, 1977. Viability of *Rhizobium trifolii* bacteroids from clover root nodules. Plant Sci. Lett. 10:299–304.

Haack, A. 1964. Über den Einfluss der Knöllchenbakterien auf die Wurzelhaare von Leguminosen und Nichtleguminosen. Zentbl. Bakt. Parasitkde (Abt II) 117: 343–366.

Hamblin, J., and S. P. Kent. 1973. Possible role of phytohaemagglutination in *Phaseolus vulgaris* L. Nature 245:28–30.

Hardy, R. W. F., and W. S. Silver (eds.). 1977. A Treatise on Dinitrogen Fixation, Section III, Biology. John Wiley & Sons, Inc., New York.

Heumann, W., W. Kamberger, A. Pühler, A. Rosch, R. Springer, and H. J. Burkardt. 1976. Conjugation and transduction in *Rhizobium lupini*. In: W. E. Newton and C. J. Nyman (eds.), Proceedings of the First International Symposium on Nitrogen Fixation, pp. 383–390. Washington State University Press, Pullman.

Higashi, S. 1967. Transfer of clover infectivity of *Rhizobium trifolii* to *Rhizobium phaseoli* as mediated by an episomic factor. J. Gen. Appl. Microbiol. 13:391–403.

Higashi, S. 1969. Electron microscopic studies on the infection thread developing in the root hair of *Trifolium repens* L. infected with *Rhizobium trifolii*. J. Gen. Appl. Microbiol. 12:147–156.

Holl, F. B., and T. A. LaRue. 1976. Genetics of legume plant hosts. In: W. E. Newton and C. J. Nyman (eds.), Proceedings of the First International Symposium on Nitrogen Fixation, pp. 391–399. Washington State University Press, Pullman.

Hornez, J. P., B. Courtois, D. Défives, and J.-C. Derieux. 1974. Étude des membranes internes dans les bactéroides de *Rhizobium meliloti* au sein des nodules de luzerne (*Medicago sativa*). C. R. Acad. Sci. [D] (Paris) 278:157–160.

Hubbell, D. H., V. M. Morales, and M. Umali-Garcia. 1978. Pectolytic enzymes in *Rhizobium*. Appl. Environ. Microbiol. 35:210–213.

Humphrey, B. A., and J. M. Vincent. 1975. Specific and shared antigens in *Rhizobium meliloti*. Microbios 13:71–76.

Johnston, A. W. B., and J. E. Beringer. 1975. Identifications of the *Rhizobium* strains in pea root nodules using genetic markers. J. Gen Microbiol. 87:343–350.

Johnston, A. W. B., and J. E. Beringer. 1976a. Mixed inoculations with effective and ineffective strains of *Rhizobium leguminosarum*. J. Appl. Bacteriol. 40:375–380.

Johnston, A. W. B., and J. E. Beringer. 1976b. Pea root nodules containing more than one *Rhizobium* species. Nature 263:502–504.

Johnston, A. W. B., and J. E. Beringer. 1977. Chromosomal recombination between Rhizobium species. Nature 267:611–613.

Johnston, A. W. B., S. M. Setchell, and J. E. Beringer. 1978. Interspecific crosses between *Rhizobium leguminosarum* and *R. meliloti:* Formation of haploid recombinants and R primes. J. Gen. Microbiol. 104:209–218.

Kondorosi, A., G. B. Kiss, T. Forrai, E. Vincze, and Z. Banfalvi. 1977a. Circular linkage map of *Rhizobium meliloti* chromosome. Nature 268:525–527.

Kondorosi, A., Z. Svab, G. B. Kiss, and R. A. Dixon. 1977b. Ammonia assimilation and nitrogen fixation in *Rhizobium meliloti*. Molec. Gen. Genet. 157:221–226.

Kowalczuk, E., and Z. Lorkiewicz. 1977. Transfer of R1drd19 plasmid from *Escherichia coli* to *Rhizobium trifolii* by conjugation. Acta Microbiol. Polon. 26:9–18.

Kurz, W. G. W., and T. A. LaRue. 1975. Nitrogenase activity in rhizobia in absence of plant host. Nature 256:407–409.

Labandera, C. A., and J. M. Vincent. 1975. Competition between an introduced strain and native Uruguayan strains of *Rhizobium trifolii*. Plant Soil 42:327–347.

Law, I. J., and B. W. Strijdom. 1977. Some observations on plant lectins and *Rhizobium* specificity. Soil Biol. Biochem. 9:79–84.

Libbenga, K. R., and R. J. Bogers. 1974. Root-nodule morphogenesis. In: A. Quispel (ed.), The Biology of Nitrogen Fixation. North-Holland Research Monographs, Frontiers of Biology, Vol. 33, pp. 430–472. North-Holland Publishing Company, Amsterdam.

Lowther, W. L., and J. F. Loneragan. 1970. Calcium in the nodulation of legumes. In: M. J. Norman (ed.), Proceedings of the XIth International Grassland Congress, pp. 446–450. University of Queensland Press, Brisbane.

Ludwig, R. A., and E. A. Signer. 1977. Glutamine synthetase and control of nitrogen fixation in *Rhizobium*. Nature 267:245–248.

Macgregor, A. N., and M. Alexander. 1972. Comparison of nodulating and non-nodulating strains of *Rhizobium trifolii*. Plant Soil 36:129–139.

MacKenzie, C. R., and D. C. Jordan. 1972. Cell-wall composition and viomycin resistance in *Rhizobium meliloti*. Can. J. Microbiol. 18:1168–1170.

MacKenzie, C. R., and D. C. Jordan. 1974. Ultrastructure of root nodules formed by ineffective strains of *Rhizobium meliloti*. Can. J. Microbiol. 20:755–758.

Maier, R. J., and W. J. Brill. 1976. Ineffective and non-nodulating mutant strains of *Rhizobium japonicum*. J. Bacteriol. 127:763–769.

Marques Pinto, C., P. Y. Yao, and J. M. Vincent. 1974. Nodulating competitiveness amongst strains of *Rhizobium meliloti* and *R. trifolii*. Aust. J. Agric. Res. 25:317–329.

McComb, J. A., J. Elliott, and M. J. Dilworth. 1975. Acetylene reduction by *Rhizobium* in pure culture. Nature 256:409–410.

Meade, H. M., and E. R. Signer. 1977. Genetic mapping of *Rhizobium meliloti*. Proc. Natl. Acad. Sci. USA 74:2076–2078.

Munns, D. N. 1968. Nodulation of *Medicago sativa* in solution culture. I. Acid-sensitive steps. Plant Soil 28:129–146.

Nadler, K. D., and V. J. Avissar. 1977. Heme synthesis in soybean root nodules. Plant Physiol. 60:433–436.

Napoli, C. A., and D. H. Hubbell. 1975. Ultrastructure of *Rhizobium*-induced infection threads in clover root hairs. Appl. Microbiol. 30:1003–1009.

Newcomb, W. 1976. A correlated light and electron microscopic study of symbiotic growth and differentiation in *Pisum sativum* root nodules. Can. J. Bot. 54:2163–2186.
Newton, W. E., and C. J. Nyman (eds.). 1976. Proceedings of the First International Symposium on Nitrogen Fixation. Washington State University Press, Pullman.
Newton, W. E., J. R. Postgate, and C. Rodriguez-Barrueco (eds.). 1977. Recent Developments in Nitrogen Fixation. Academic Press, London.
Nuti, M. P., A. M. Ledeboer, A. A. Lepidi, and R. A. Schilperoort. 1977. Large plasmids in different *Rhizobium* species. J. Gen. Microbiol. 100:241–248.
Nutman, P. S. 1949. Nuclear and cytoplasmic inheritance of resistance to infection by nodule bacteria in red clover. Heredity 3:263–271.
Nutman, P. S. 1959. Some observations on root-hair infection by nodule bacteria. J. Exp. Bot. 10:250–263.
Nutman, P. S. (ed.). 1976. Symbiotic Nitrogen Fixation in Plants. International Biological Programme 7. Cambridge University Press, London.
Olivares, J., E. Montoya, and A. Palomares. 1977. Some effects derived from the presence of extrachromosomal DNA in *Rhizobium meliloti*. In: W. E. Newton, J. R. Postgate, and C. Rodriguez-Barrueco (eds.), Recent Developments in Nitrogen Fixation, pp. 375–385. Academic Press, London.
Paau, A. S., D. Lee, and J. R. Cowles. 1977. Comparison of nucleic acid content in populations of free-living and symbiotic *Rhizobium meliloti* by flow microfluorimetry. J. Bacteriol. 129:1156–1158.
Pagan, J. D., J. J. Child, W. R. Scowcroft, and A. H. Gibson. 1975. Nitrogen fixation by *Rhizobium* cultured on a defined medium. Nature 256:406–407.
Pagan, J. D., W. R. Scowcroft, W. F. Dudman, and A. H. Gibson. 1977. Nitrogen fixation in nitrate-reductase-deficient mutants of cultured rhizobia. J. Bacteriol. 129:718–723.
Pankhurst, C. E. 1974. Ineffective *Rhizobium trifolii* mutants examined by immune-diffusion, gel electrophoresis and electron microscopy. J. Gen. Microbiol. 82:405–413.
Pankhurst, C. E. 1977. Symbiotic effectiveness of antibiotic-resistant mutants of fast and slow growing strains of *Rhizobium* nodulating *Lotus* species. Can J. Microbiol. 23:1026–1033.
Pankhurst, C. E., and A. H. Gibson. 1973. Rhizobium strain influence on disruption of clover nodule development at high root temperature. J. Gen. Microbiol. 74:219–231.
Pankhurst, C. E., E. A. Schwinghamer, and F. J. Bergersen. 1972. The structure and acetylene-reducing activity of root nodules formed by a riboflavin requiring mutant of *Rhizobium trifolii*. J. Gen. Microbiol. 70:161–177.
Pate, J. S. 1977. Functional biology of dinitrogen fixation by legumes. In: R. W. F. Hardy and W. S. Silver (eds.), A Treatise on Dinitrogen Fixation. Section III, Biology, pp. 473–517. John Wiley & Sons, Inc., New York.
Planqué, K., and J. W. Kijne. 1977. Binding of pea lectins to a glycan type polysaccharide in the cell walls of *Rhizobium leguminosarum*. FEBS Lett. 73:64–66.
Quispel, A. (ed.). 1974. The Biology of Nitrogen Fixation. North-Holland Research Monographs, Frontiers of Biology, Vol. 33. North-Holland Publishing Company, Amsterdam.
Rigaud, J., and A. Puppo. 1977. Effect of nitrite upon leghemoglobin and interaction with nitrogen fixation. Biochim. Biophys. Acta 497:702–706.
Rougé, P., and L. Labroue. 1977. On the role of phytohemagglutinins in relation to specific attachment of infectious *Rhizobium leguminosarum* strains to peas. C. R. Acad. Sci. [D] (Paris) 284:2423–2426.

Roughley, R. J. 1970. The influence of root temperature, *Rhizobium* strain and host selection on the structure and nitrogen-fixing efficiency of the root nodules of *Trifolium subterraneum*. Ann. Bot. 34:631–646.

Roughley, R. J., P. J. Dart, and J. M. Day. 1976. The structure and development of *Trifolium subterraneum* L root nodules. II. In plants grown at suboptimal root temperatures. J. Exp. Bot. 27:441–450.

Rovira, A. D. 1961. Rhizobium numbers in the rhizospheres of red clover and paspalum in relation to soil treatment and the numbers of bacteria and fungi. Aust. J. Agric. Res. 12:77–83.

Sahlman, K., and G. Fåhraeus. 1963. An electron microscope study of root-hair infection by *Rhizobium*. J. Gen. Microbiol. 33:425–427.

Sanders, R. E., R. W. Carlson, and P. Albersheim. 1978. A *Rhizobium* mutant incapable of nodulation and normal polysaccharide excretion. Nature 271: 240–242.

Schubert, K. R., and H. J. Evans. 1977. The relation of hydrogen reactions to nitrogen fixation in nodulated symbionts. In: W. E. Newton, J. R. Postgate, and C. Rodriguez-Barrueco (eds.), Recent Developments in Nitrogen Fixation, pp. 469–485. Academic Press, London.

Schwinghamer, E. A. 1977. Genetic aspects of nodulation and dinitrogen fixation by legumes: The microsymbiont. In: R. W. F. Hardy and W. S. Silver (eds.), A Treatise on Dinitrogen Fixation, Section III, Biology, pp. 577–622. John Wiley & Sons, Inc., New York.

Sik, T., and I. Barabas. 1977. The correlation of nitrate reductase and nitrogenase in *Rhizobium* symbiosis. In: W. E. Newton, J. R. Postgate, and C. Rodriguez-Barrueco (eds.), Recent Developments in Nitrogen Fixation, pp. 365–373. Academic Press, London.

Sik, T., A. Kondorosi, I. Barabas, and Z. Svab. 1976. Nitrate reductase and effectiveness in *Rhizobium*. In: W. E. Newton and C. J. Nyman (eds.), Proceedings of the First International Symposium on Nitrogen Fixation, pp. 374–383. Washington State University Press, Pullman.

Sik, T., and L. Orosz. 1971. Chemistry and genetics of *Rhizobium meliloti* phage 16-3. Plant Soil, Special Vol., pp. 57–62.

Skotnicki, M. L., and B. G. Rolfe. 1978. Transfer of nitrogen fixation genes from a bacterium with the characteristics of both *Rhizobium* and *Agrobacterium*. J. Bacteriol. 133:518–526.

Sprent, J. I. 1976. Nitrogen fixation by legumes subjected to water and light stresses. In: P. S. Nutman (ed.), Symbiotic Nitrogen Fixation in Plants. International Biological Programme 7, pp. 405–420. Cambridge University Press, London.

Sutton, W. D., N. M. Jepsen, and B. D. Shaw. 1977. Changes in the number, viability and amino-acid incorporating activity of *Rhizobium* bacteroids during lupin nodule development. Plant Physiol. 59:741–744.

Thornton, H. G. 1936. The action of sodium nitrate upon the infection of lucerne root-hairs by nodule bacteria. Proc. R. Soc. Lond. [Biol.] 119:474–491.

Trinick, M. J. 1973. Symbiosis between *Rhizobium* and the non-legume, *Trema aspera*. Nature 244:459–460.

Trinick, M. J. 1976. *Rhizobium* symbiosis with a non-legume. In: W. E. Newton and C. J. Nyman (eds.), Proceedings of the First International Symposium on Nitrogen Fixation, pp. 507–517. Washington State University Press, Pullman.

Truchet, G., and J. Dénarié. 1973a. Structure et activité réductrice d'acetylene des nodules de Luzerne (*Medicago sativa* L) induits par des mutants de *Rhizobium meliloti* auxotrophes pour l'adenine et pour l'uracile. C. R. Acad. Sci. [D] (Paris) 277:841–844.

Truchet, G, and J. Dénarié. 1973b. Ultrastructure et activité réductrice d'acetylene des nodosités de Luzerne (*Medicago sativa* L) induites par des souches de *Rhizobium meliloti* auxotrophes pour la leucine. C. R. Acad. Sci. [D] (Paris) 277:925–928.

Tsien, H. C., P. S. Cain, and E. L. Schmidt 1977. Viability of *Rhizobium* bacteroids. Appl. Environ. Microbiol. 34:854–856.

Upchurch, R. G., and G. H. Elkan. 1978. Ammonia assimilation in *Rhizobium japonicum* colonial derivatives differing in nitrogen-fixing ability. J. Gen. Microbiol. 104:219–225.

Van Brussel, A. A. N., K. Planqué, and A. Quispel. 1977. The wall of *Rhizobium leguminosarum* in bacteroid and free-living forms. J. Gen. Microbiol. 101:51–56.

Van Egeraat, A. W. S. M. 1975. The possible role of homoserine in the development of *Rhizobium leguminosarum* in the rhizosphere of pea seedlings. Plant Soil 42:381–386.

Van Rensburg, H. J., and B. W. Strijdom. 1971. Stability of infectivity in strains of *Rhizobium japonicum*. Phytophylactica 3:125–130.

Van Rensburg, H. J., and B. W. Strijdom. 1972a. A bacterial contaminant in nodules of leguminous plants. Phytophylactica 4:1–8.

Van Rensburg, H. J., and B. W. Strijdom. 1972b. Information on the mode of entry of a bacterial contaminant into nodules of some leguminous plants. Phytophylactica 4:73–78.

Vincent, J. M. 1954. The root-nodule bacteria of pasture legumes. Proc. Linn. Soc. New South Wales 79:iv-xxxii.

Vincent, J. M. 1965. Environmental factors in the fixation of nitrogen by the legume. In: W. V. Bartholemew and F. E. Clark (eds.), Soil Nitrogen, pp. 384–435. American Society of Agronomy, Madison, Wisc.

Vincent, J. M., and L. M. Waters 1953. The influence of host on competition amongst clover root-nodule bacteria. J. Gen. Microbiol. 9:357–370.

Vincent, J. M., A. S. Whitney, and J. Bose (eds.). 1977. Exploiting the Legume-*Rhizobium* Symbiosis in Tropical Agriculture. College of Tropical Agriculture, Misc. Pub. 145. University of Hawaii.

Williams, L. F., and D. L. Lynch. 1954. Inheritance of a non-nodulating character in the soybean. Agron. J. 46:28–29.

Wilson, M. H. M., B. A. Humphrey, and J. M. Vincent. 1975. Loss of agglutinating specificity in stock cultures of *Rhizobium meliloti*. Arch. Microbiol. 103:151–154.

Yao, P. Y., and J. M. Vincent. 1969. Host specificity in the root hair "curling factor" of *Rhizobium* spp. Aust. J. Biol. Sci. 22:413–423.

Yao, P. Y., and J. M. Vincent. 1976. Factors responsible for the curling and branching of clover root hairs by *Rhizobium*. Plant Soil 45:1–16.

Yu, K. K.-Y., and D. C. Jordan. 1971. Cation content and cation exchange capacity of intact cells and cell envelopes of viomycin-sensitive and -resistant strains of *Rhizobium meliloti*. Can. J. Microbiol. 17:1283–1286.

Zurkowski, W., and Z. Lorkiewicz. 1977. Bidirectional replication of the chromosome in *Rhizobium trifolii*. Mol. Gen. Genet. 156:215–219.

Nitrogen Fixation, Volume II
Edited by W. E. Newton and W. H. Orme-Johnson

Inoculation of Legumes With *Rhizobium* in Competition With Naturalized Strains

G. E. Ham

A more efficient nitrogen-fixing symbiosis could be one means of obtaining efficiency and stability of soybean yields. Many strains of rhizobia in the soil fix insufficient nitrogen for maximum yields. Nodule formation of soybean roots is not a sure indicator of nitrogen fixation. Some rhizobia induce nodule formation and subsist within the roots without adding any nitrogen to the plant and are called ineffective. Other rhizobia are very effective and the nodules supply enough nitrogen for excellent plant growth. Many rhizobia, however, are intermediate in nitrogen fixation. The most economical way to provide additional nitrogen for maximum soybean yields is through the use of more effective strains of *Rhizobium*.

Inoculation is not simply a matter of providing compatible, effective rhizobia for soybeans in a soil that has no soybean rhizobia present. Today, inoculation must meet the challenge of providing superior strains and of inoculating in a manner that will establish the inoculated strains in the soybean nodule on soils containing other naturalized strains of rhizobia. Competition among strains of rhizobia occurs in broth and peat (Marshall, 1956; Vincent, Thompson, and Donovan, 1962; Sherwood and Masterson, 1974) and in the rhizosphere of the legume (Nichol and Thornton, 1941; Read, 1953; Waters, 1956). Such competition can result in rapid changes in the proportions of strains in a common environment. Competition for available infection sites on roots also occurs (Johnson, Means, and Weber, 1965; Brockwell and Dudman, 1968; Caldwell, 1969) and is known to be largely independent of the relative numbers of the bacterial strains to which the roots are exposed. Dominance of a given strain of rhizobia in cultures or on roots appears to be independent of nitrogen-fixing ability.

IDENTIFICATION OF STRAINS

In order to study *Rhizobium* competition in the soil when other microorganisms and usually other rhizobia are present, the strains must be distinguishable. The availability of serological techniques for strain identification (Read, 1953; Marshall, 1956; Date and Decker, 1965; Schmidt, Bankole, and Bohlool, 1968) and other traits, such as the chlorotic factor characteristic of some soybean–*Rhizobium japonicum* interactions (Johnson, Means, and Weber, 1965), permit the success and persistence of inoculant strains to be determined with considerable precision, provided that appropriate controls are used. More recently, the fluorescent antibody technique has been adapted for use in natural habitats by Schmidt, Bankole, and Bohlool (1968). The technique is useful for direct autecological study of microorganisms in natural environments. Because *Rhizobium* detection is strain specific, this technique can be used in the normal nonsterile soil environment.

Serology has been used extensively to identify different strains of *Rhizobium* and to measure the survival of inoculant strains, competition among inoculant strains, and the nature of the indigenous soil rhizobia. Rhizobia need not be isolated from the nodules for identification because antisera may be reacted with crushed nodule suspensions (Means, Johnson, and Date, 1964). Crushed nodule suspension tests have allowed for rapid analysis of nodules from large scale field experiments of competition between strains of rhizobia (Johnson and Means, 1963; Damirgi, Frederick, and Anderson, 1967; Caldwell and Hartwig, 1970; Ham, Cardwell, and Johnson, 1971; Ham, Frederick, and Anderson, 1971; Kapusta and Rouwenhorst, 1973). Other methods of distinguishing *Rhizobium* include phage typing or the use of mutant markers. More recently, enzyme polymorphism has been suggested as a recognition test (Myron, McAdam, and Portlock, 1978).

COMPETITION STUDIES

Competition within Indigenous Soil Populations

Serological results indicated that different populations of rhizobia exist in the nodules of soybeans grown in different soils from Iowa, Mississippi, South Carolina, and Maryland (Johnson and Means, 1963). In five of six soils, 49% to 82% of the isolates were of one or two predominant groups, with the predominant groups differing among soils. The same distribution of serogroups was obtained from plants in a given field and plants grown in the greenhouse in a sample of soil from the same field. The distribution of serogroups in the nodules of the plants grown in the same soil in two successive years was very similar. Caldwell and Hartwig (1970) reported considerable variability among locations in the serological distribution of rhizobia in nodules on field-grown soybeans in Mississippi, North Carolina, South Carolina, and Florida. In an area where soybeans were grown for many years, a population equilibrium was reached and fields in that area had a similar distribution of *R. japonicum*.

Damirgi, Frederick, and Anderson (1967) reported that serogroups 3, 31, and 123 were present in soybean nodules in all areas of Iowa that were tested; however, the frequency of these serogroups varied among soil types. Serogroup 123 was the dominant *R. japonicum* serogroup found, occupying 41% of the nodules in northwestern Iowa, 46% in southwest, 55% in south central, and 63% in east central. Variability between soil sites on the same soil type was less than the variability between soil types. Serogroups 71a, 110, and 117 were found infrequently in all areas, whereas serogroup 135 was predominant in alkaline soils.

Ham, Frederick, and Anderson (1971) studied serogroups of *R. japonicum* present in the nodules of soybeans and their relationship to soil properties on 613 nodule samples obtained at 75 Iowa locations. Serogroup 123 was dominant in both years (67.6% and 62.6% of all nodules sampled) and was found in all areas except where soil pH was above 7.5, in which case serogroup 135 dominated. Soil pH accounted for 81% of the variation in the occurrence of serogroup 135.

Different planting dates influenced the occurrence of serogroups in soybean nodules. The abundance of serogroup c3 was greater in the late planting dates in 2 years and serogroup c2 increased in the third year (Caldwell and Weber, 1970). Serogroup 110 and the composite serogroups 122 to 125 were more frequent in the earliest planted soybeans. The serological distribution also differed when nodules were sampled from plants at different stages of development (Caldwell and Weber, 1970; Weber et al., 1971).

Serogroup recovery was related to temperature (Vincent, 1965). Strains infrequently recovered at 10°C and 20°C became dominant at 30°C, whereas strains forming most of the nodules at 10°C and 20°C formed few nodules at 30°C.

Competition among Introduced Populations or between Introduced and Indigenous Soil Populations

The most difficult situation in the use of rhizobial inocula occurs when it becomes necessary to establish and maintain a strain in a soil containing the same rhizobial species. This establishment is particularly difficult if the natural population is already large at planting time.

Caldwell (1969) studied the competition among three strains of *R. japonicum* introduced on the seed of a single genotype of soybeans into a field free of the organisms. The three strains differed in their competitive ability. Strain 110 produced 98% of the nodules when applied in combination with strain 38 and 76% when in combination with strain 76. Strain 76 produced 95% of the nodules when applied in combination with strain 38. Treatments containing strain 110 alone or in combination had the highest seed yield and the highest protein percentage. The application of strains 38, 76, and 110, singly and in combination, produced from 98% to 100% of the nodules. The remaining nodules were serologically unrelated to the inocula or belonged to the "c1" serogroup found locally. The few indigenous strains did not affect nodulation much in the first year, but occupied more nodule sites in the second and third years of the test, indicating either lack of persistence of applied strains or better competitive ability of indigenous strains (Weber et al., 1971).

Abel and Erdman (1964) inoculated soybeans with 21 strains of *R. japonicum* in soil where soybeans had not been grown before and reported yields of 1564 to 2854 kg of seed per ha. The yield of the uninoculated control was 1369 kg per ha. When the same strains were used as inoculants on seed planted in soil where nodulated soybeans had been grown previously, the resulting seed yields were not significantly different from the control. The results indicated that rhizobia in the soil produced a significant proportion of the nodules. Johnson, Means, and Weber (1965) demonstrated an average of 5% of nodules produced by strains of *R. japonicum* applied at the standard rate of inoculation. The remaining 95% of the nodules were produced by rhizobia present in the soil prior to planting. Caldwell and Vest (1970) evaluated 20 strains and two commercial preparations of *R. japonicum* on five soybean varieties. Yields of inoculated plants were increased significantly over uninoculated plants in three rhizobia-free soils. Introduced strains did not increase seed yield when the soil contained *R. japonicum*. Only 5% to 10% of the nodules were produced by the inoculum on the seed.

Ham, Cardwell, and Johnson (1971) reported that soybean seed yields and seed protein percentage were not significantly increased by inoculating soybean seeds at planting time with three of the more effective strains from inoculation studies on rhizobia-free soils. The uninoculated controls were adequately nodulated by rhizobia in the soil from previous soybean crops. Serotyping of nodules indicated a substantial range of recovery (0% to 17%) depending of the strain and location. The recovery of strain 138 was generally higher than the recovery of strains 110 and 126. Kapusta and Rouwenhorst (1973) found that recovery of *R. japonicum* strain 138 in nodules increased from 18% in the uninoculated plots to 60% in plots receiving 15 billion rhizobia per cm of row. The control contained between 30,000 and 300,000 *R. japonicum* per g of soil and the predominant naturalized serogroup was 123 (63% of nodules on uninoculated plants).

Weaver and Frederick (1974) determined the quantitative relationship between soil and inoculum *R. japonicum* strain 138 in the field. If the inoculum rhizobia are to form 50% or more of the nodules, an inoculum rate of at least 1000 times the soil population must be used. Perhaps with different inoculation techniques or more competitive rhizobia, the ratio of inoculum soil rhizobia could be reduced.

Johnston and Beringer (1976) reported that an ineffective strain of *Rhizobium leguminosarum* was more competitive and influenced the nodulation pattern by an effective strain on the same root system.

Roughley, Blowes, and Herridge (1976) reported naturalized populations of *Rhizobium trifolii* at five sites in Australia ranging from 4×10^6 rhizobia/g to no detectable rhizobia. There were marked differences in competitive ability among the introduced strains and these differences were modified by the host cultivar and the site. At locations where rhizobia were abundant at sowing, they formed most of the nodules regardless of the inoculum used. Marques Pinto, Yao, and Vincent (1974) reported that competitive superiority was not necessarily related to speed of nodule formation or number of nodules produced when the strains were tested individually. The strain of *Rhizobium* capable of forming the most nodules when competing on equal numerical terms was more compatible with the host than was the less successful competitor. The nature of the extra compatibility was not apparent.

EFFECT OF HOST GENOTYPE ON COMPETITION

Caldwell and Vest (1968) showed that differences exist among plant genotypes in their acceptance of *R. japonicum* strains from a mixed, naturalized population in the soil. Specific interactions were noted between soybean genotypes and *R. japonicum* serogroups in nodule formation. Closely related soybean genotypes had similar distributions of rhizobia in their nodules. Roughley, Blowes, and Herridge (1976) reported that *R. trifolii* strain W495 was outstanding on Woogenellup cultivar as compared to Mt. Barker cultivar.

EFFECT OF SOIL MICROORGANISMS ON RHIZOBIA

Damirgi and Johnson (1966) reported that 20 of the 24 isolates of actinomycetes produced no inhibition of eight *R. japonicum* strains. One isolate inhibited only one strain, two other isolates showed slight inhibition on two strains of rhizobia, and one isolate inhibited all eight *R. japonicum* strains tested. When actinomycetes were introduced at planting, the number of nodules formed by strains 76 and 110 was reduced by 40% and 31%, respectively.

Smith and Miller (1974) reported that eight of nine rhizosphere bacterial isolates inhibited *R. japonicum* on agar. However, nodulation was not affected by rhizosphere bacteria when plants were grown in vermiculite or in sterile soil. They suggest that the rhizosphere of the soybean plant not only influences the types and numbers of bacteria present in the rhizosphere but also the ability of the rhizobia to compete and function as free-living bacteria among this active population. The plant influence on rhizobia in the rhizosphere is of prime importance and may actually mask any potential effects of a rhizobial-inhibiting bacterium.

EFFECT OF SOIL ENVIRONMENTAL FACTORS ON RHIZOBIA

For a large part of their existence, rhizobia survive as free-living bacteria before the process of nodule formation is possible. As free-living soil bacteria, rhizobia are affected by pH (Holding and King, 1963; Jones, 1963; Vincent, 1958, 1965; Munns, 1968), temperature (Marshall, 1964; Marques Pinto, Yao, and Vincent, 1974; Roughley, Blowes, and Herridge, 1976), desiccation (Vincent, Thompson, and Donovan, 1962; Foulds, 1971; Hamdi, 1971), soil type (Johnson and Means, 1963; Damirgi, Frederick, and Anderson, 1971; Ham, Frederick, and Anderson, 1971), and soil moisture (Sherwood and Masterson, 1974).

GENERAL CONSIDERATIONS

One of the limitations to developing improved inoculation procedures for soybeans is the lack of information on why one strain of *Rhizobium* succeeds in a given situation and another fails. Some strains have a much greater likelihood of success as inoculum than others. *Rhizobium* added as inoculum must compete with the microflora of the soil and other rhizobia for available substrate necessary for growth and survival in the soil. Also, *Rhizobium* must compete with other rhizobium for nodule sites. Strains that have adapted successfully to the soil environment and compete very favorably with introduced rhizobia need to be examined in order to determine what strain properties lead to widespread distribution.

Establishing a desired *Rhizobium* strain in the nodules of the legume cultivar one chooses to plant in a given field is an important problem now and in the future. Unless a way is found to enhance the competition of applied strains with those in the soil, more effective strain-variety combinations are academic and the selection of better legume cultivars could be hampered because their performance could be affected by soil *Rhizobium* of unknown effectiveness. The naturalized population of *Rhizobium* in breeding nurseries is unknown in many cases. One can only speculate about how many genotypes of soybeans have been chosen on the basis of their efficient symbiosis with a naturalized population at a particular location, and then have failed to meet expectations when tested in other areas where the rhizobial populations are different.

Competition between infective strains of root nodule bacteria during the nodulation of the host plant is a complex phenomenon that occurs between strains of the natural soil population, between naturalized strains and inoculum strains, and between strains applied in a mixed inoculum. The relative success of different strains of *Rhizobium* is probably the result of many factors interacting to affect rhizobial survival and root colonization as well as competition for nodule sites. Such factors are obviously of great importance and their role must be defined and clarified.

REFERENCES

Abel, G. H., and L. H. Erdman. 1964. Response of "Lee" soybeans to different strains of *Rhizobium japonicum*. Argon. J. 56:423–424.

Brockwell, J., and W. F. Dudman. 1968. Ecological studies of the root-nodule bacteria introduced into field environments. II. Initial competition between seed inocula in the nodulation of *Trifolium subterraneum L.* seedlings. Aust. J. Agric. Res. 19:749–757.

Caldwell, B. E. 1969. Initial competition of root-nodule bacteria on soybeans in a field environment. Agron. J. 61:813–815.

Caldwell, B. E., and E. E. Hartwig. 1970. Serological distribution of soybean root nodule bacteria in soils of southeastern USA. Agron. J. 62:621–622.

Caldwell, B. E., and G. Vest. 1968. Nodulation interactions between soybean genotypes and serogroups of *Rhizobium japonicum*. Crop Sci. 8:680–682.

Caldwell, B. E., and G. Vest. 1970. Effect of *Rhizobium japonicum* on soybean seed yields. Crop Sci. 10:19–21.

Caldwell, B. E., and D. F. Weber. 1970. Distribution of *Rhizobium japonicum* serogroups in soybean nodules as affected by planting dates. Agron. J. 62:12–14.

Damirgi, S. M., L. R. Frederick, and I. C. Anderson. 1967. Serogroups of *Rhizobium japonicum* in soybean nodules as affected by soil types. Agron. J. 59:10–12.

Damirgi, S. M., and H. W. Johnson. 1966. Effect of soil actinomycetes on strains of *Rhizobium japonicum*. Agron. J. 58:223–224.

Date, R. A., and A. M. Decker. 1965. Minimal antigenic constitution of 28 strains of *Rhizobium japonicum*. Can. J. Microbiol. 11:1–8.

Foulds, W. 1971. Effect of drought on three species of *Rhizobium*. Plant Soil 35:665–667.

Ham, G. E., V. B. Cardwell, and H. W. Johnson. 1971. Evaluation of *Rhizobium japonicum* inoculants in soils containing naturalized populations of rhizobia. Agron. J. 63:301–303.

Ham, G. E., L. R. Frederick, and I. C. Anderson. 1971. Serogroups of *Rhizobium japonicum* in soybean nodules sampled in Iowa. Agron. J. 63:69–72.

Hamdi, Y. A. 1971. Soil water tension and the movement of rhizobia. Soil Biol. Biochem. 3:121–126.

Holding, A. J., and J. King. 1963. The effectiveness of indigenous populations of *Rhizobium trifolii* in relation to soil factors. Plant Soil 18:191–198.

Johnson, H. W., and U. M. Means. 1963. Serological groups of *Rhizobium japonicum* recovered from nodules of soybeans (*Glycine max*) in field soils. Agron. J. 55:269–271.

Johnson, H. W., U. M. Means, and C. R. Weber. 1965. Competition for nodule sites between strains of *Rhizobium japonicum*. Agron. J. 57:179–185.

Johnston, A. W. B., and J. E. Beringer. 1976. Mixed inoculations with effective and ineffective strains of *Rhizobium leguminosarum*. J. Appl. Bact. 40:375–380.

Jones, D. G. 1963. Symbiotic variation of *Rhizobium trifolii* with S-100 Norwork White Clover [Trifolium repens (L.)]. J. Sci. Food Agr. 14:740.

Kapusta, G., and D. L. Rouwenhorst. 1973. Influence of inoculum size on *Rhizobium japonicum* serogroup distribution frequency in soybean nodules. Agron. J. 65:916–919.

Marques Pinto, C., P. Y. Yao, and J. M. Vincent. 1974. Nodulating competitiveness amongst strains of *Rhizobium meliloti* and *R. trifolii*. Aust. J. Agric. Res. 25:317–29.

Marshall, K. C. 1956. Competition between strains of *Rhizobium trifolii* in peat and broth culture. J. Aust. Inst. Agric. Sci. 22:137–140.

Marshall, K. C. 1964. Survival of root nodule bacteria in dry soils exposed to high temperatures. Aust. J. Agric. Res. 15:273–281.

Means, U. M., H. W. Johnson, and R. A. Date. 1964. Quick serological method of classifying strains of *Rhizobium japonicum* in nodules. J. Bacteriol. 87:547–553.

Munns, D. N. 1968. Nodulation of *Medicago sativa* in solution culture. I. Acid sensitive steps. Plant Soil 28:129–146.

Myron, L. R., N. J. McAdam, and P. Portlock. 1978. Enzyme polymorphism as an aid to identification of *Rhizobium* strains. Soil Biol. Biochem. 10:79–80.

Nicol, H., and H. G. Thornton. 1941. Competition between related strains of nodule bacteria and its influence on infection of the legume host. Proc. R. Soc. Lond. [Biol.] 130:32–59.

Read, M. P. 1953. The establishment of serologically identifiable strains of *Rhizobium trifolii* in field soils in competition with the native microflora. J. Gen. Microbiol. 9:1–14.

Roughley, R. J., W. M. Blowes, and D. F. Herridge. 1976. Nodulation of *Trifolium subterraneum* by introduced rhizobia in competition with naturalized strains. Soil Biol. Biochem. 8:403–407.

Schmidt, E. L., R. O. Bankole, and B. B. Bohlool. 1968. Fluorescent antibody approach to study of rhizobia in soil. J. Bacteriol. 95:1987–1992.

Sherwood, M. T., and C. L. Masterson. 1974. Importance of using the correct test host in assessing the effectiveness of indigenous populations of *Rhizobium trifolii*. Ire. J. Agric. Res. 13:101–108.

Smith, R. S., and R. H. Miller. 1974. Interactions between *Rhizobium japonicum* and soybean rhizosphere bacteria. Agron. J. 66:564–567.

Vincent, J. M. 1958. Survival of the root nodule bacteria. In: Hollswroth (ed.), Nutrition of the Legumes, pp. 108–123. Butterworths, London.

Vincent, J. M. 1965. Environmental factors in the fixation of nitrogen by the legume. In: W. V. Bartholomew and F. E. Clark (eds.), Soil Nitrogen. Agronomy, Vol. 10, pp. 384–385. American Society of Agronomy, Madison, Wisconsin.

Vincent, J. M., J. A. Thompson, and K. O. Donovan. 1962. Death of the root nodule bacteria on drying. Aust. J. Agric. Res. 13:258–270.

Waters, L. M. 1956. Nodulation tests with mixed inoculants. J. Aust. Inst. Agric. Sci. 22:141–142.

Weaver, R. W., and L. R. Frederick. 1974. Effect of inoculum rate on competitive nodulation of *Glycine max*. L. Merrill. II. Field studies. Agron. J. 66:233–236.

Weber, D. F., B. E. Caldwell, C. Sloger, and H. G. West. 1971. Some USDA studies on the soybean-*Rhizobium* symbiosis. Plant Soil, Special Vol., pp. 293–304.

Nitrogen Fixation, Volume II
Edited by W. E. Newton and W. H. Orme-Johnson

Detection, Isolation, and Characterization of Large Plasmids in *Rhizobium*

R. K. Prakash, P. J. J. Hooykaas,
A. M. Ledeboer, J. W. Kijne, R. A. Schilperoort,
M. P. Nuti, A. A. Lepidi, F. Casse, C. Boucher,
J. S. Julliot, and J. Dénarié

An understanding of the genetic control and molecular basis of the reactions involved in the establishment of effective nodules by rhizobia on legume roots should lead to the improvement of nitrogen fixation in legumes and provide a background of knowledge for the assessment of new symbiotic associations. Although the mechanism of initiation of symbiosis is still a matter for speculation, there is much evidence that suggests that the recognition step between bacterium and legume host is mediated by a specific interaction between bacterial surface polysaccharides and plant lectins (Dazzo and Hubbell, 1975; Wolpert and Albersheim, 1976; Planqué and Kijne, 1977). This accounts for one of the so-called symbiotic properties of rhizobia, host specificity (Hsp). Infectiveness (Nod) and effectiveness (Fix) have long been recognized to be unstable characteristics in some strains. This instability (see Beringer, 1976) has led to the hypothesis that these properties might be plasmid borne.

Relatively little has been published about plasmids in rhizobia. DNA sedimentation profiles in sucrose or CsCl/EtBr gradients have been

Part of this work was supported by contracts 0441 and 0414 from the Plant Protein Program of the Commission of the European Communities and grant 78-7-669 from the Délégation Générale à La Recherche Scientifique et Technique.

reported by Klein et al. (1975), Tshitenge et al. (1975), Dunican, O'Gara, and Tierney (1976), and Olivares, Montoya, and Palomares (1977), who studied cleared lysates of different *Rhizobium* species. Sutton (1974) and Zurkowski and Lorkiewicz (1976) characterized their isolated DNAs by renaturation kinetics and electron microscopy, respectively. The calculated molecular weight (MW) of the satellite DNAs ranged from 28 to 64 $\times$ 10^6.

It has been shown previously that other members of the Rhizobiaceae, namely *Agrobacterium tumefaciens*, *Agrobacterium rhizogenes*, *Agrobacterium radiobacter*, and *Agrobacterium rubi* (Zaenen et al., 1974; Currier and Nester, 1976b; Sciaky, Montoya, and Chilton, 1978), harbor large plasmids of MW ranging from about 100 to 160 $\times$ 10^6. *A. tumefaciens* and *A. rubi* are able to induce tumors on dicotyledonous plants, and *A. rhizogenes* provokes root proliferation at wound sites. Several authors have shown that the large plasmid (Ti) confers oncogenicity to *A. tumefaciens* (Van Larebeke et al., 1974; Watson et al., 1975). It was of interest to know whether *Rhizobium* species, which are able to induce cell multiplication in their host plants, also harbor large plasmids.

New procedures have been developed for the isolation and characterization of the large *Agrobacterium* plasmids (Zaenen et al., 1974; Ledeboer et al., 1976; Currier and Nester, 1976a). By sedimentation analysis of lysates on alkaline sucrose gradients and by dye buoyant density centrifugation, large naturally occurring cryptic plasmids have been detected in several *Rhizobium* species (Nuti et al., 1977; Ledeboer, 1978). These plasmids were overlooked in earlier studies because cleared lysates were always used. No findings correlating symbiotic properties with the presence of plasmids in *Rhizobium* have yet been published. Assessment of their ecological and genetical significance can be based on: the detection of such plasmids in a large number of different strains; studies of the purified DNA molecules to detect possible homologies; and genetic studies, such as plasmid curing and transfer.

In this paper, we briefly describe: 1) methods used for the detection and isolation of large DNA molecules in *Rhizobium*; 2) a screening for plasmids in 35 *Rhizobium* strains chosen because of their various geographical origin, previous genetic and physiological studies, or agricultural use for inoculation; 3) biophysical characterization of some plasmids (reassociation kinetics, agarose gel electrophoresis of CCC DNA, or restriction enzyme digest fingerprints); 4) preliminary attempts at curing; and 5) the transfer of Ti plasmids between *Agrobacterium* and *Rhizobium*.

METHODS FOR THE DETECTION AND ISOLATION OF LARGE CCC DNA MOLECULES IN *RHIZOBIUM*

Although the use of the cleared lysate procedure (Clewell and Helinski, 1969) has been successful for the detection and isolation of low molecular weight plasmids in *Rhizobium* (Dunican, O'Gara, and Tierney, 1976; Olivares, Montoya, and Palomares, 1977), difficulties have been encountered in applying this technique to the detection and isolation of large CCC DNA molecules in *A. tumefaciens* (Zaenen et al., 1974; Currier and Nester, 1976a; Ledeboer et al., 1976) and *Rhizobium* (Nuti et al., 1977; Ledeboer, 1978). The use of strongly polar detergents, such as sodium dodecyl sulfate or sarcosinate, which allows DNA dissociation from membrane complexes (Zaenen et al., 1974; Ledeboer et al., 1976), allowed reproducible recovery of such molecules in both *Agrobacterium* and *Rhizobium*. Lysis was found to be more efficient when *Agrobacterium* was preincubated in the presence of carbenicillin (Ledeboer et al., 1976) or when dodecylamine was added to the lysis mixture (Zaenen et al., 1974). These lysis procedures (Zaenen et al., 1974; Ledeboer et al., 1976), followed by velocity sedimentation, allowed detection of large plasmids in both *Agrobacterium* and *Rhizobium*. As shown by Ledeboer et al. (1976) and Nuti et al. (1977), the plasmid nature of isolated DNA could be further demonstrated by reassociation kinetics analysis (Britten and Kohne, 1968) and molecular weights could be estimated by the same technique (Dons, 1975).

Reassociation kinetics is an assay of molecular homogeneity and shows the plasmid nature of a DNA fraction even if the molecules studied are nicked (linear or open circular). This property is very useful with very large molecules that are difficult to isolate and store as CCC DNA, but this procedure is essentially analytical: the isolated DNA is denatured and cannot be used for further physical or biochemical characterization.

Currier and Nester (1976a) developed a preparative procedure for the isolation on a large scale of CCC DNA molecules of MW up to 160×10^6 from *Agrobacterium*. The main characteristics of their method are: neutralization of the sheared lysate immediately after alkaline denaturation to prevent nicking and irreversible denaturation of plasmid DNA; and removal of most of the chromosomal single-stranded DNA by phenol treatment in the presence of 3% NaCl. This procedure has been used recently for *Rhizobium* strains (F. White and E. W. Nester, unpublished data). Slight modifications (R. K. Prakash et al., unpublished data) allow the isolation of plasmids of MW higher than 180×10^6 from *Rhizobium* and the recovery of up to 25–50 μg of plasmid DNA per liter of culture. The modifications consist mainly of: using early log phase cultures and larger initial culture volumes (up to 4 liters); shearing of lysates for only 5–15 sec; and alcohol precipitation and dialysis steps to 1–2 hr each. Flocks formation in the CsCl/EtBr gradients was avoided.

Casse et al. (1979) devised an analytical method by modifying and simplifying the Currier and Nester (1976) procedure for DNA isolation, followed by direct agarose gel electrophoresis of the crude extract. Lysis was performed directly in buffer at pH 12.3 to allow an efficient denaturation of chromosomal DNA while the initial shearing step was avoided. Chloroform extraction was eliminated and during ethanol precipitation magnesium phosphate was replaced by sodium acetate to facilitate the resolubilization of DNA. Agarose gel electrophoresis was then performed according to Meyers et al. (1976). These authors showed that a plot of the relative migration (RM) of CCC DNA through the gel versus the log of plasmid MW (determined by electron microscopy) provided a linear plot for plasmids ranging from 1.9×10^6 to 93×10^6. Similarly, linearity of the curve obtained for plasmids of higher MW ranging from 90 to 140×10^6 was shown to be highly significant by calculation of the linear correlation coefficient (Casse et al., 1979). The equation of the regression line obtained ($\log_{MW} = -2.36 \log_{RM} + 4.32$) was then used to calculate the MW of numerous large *Rhizobium* plasmids (see Table 3).

For a rapid screening and preliminary characterization of large plasmids, the agarose gel electrophoresis technique is very efficient:

1. As many as 20 different strains can be screened for large plasmids by one person within 2 days.
2. Plasmid DNA of high MW can easily be detected; e.g., in the strain *A. tumefaciens* C58-C9, cured of the pTi-C58 plasmid (120×10^6), a large cryptic plasmid not found in previous studies, with a MW greater than 200×10^6, was detected (Figure 1, lane A).
3. Measurement of the relative mobility in the gel under well defined conditions allows an estimation of MW that compares favorably with contour length measurement (Meyers et al., 1976; Casse et al., 1979).
4. It provides a simple method to detect different plasmids in one given strain; for example, *A. tumefaciens* 0362 was shown to carry three different plasmids (Figure 1, lane B). Detection of all the different plasmids in a given strain is important for further genetical and biochemical studies.
5. It can separate plasmids with closely similar MW; e.g., two plasmids of around 125×10^6 were detected in *A. tumefaciens* B6-806 by Genetello et al. (1977), Ledeboer (1978), and Sciaky et al. (1978)—a Ti plasmid and an additional cryptic plasmid. Even though these plasmids could not be distinguished by electron microscopy (Sciaky et al., 1978), they were separated by gel electrophoresis (Figure 1, lane C) within 4 hr.

SCREENING FOR PLASMIDS AND ESTIMATION OF MOLECULAR WEIGHTS

Screening for the presence of plasmids and estimation of their MW for all the strains listed in Table 1 were performed by reassociation kinetics, electron microscopy, or agarose gel electrophoresis. To conform to the nomenclature used by Sciaky et al. (1978) for large cryptic plasmids from

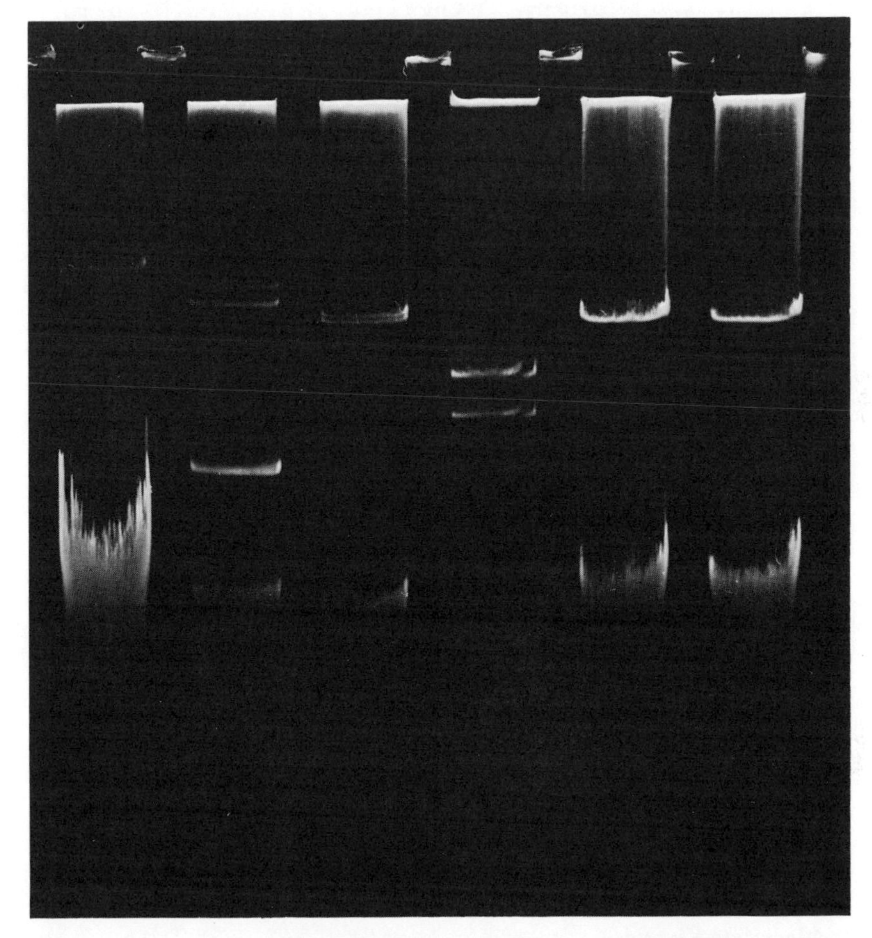

Figure 1. Agarose gel electrophoresis of ethanol-precipitated plasmid DNA from crude lysates of *Agrobacterium* and *Rhizobium* strains. (A), large cryptic plasmid of *A. tumefaciens* C58-C9 cured of pTi C58 plasmid by heat treatment; (B), *A. tumefaciens* 0362, a nopaline-utilizing strain showing three plasmids; (C), *A. tumefaciens* 806 carrying two plasmids of similar MW (around 125×10^6); (D), purified CCC DNA molecules of RP4 (36×10^6) and pGMI 165 C (46×10^6) (Julliot and Boistard, 1979); (E) and (F), *R. meliloti* U45 and 102F51.

strains of *A. tumefaciens*, we designated the naturally occurring plasmids of *Rhizobium leguminosarum*, *Rhizobium meliloti*, and *Rhizobium trifolii* as pRle, pRme, and pRtr, respectively, followed by the number of the strain in which the plasmid was found and serial letters (a, b, c . . .) in case of multiple plasmids, as shown in Table 3.

Table 1. Bacterial strains used and their relevant properties

Strains[a]

Wild-type isolates

R. leguminosarum RCR1001, RYPK/A171, RCR1016 (Nod^-), RCR1018, RCR1039

R. trifolii RCR5; RCR0402 (Nod^-)

R. phaseoli RCR3605

R. japonicum RCR3407

R. lupini RCR3210

R. 'dolichos' RDI[b]

R. meliloti[c] RCR2001, RCR2011, Rf22, Rm12, V7, 54032, I1, S14, 3DoA20a, S26, S33, 102F28, 102F51, B251, U45, U54, Sa10, Lb1, Ls 2a, Ve8, Rm41, L5-30, 311, A145, 1322, Rm11

Mutants isolated in the laboratory

R. trifolii LPR 5002, a $Rif^R Str^R$ derivative of RCR5

R. leguminosarum LPR 1705, an auxotrophic mutant of RYPK/A171 (Dr. J. Kijne, Bot. Lab., University of Leiden, The Netherlands)

R. leguminosarum LPR 1107 (=JB897), a mutant of JB300 that is Trp^- Phe^- Str^r (Dr. J. Beringer, John Innes Institute, Norwich, England)

R. leguminosarum LPR 180, a "rough" derivative of LPR 1705

R. leguminosarum LPR 115 a "rough" derivative of JB897

Properties

Bacteria were tested for symbiotic properties on: *Pisum sativum* (*R. leguminosarum*), *Trifolium pratense* (*R. trifolii*), *Phaseolus vulgaris* (*R. phaseoli*), *Medicago sativa* (*R. meliloti*), *Glycine max.* (*R. japonicum*), *Lupinus luteus* (*R. lupini*), and *Lablab purpureus* (*R. "dolichos"*). They were Nod^+ Fix^+ unless otherwise stated. Agrobacteria were tested for tumorigenicity on *Pisum sativum*, *Kalanchöe daigremontiana*, *Nicotiana tabaccum*, and *Helianthus annuus*.

[a] LPR = Leiden-Pisa Rhizobium Collection; RCR = Rothamsted Collection of Rhizobia, Harpenden, U.K. (Prof. P. S. Nutman); RYPK = Rÿksdienst voor de Ysselmeerpolders, Kampen, The Netherlands (Dr. D. van Schreven); JB = Collection of Dr. J. Beringer, John Innes Institute, Norwich, U.K.

[b] Isolated in Afgoye (somalia).

[c] Casse et al. (1979).

Reassociation Kinetics

By using this procedure, large plasmids were identified in all the strains reported in Table 2, regardless of whether or not they were infective. Molecular weights of the plasmids (Table 2) range between 0.7 and 1.6 $\times$ 10^8 for all except three strains, *R. leguminosarum* LPR 1107 and LPR 1705 and *R. trifolii* RCR5, in which the presence of a plasmid of MW from 3.2 to 4.0 $\times$ 10^8 is suggested. However, this result could be due to the presence of at least two different plasmids of lower but similar MW. Results suggested the presence of more than one plasmid in some of the strains examined. In *R. leguminosarum* LPR 1705 and LPR 1107, two separate peaks were observed in alkaline sucrose gradients (Figure 2) and independently checked for their molecular complexity (Figures 3 and 4). In *R. trifolii* RCR5, on the basis of the comparison of sedimentation profiles and reassociation kinetics, the presence of more than one plasmid was indicated (Nuti et al., 1977).

Electron Microscopy

Contour length measurements, using the procedure described by Zaenen et al. (1974), were performed on *R. leguminosarum* RCR1001 and *R. meliloti* strains L5–50, Rm 12, Rm 41, and 1322 after CCC DNA isolation according to Currier and Nester (1976a). Results are given in Figure 5 and Table 3.

Agarose Gel Electrophoresis

This method was first used to identify plasmids in 25 effective *R. meliloti* strains (Table 1). The geographical origin and characteristics of these strains are described elsewhere (Casse et al., 1979). Large plasmids were found in 22 strains, of which 8 carried more than one plasmid (Figures 1 and 6; Table 3). A calibration curve was drawn by plotting the log of relative migration through the gel versus the log of plasmid molecular weight as determined by contour length measurements by electron microscopy for the plasmids of four *R. meliloti* strains, L5-30, 12, 1322, and 41. These strains were chosen because they carried only one plasmid each with a different relative mobility (see Figure 6) and because of previous genetic studies on strain L5-30 (Dénarié, Truchet, and Bergeron, 1976) and strain 41 (Kondorosi et al., 1977). Molecular weights of the 28 plasmids studied

Table 2. Molecular weights of *Rhizobium* plasmids estimated from C_0t values

Strain	Species	C_0t (1/2) plasmid(s)	Estimated MW ($\times 10^8$)
RCR1001	*R. leguminosarum*	0.07	1.3[a]
LPR1705	*R. leguminosarum*	0.16; 0.06	3.2[b]; 1.2
LPR 1107	*R. leguminosarum*	0.18; 0.07	3.6[b]; 1.3
RCR1016	*R. leguminosarum* (Nod⁻)	0.04	0.7–1.0
RCR5	*R. trifolii*	0.20	3.5–4.0[b]
RCR0402	*R. trifolii* (Nod⁻)	0.08	1.4
RCR3605	*R. phaseoli*	0.07	1.3
RCR2001	*R. meliloti*	0.06	1.2
RCR3407	*R. japonicum*	0.07	1.3
RD1	*R. "dolichos"*	0.09	1.6
LBA601	*A. tumefaciens*	0.06	1.2

[a] A peak of molecular complexity comparable to the larger plasmids of other *R. leguminosarum* strains has sometimes been observed.

[b] The presence of more than one plasmid is suggested on the basis of the sedimentation properties.

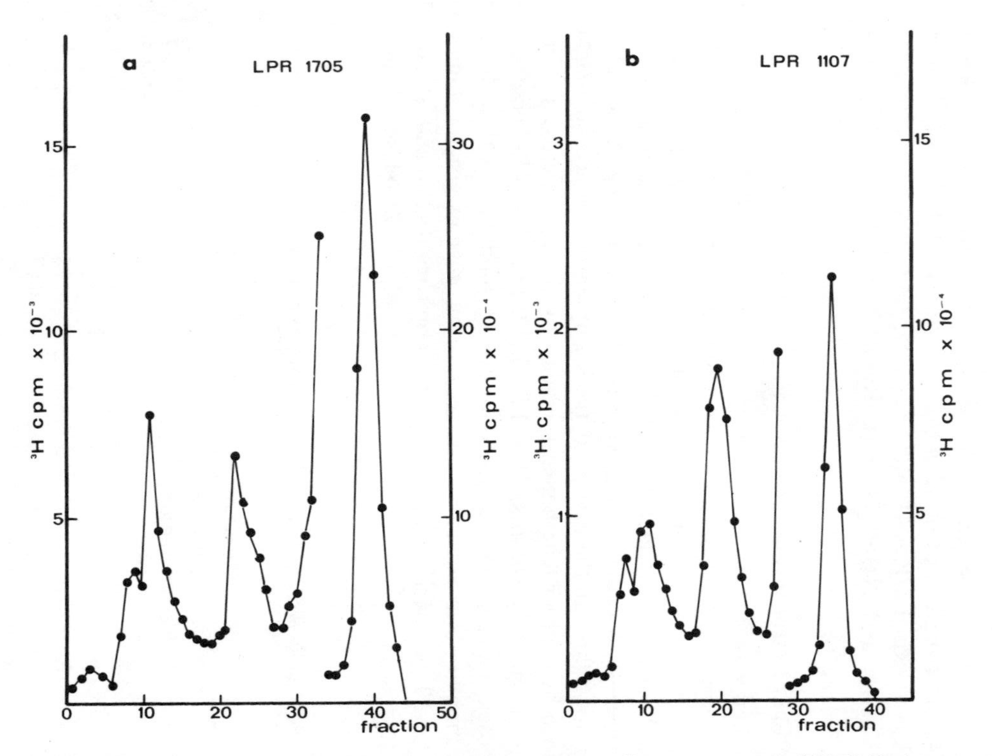

Figure 2. Sedimentation profiles of *R. leguminosarum* lysates. Alkaline sucrose gradients were run in SW 61 Beckman rotor for 17 min at 37,000 rpm and 6°C, and then processed as described by Nuti et al. (1977).

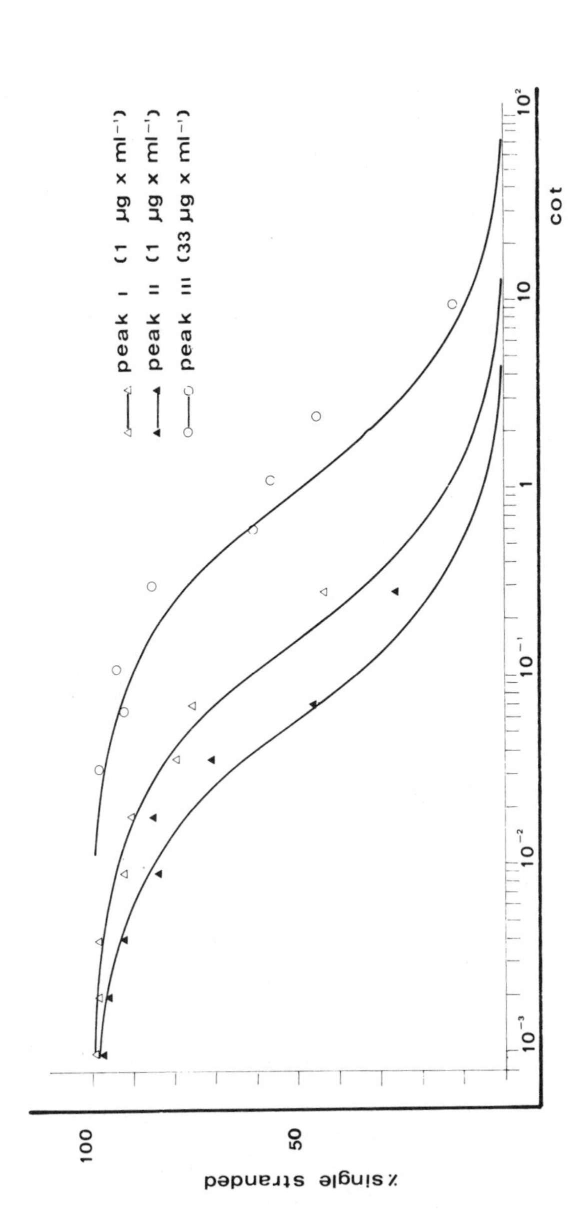

Figure 3. Reassociation kinetics of the two plasmid peaks and chromosomal peak of *R. leguminosarum* LPR 1705 lysate separated as in Figure 2. On the basis of the sedimentation properties, it is suggested that two plasmids having similar MW are present in peak I.

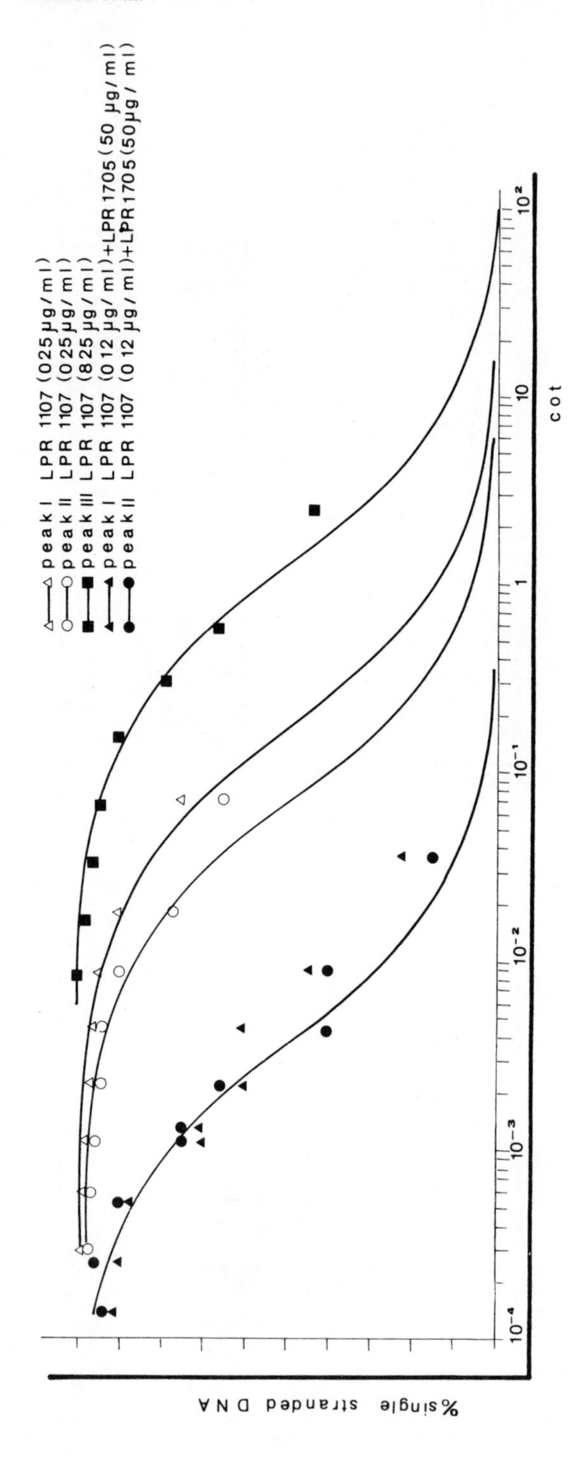

Figure 4. Reassociation kinetics of the plasmids and chromosomal peaks of *R. leguminosarum* LPR 1107. On the basis of the sedimentation profile, it is suggested that two plasmids having similar MW are present in peak I. In order to detect the homology between the different *R. leguminosarum*, the same peaks of LPR 1107 were reassociated in the presence of an excess of total plasmid DNA of LPR 1705.

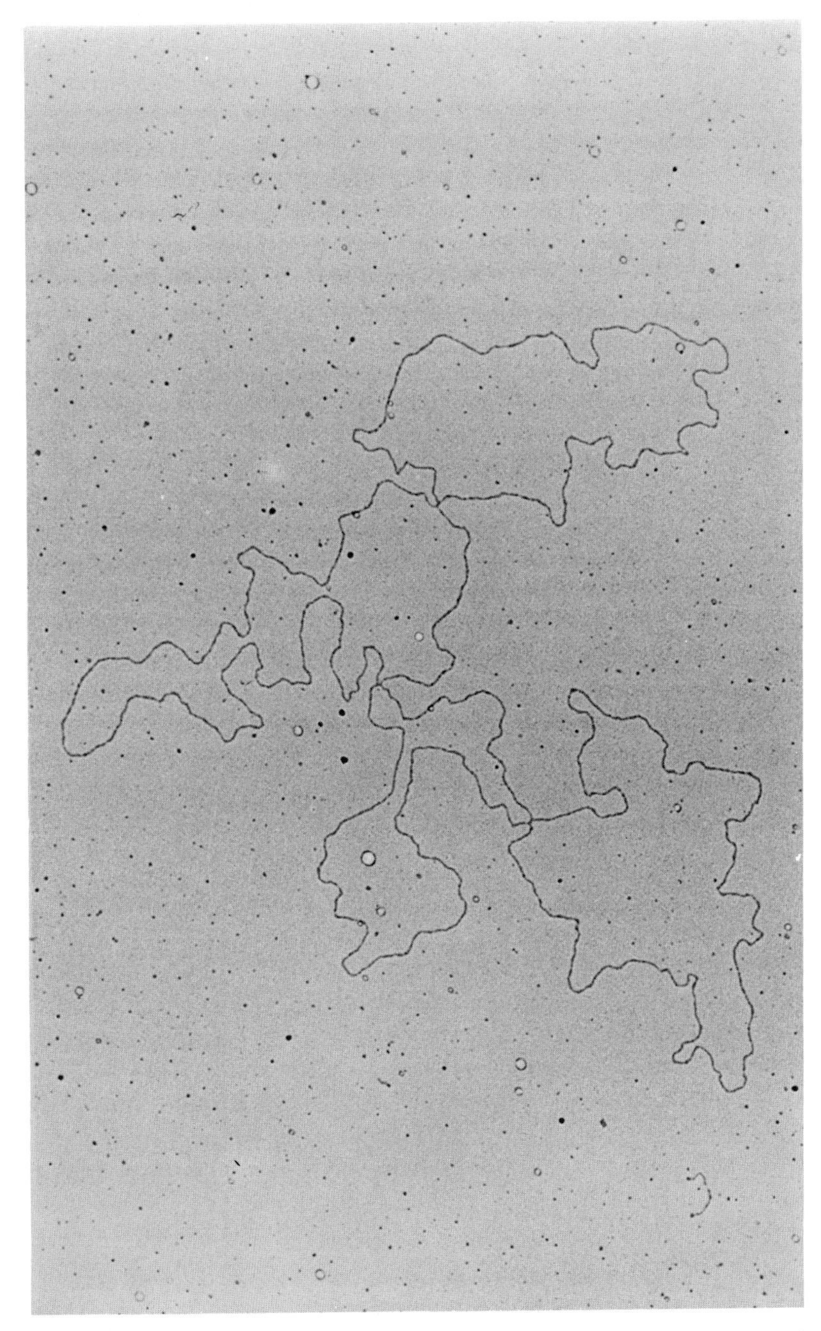

Figure 5. Electron micrograph of an open circular molecule of the plasmid pRme L5-30. Contour length 43.74 $\mu \pm 1.14$.

Table 3. Molecular weight determinations of plasmids by agarose gel electrophoresis and contour length measurements

		Electron microscopy		Gel electrophoresis	
Strains	Plasmids designations	Contour length (μ)	Mol. wt ($\times 10^6$)	Relative mobility	Mol. wt ($\times 10^6$)
R. meliloti					
L5-30	pRme-L5-30	43.74 ± 1.14	91 ± 3	10.00 ± 0.00	91 ± 3
	RP4			13.51 ± 0.11	
102F51	pRme-102F51			9.54 ± 0.10	101 ± 3
U45	pRme-U45			9.52 ± 0.10	102 ± 3
B251	pRme-B251			9.44 ± 0.23	104 ± 4
U54	pRme-U54			8.08 ± 0.10	151 ± 3[a]
1322	pRme-1322	59.21 ± 1.09	123 ± 3	8.95 ± 0.10	118 ± 3
311	pRme-311 a			16.13 ± 0.19	
	pRme-311 b			8.97 ± 0.43	117 ± 7
41	pRme-41	67.42 ± 2.51	140 ± 5	8.31 ± 0.13	140 ± 4
Ls2a	pRme-Ls2a a			9.93 ± 0.08	92 ± 3
	pRme-Ls2a b			9.57 ± 0.08	101 ± 3
102F28	pRme-102F28 a			10.83 ± 0.21	75 ± 4
	pRme-102F28 b			8.95 ± 0.08	118 ± 4
3DoA20a	pRme-3DoA20a			9.74 ± 0.21	98 ± 4
S33	pRme-S33			9.42 ± 0.25	105 ± 4
Balsac	pRme-Bal a			9.77 ± 0.10	96 ± 3
	pRme-Bal b			9.31 ± 0.10	107 ± 3
S14	pRme-S14 a			9.78 ± 0.11	96 ± 3
	pRme-S14 b			8.58 ± 0.15	131 ± 4

V7	pRme-V7			7.49 ± 0.28	>179[a]
54032	pRme-54032			9.4 ± 0.24	105 ± 5
I1	pRme-I1 a			9.82 ± 0.30	95 ± 4
	pRme-I1 b			9.45 ± 0.18	104 ± 3
12	pRme-12	51.93 ± 2.50	108 ± 6	9.33 ± 0.13	107 ± 3
Rf 22	pRme-Rf 22 a			9.83 ± 0.16	94 ± 3
	pRme-Rf 22 b			8.96 ± 0.20	118 ± 4
Ve8	pRme-Ve8 a			9.82 ± 0.13	95 ± 3
	pRme-Ve8 b			9.45 ± 0.12	104 ± 3
A. tumefaciens					
C58	pAt-C58			7.08 ± 0.21	>206[a]
	pTi-C58			8.93 ± 0.12	119 ± 3
R. leguminosarum					
1001	pRle-1001	65.8 ± 1.5	136 ± 2		
A171	pRle-A171 a				110 ± 4
	pRle-A171 b				>194[a]
	pRle-A171 c				>222[a]
LPR 115	pRle-LPR 115 a				99 ± 3
	pRle-LPR 115 b				>164[a]
	pRle-LPR 115 c				>203[a]
R. trifolii					
RCR5	pRtr-RCR 5				>181[a]

[a] MW estimated from the regression straight line but outside the interval in which linearity was demonstrated.

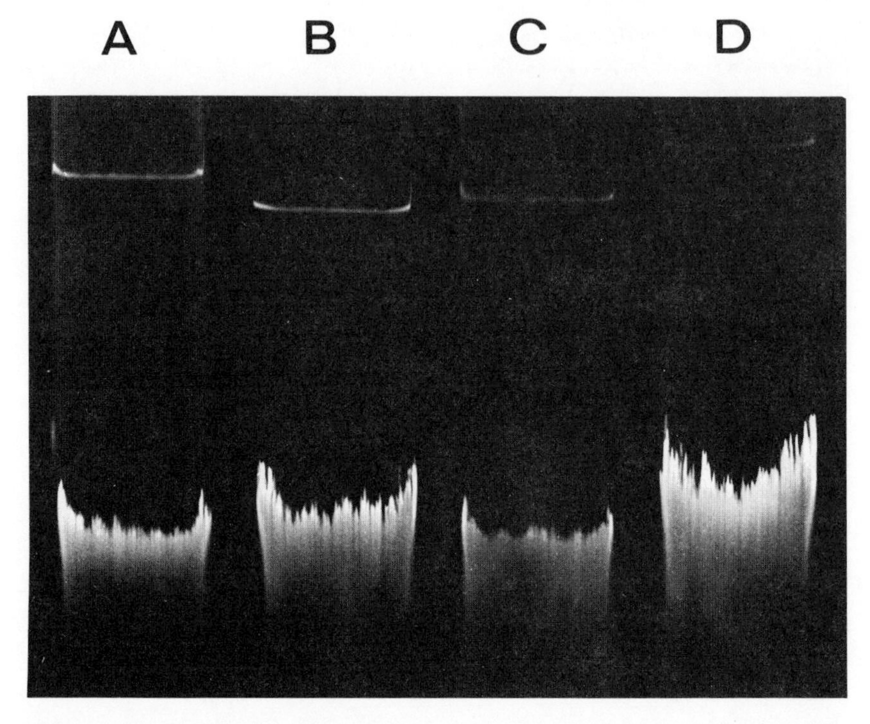

Figure 6. Agarose gel electrophoresis of ethanol-precipitated plasmid DNA from crude lysates of *Rhizobium* and *Agrobacterium* strains. (A), *R. meliloti* U54; (B), *R. meliloti* 12; (C), *R. meliloti* 1322; (D), *A. tumefaciens* C58-C9.

them having a MW of more than 190 × 10^6 (Table 3), explaining the high ranged from about 90 to more than 180 × 10^6 with the exceptions of a small plasmid of MW lower than 30 × 10^6 in strain 311 and a plasmid of 75 × 10^6 in 102F28 (Table 3). No plasmid was detected by either this procedure or CsCl/EtBr gradient centrifugation in the three *R. meliloti* strains RCR 2011, A145, and S26. The failure to detect supercoiled DNA in these strains may be due to method limitation or it may reflect an actual absence of extrachromosomal DNA. The former suggestion gains support from work with alkaline sucrose gradients followed by reassociation kinetics, which showed the presence of a very large plasmid of MW 260 × 10^6 in strain 2011. The plasmid's size has been confirmed by contour length measurement (A. Pühler, unpublished data).

The method was then used on *R. leguminosarum* strains LPR 1705 and LPR 115 and *R. trifolii* RCR5 in which renaturation kinetics suggested the presence of multiple plasmids, including unusually large ones. The results were in agreement with those obtained by reassociation kinetics. LPR 1705 and LPR 115 (derived from JB897) carry at least three plasmids, two of

C_0t values (see Figure 9, lanes C and D; Table 3). However, in *R. trifolii* RCR5, only one large plasmid of more than 1.8×10^8 was found (see Figure 9, lane A; Table 3).

COMPARATIVE STUDIES OF PLASMIDS

The separate labeled plasmids of *R. leguminosarum* LPR 1107 were renatured with unlabeled nonseparated plasmids of LPR 1705. The data indicate that the larger plasmids had about 80% homology with the plasmid of LPR 1705, and the smaller even more (Figure 4). These large plasmids were further characterized by digestion with the restriction endonucleases SmaI and EcoRI. A large number of fragments were detected (Figure 7). A comparison of fragmentation patterns (Figure 7) indicates that the two *R. leguminosarum* plasmids share many common bands, as might be expected from the homology studies (Figure 4). Since each of these two strains carries at least three plasmids, the homology between one class of plasmids can be partly masked on the restriction pattern by nonhomologous plasmids. This drawback was recently pointed out by Sciaky et al. (1978) in a comparative study of *A. tumefaciens* plasmids. Further biochemical comparison of plasmid DNAs should preferably be done with *Rhizobium* strains carrying only one plasmid or with purified DNA of each plasmid.

ATTEMPTS TO "CURE" *RHIZOBIA* OF THEIR LARGE PLASMIDS

A. tumefaciens strains have been cured of their Ti plasmids by culturing them at high temperatures (37°C). Plasmid-cured strains were not oncogenic (Van Larebeke et al., 1974). MacGregor and Alexander (1971) and Sanders, Carlson, and Albersheim (1978) correlated the loss of infectiveness in *Rhizobium* with the inability to form mucous colonies on yeast extract–mannitol salts media. Attempts were made to isolate "rough" mutants from Nod^+ Fix^+ *R. leguminosarum* LPR 1705 by heat treatment. After incubation at 33°C for 2 weeks, 10 rough derivatives were picked and checked for infectiveness on *Pisum sativum*. All rough strains were Nod^-. One of these derivatives (LPR 180) was further characterized. Like the parental strain, LPR 180 does not grow on nutrient broth and requires yeast extract in the medium for full growth. Chromosomal DNA of LPR 180 and LPR 1705 were 100% homologous according to reassociation kinetics analysis. The absence of the 111×10^6 plasmid in LPR 180 was shown by dye-buoyant density centrifugation followed by restriction enzyme fingerprinting (Figure 8) and by agarose gel electrophoresis (Figure 9, lanes C and D). However, the two larger plasmids characteristic for LPR 1705 were still present (Figure 9).

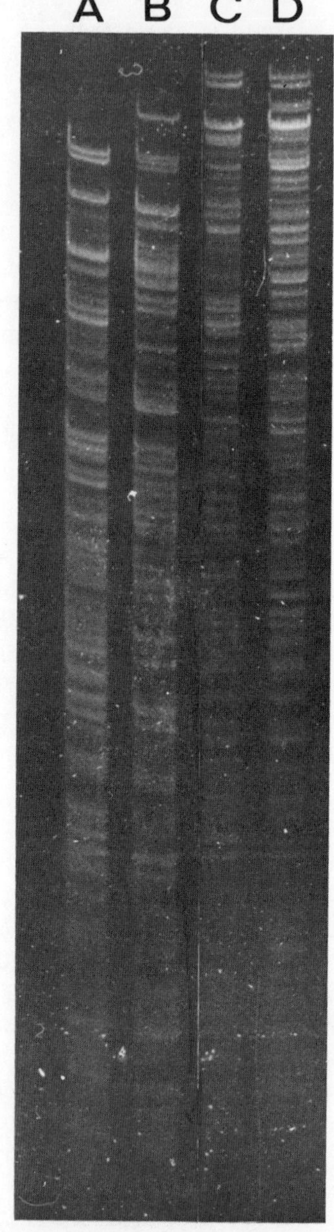

Figure 7. EcoR1 and Sma1 digest fingerprints of plasmid DNA of *R. leguminosarum*. Lanes A and B, Sma1 digests of LPR 1705 and LPR 1107 plasmids, respectively. Lanes C and D, EcoR1 digests of LPR 1705 and LPR 1107 plasmids, respectively.

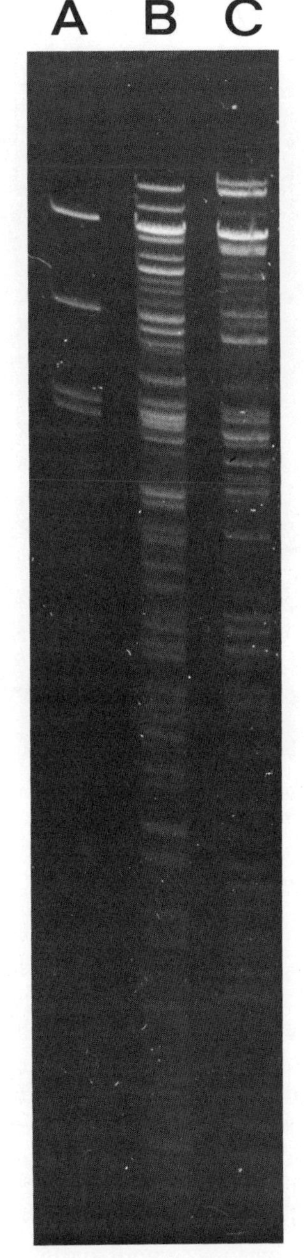

Figure 8. SmaI digest fingerprints of *R. leguminosarum* plasmids of: (A), λ phage; (B), LPR 180 Nod$^-$; (C), LPR 1705 Nod$^+$, Fix$^+$.

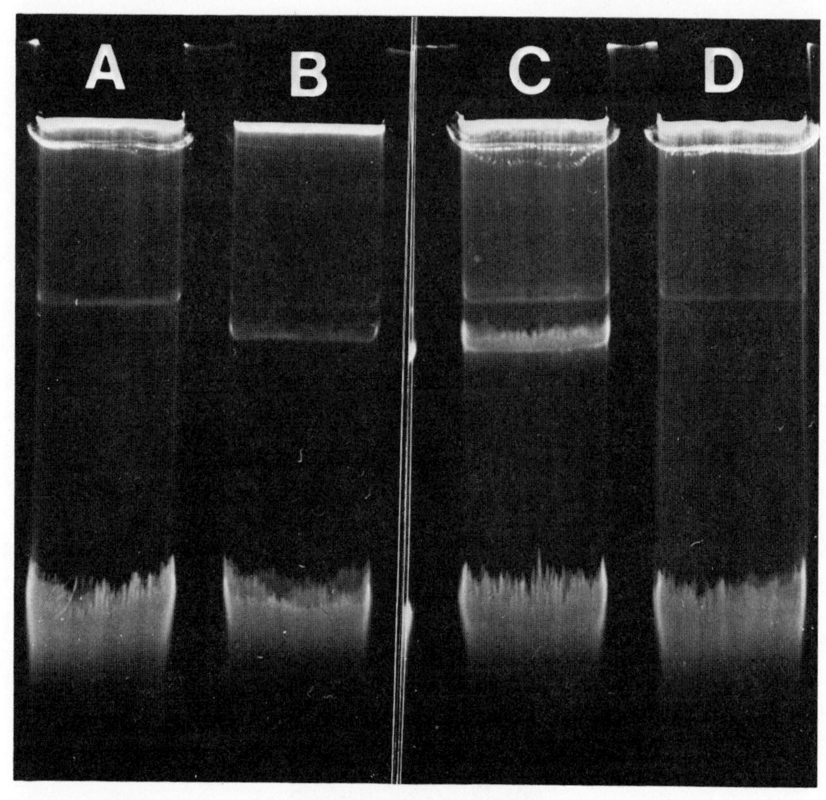

Figure 9. Agarose gel electrophoresis of ethanol-precipitated plasmid DNA from crude lysates of *Rhizobium* strains. (A), *R. trifolii* RCR 5; (B), *R. trifolii* LPR 511, transconjugant of *R. trifolii* RCR5 that has received an octopine Ti plasmid from *A. tumefaciens* LBA 661; (C), *R. leguminosarum* LPR 1705; (D), *R. leguminosarum* LPR 180, a non-nodulating derivative of LPR 1705 cured of its pRle 1705a plasmid.

Phages were isolated that lysed LPR 1705, namely, LPB 1 and LPB 51. Whereas phage LPB 1 gave as many plaques on LPR 180 as on the parental strain, the efficiency of plaque formation of LPB 51 on LPR 180 was severely reduced (about 10^4 lower e.o.p.). The low efficiency of plaque formation by LPB 51 on LPR 180 is probably due to the inability of the phage to attach to this bacterium. Although LPB 51 attaches very efficiently to all *R. trifolii* and *R. leguminosarum* strains tested (including wild-type isolates on which it does not form plaques), it does not attach detectably to LPR 180 or to *R. meliloti* strains. This suggests that a phage receptor site that is common to *R. trifolii* and *R. leguminosarum* strains is absent in LPR 180. An alternative explanation could be that no stable attachment is possible in the absence of a normal amount of exopolysaccharides.

Studies of LPR 180 surface polysaccharide composition showed that this strain contains a lectin-agglutinable fraction, although it was noticed that unhydrolyzed LPS and viable bacteria were not agglutinated (Van Der Schaal et al., unpublished data).

The loss of the smooth colony type does not necessarily mean loss of infectiveness. At least three derivatives of another *R. leguminosarum* strain, LPR 1107 isolated as "rough" mutants by selection for phage resistance to LPB 51, were fully infective and effective on *P. sativum*. The markers of the parental strain (*phe*, *trp*, *str*), except for phage sensitivity, were retained.

TRANSFER OF Ti PLASMIDS BETWEEN *AGROBACTERIUM* AND *RHIZOBIUM* STRAINS

Ti plasmids are large plasmids that confer on *Agrobacterium* strains the ability to induce tumors on dicotyledonous plants. Ti plasmids carry genes that determine whether tumors induced on *Kalanchöe daigremomtiana* will have a rough or a smooth surface and whether or not they will contain unusual amino acids, such as octopine (N^2-(D-carboxyethyl)-L-arginine) or nopaline (N^2-(1,3-dicarboxypropyl)-L-arginine). The plasmids enable the bacteria to utilize these unusual amino acids. Ti plasmids can be introduced by conjugation into strains lacking the plasmid. They can be mobilized by RP4 (Chilton et al., 1976; Hooykaas et al., 1977; Van Larebeke et al., 1977) and other R plasmids of the incompatibility group P-1 (Hooykaas, Roobol, and Schilperoort, 1979). Ti^+ transconjugants can be selected by their ability to form normal colonies on solid media containing octopine or nopaline as the sole nitrogen source, provided that the nitrogen-containing compounds in agar are removed by extensive washing. Hooykaas, Roobol, and Schilperoort (1979) have developed an indicator medium on which Ti plasmid–containing strains form orange-yellow colonies, whereas the colonies of the plasmid-lacking strains remain translucent. In these experiments donors were counterselected by use of an antibiotic to which the recipient is resistant. Recently, it was found that Ti plasmids can act as conjugative plasmids themselves; however, conjugation only takes place when donor and recipient bacteria are incubated together on a solid minimal medium containing octopine for octopine Ti plasmids (Kerr, Manigault, and Tempé, 1977; Genetello et al., 1977; Petit et al., 1978; Hooykaas, Roobol, and Schilperoort, 1979). Mutants can be isolated that carry plasmids derepressed for transfer (Petit et al., 1978; Hooykaas, Roobol, and Schilperoort, 1979). We tried to introduce Ti plasmids into *Rhizobium:* 1) to determine if these plasmids are of the same incompatibility group as *Rhizobium* plasmids, which could provide a method to cure *Rhizobium* of their own plasmids; and 2) to compare the expression of Ti in *Agrobacterium* and *Rhizobium* in order to estimate the physiological similarities of

Table 4. Characteristics of *Rhizobium* strains harboring Ti plasmids as compared with reference strains[a]

	RCR 5	LBA 57	LPR 501	LBA 643	LBA 661	LPR 511	LBA 672	LPR 518	LBA 8374	LPR 519
Ti plasmid	−	+	+	+	+	+	+	+	+	+
Tumor induction on:										
Kalanchoë daigremontiana	−	+	+	+	+	+	+	+	+	+
Pisum sativum	−	+	−	+	+	−	+	+	+	NT
Tumor morphology on *K. daigremontiana*		rough	rough	rough	rough	rough	smooth	smooth	smooth	
Synthesis of guanidines in tumors		octopine	octopine	octopine	octopine	octopine	nopaline + octopine	NT	nopaline	NT
Breakdown of octopine	−	+	+	+	+	+	+	+	+	+
Breakdown of nopaline	−	−	−	−	−	−	+	+	+	+
Sensitivity to:										
Phage LPB 1	+	−	+	−	−	+	−	+	−	+
Phage LPB 51	+	−	+	−	−	+	−	+	−	+
Nodulation of										
Trifolium pratense	+	−	+	−	−	+	−	+	+	NT
Nitrogen fixation in nodules	+	−	+	−	−	+	−	+	−	NT

[a] Origin of strains: LBA 57 (*A. tumefaciens* B6S3 Met^-Ilv^-(RP4)); LPR 501 (RCR5 harboring Ti(B6) from LBA 57); LBA 643 *A. tumefaciens* C58 cured, harboring the Ti plasmid from LPR 501); LBA 661 (C58 cured carrying Ti(B6)); LPR 511 (LPR 5002 with Ti(B6)); LBA 672 (*A. tumefaciens* with a cointegrate plasmid of Ti(B6) and Ti(C58)); LPR 518 (LPR 5002 with the plasmid of LBA672); LBA 8374 (T37 with a constitutive nopaline plasmid); LPR 519 (LPR 5002 carrying the Ti plasmid of LBA 8374).

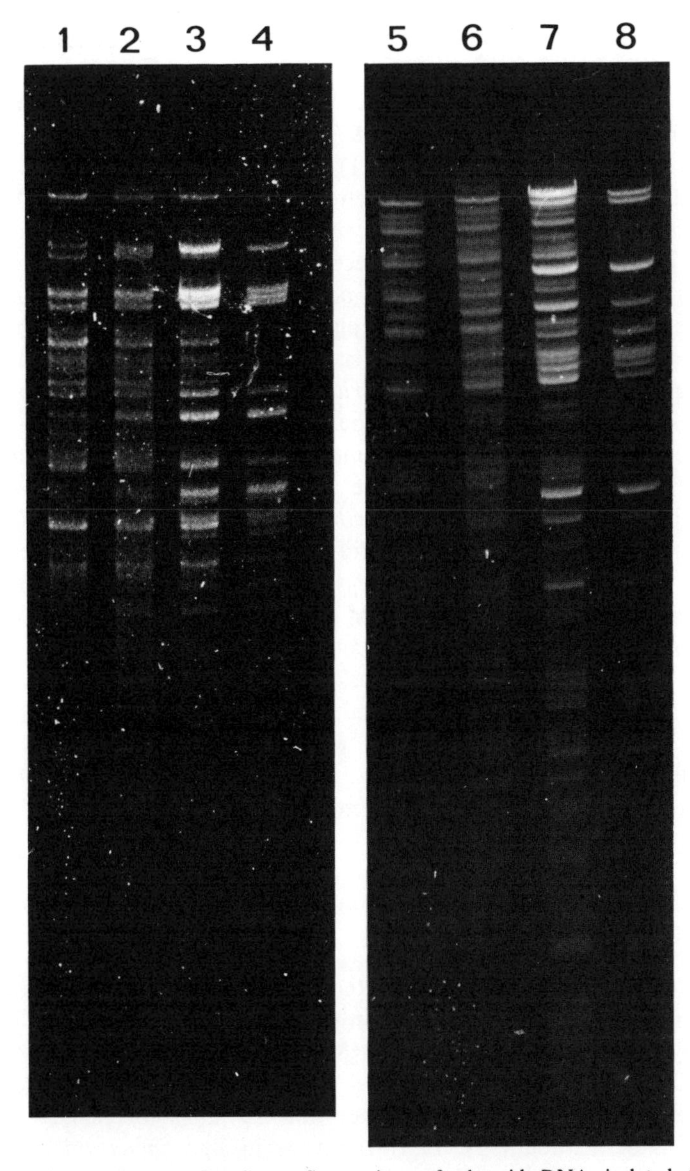

Figure 10. SmaI and EcoRI digest fingerprints of plasmid DNA isolated from the exconjugant *R. trifolii* LPR 511 compared with plasmid DNAs of the parental strains *R. trifolii* LPR 5002 and *A. tumefaciens* LBA 661. (1). SmaI, LPR 5002 (CCC band); (2), SmaI, LPR 5002 (linear and OC band of the CsCl/EtBr gradient); (3), SmaI, LPR 511 (CCC band); (4), SmaI, LBA 661 (CCC band); (5), EcoRI, LPR 5002 (CCC band); (6), EcoRI, LPR 5002 (linear and OC band of the CsCl/EtBr gradient); (7), EcoRI, LPR 511 (CCC band); (8), EcoRI, LBA 661 (CCC band). Comparison of lanes (1) and (2) and of lanes (5) and (6) shows that, after DNA isolation by the Currier and Nester (1976a) procedure, the upper band of the CsCl/EtBr gradient contains large amounts of linear and OC plasmid DNA.

bacteria of these two genera, including the expression of genes controlling relationships with a host plant.

Ti plasmids were introduced into *R. trifolii* (RCR5) either by mobilization with RP_4 (Hooykaas et al., 1977) or by per se transfer (Table 4). Transfer to *R. leguminosarum* (A171) and *R. meliloti* (RCR2001) was not observed, possibly because of restriction of the incoming DNA. After the introduction of a Ti plasmid, *R. trifolii* strains become able to induce tumors. Depending on the plasmid received, *R. trifolii* (like *A. tumefaciens*) induced smooth tumors containing nopaline or rough tumors containing octopine. The Ti plasmids were maintained stably in *Rhizobium*; spontaneous loss of markers was not observed. Ti plasmids could be transferred back from *Rhizobium* to *Agrobacterium;* transfer from *Rhizobium* (in the absence of a mobilizing R factor) could take place only when the bacteria were incubated on a solid minimal medium containing octopine (for octopine Ti plasmids). Plasmids that are derepressed for transfer in *A. tumefaciens* behaved identically in *R. trifolii.* Remarkably, *R. trifolii* strains containing an octopine Ti plasmid did not induce tumors on pea epicotyls, whereas *A. tumefaciens* strains harboring the same plasmid did. *Rhizobium* strains carrying a nopaline Ti plasmid or cointegrate plasmid of an octopine and a nopaline plasmid were virulent on the pea.

On introduction of the Ti plasmid, the large plasmid that is naturally present in *R. trifolii* RCR5 remained, which indicates that these plasmids are compatible. Evidence for that was obtained from alkaline sucrose gradients, restriction fingerprinting, and by agarose gel electrophoresis of some exconjugants containing an octopine Ti plasmid (Figure 9, lanes A and B, and Figure 10).

SUMMARY

A property of the gram-negative bacteria grouped in the Rhizobiaceae is the ability to induce cell multiplication in the plant host. The presence of large, naturally occurring plasmids having MW ranging from 90×10^6 to more than 220×10^6 seems to be a general feature of such bacteria, except for a few strains of *R. meliloti*, where they may remain undetected because of limitations of methodology. In rhizobia, as in *Agrobacterium*, more than one large plasmid is often detected in a given strain and therefore, in some strains, more than 10% of the genetic information is extrachromosomal. It is worthy of note that plasmids of MW less than 90×10^6 are rarely found and that very small plasmids suitable as cloning vehicles for genetic engineering have not been detected.

Ti plasmids were transferred into a *Rhizobium* strain. Transconjugants were able to induce tumors but were still able to form effective nodules on the legume hosts. The type of induced tumors and the regulation of Ti transfer were similar in both bacteria, which showed close physiological similarities even at the regulatory level. The Ti plasmid was shown to be compatible with the *Rhizobium* plasmid in this strain. A bacterial strain has been recently reported that can induce both nodule formation on clover and crown-gall on a suitable host (Skotnicki and Rolfe, 1978).

Curing *A. tumefaciens* C58 of its Ti plasmid (120×10^6) by heat treatment has been shown to give rise to nononcogenic derivatives still harboring a very large cryptic plasmid (more than 200×10^6). Similarly, in one *Rhizobium* strain a plasmid of 110×10^6 was lost on heat treatment while the other two, having a larger size, remained. This cured *Rhizobium* strain lost its nodule-inducing ability, suggesting that this plasmid could play a role in the control of infectivity.

The analytical and preparative methods for studying of large plasmid DNAs described in this chapter proved to be reliable for detection and characterization of plasmids of MW higher than 150×10^6.

Understanding of plasmid biological significance requires further genetic studies. Genetic markers on the plasmids are needed to allow easier detection of curing and transfer. These markers could be either natural ones, such as phage sensitivity and bacteriocin sensitivity (or production), or artificially introduced ones, such as transposons. The recent developments of chromosome mapping procedures (Beringer and Hopwood, 1976; Meade and Signer, 1977; Kondorosi et al., 1977; Julliot and Boistard, 1979) for *Rhizobium* species should allow an analysis of the respective roles of the chromosome and the plasmids in the genetic control of symbiotic properties.

ACKNOWLEDGMENTS

The authors are indebted to Dr. J. E. Beringer for a critical reading of the manuscript.

REFERENCES

Beringer, J. E. 1976. Plasmid transfer in *Rhizobium*. In: W. E. Newton and C. J. Nyman (eds), Proceedings of the 1st International Symposium on Nitrogen Fixation, pp. 693–717. Washington State University Press, Pullman.

Beringer, J. E., and D. A. Hopwood. 1976. Chromosomal recombination and mapping in *Rhizobium leguminosarum*. Nature 264:291–293.

Britten, R. J., and D. E. Kohne. 1968. Repeated sequences in DNA. Science 161:529–540.

Casse, F., C. Boucher, J. S. Julliot, and J. Dénarié. 1979. Identification and characterization of large plasmids in *Rhizobium meliloti* using agarose gel electrophoresis. J. Gen. Microbiol. 113:229–242.

Chilton, M. D., S. K. Farrand, R. Levin and E. W. Nester. 1976. RP_4 promotion of transfer of a large *Agrobacterium* plasmid which confers virulence. Genetics 83:609–618.

Clewell, D. B., and D. R. Helinski. 1969. Supercoiled circular DNA-protein complex in *Escherichia coli:* Purification and induced conversion to an open circular DNA form. Proc. Natl. Acad. Sci. USA 62:1159–1166.

Currier, T. C., and E. W. Nester. 1976a. Isolation of covalently closed circular DNA of high molecular weight from bacteria. Anal. Biochem. 76:431–441.

Currier, T. C., and E. W. Nester. 1976b. Evidence for diverse types of large plasmids in tumor-inducing strains of *Agrobacterium*. J. Bacteriol. 126:157–165.

Dazzo, R. B., and B. H. Hubbell. 1975. Antigenic differences between infective and noninfective strains of *Rhizobium trifolii.* Appl. Microbiol. 30:1017–1033.

Dénarié, J., G. Truchet, and B. Bergeron. 1976. Effects of some mutations on symbiotic properties of *Rhizobium*. In: P. S. Nutman (ed.), Symbiotic Nitrogen Fixation in Plants, pp. 47–61. Cambridge University Press, London.

Dons, J. J. M. 1975. Crown-gall—a plant tumor. Ph.D. thesis, Leiden University, Leiden, The Netherlands.

Dunican, L. K., R. O'Gara, and A. B. Tierney. 1976. Plasmid control of effectiveness in *Rhizobium:* Transfer of nitrogen-fixing genes on a plasmid from *Rhizobium trifolii* to *Klebsiella aerogenes*. In: P. S. Nutman (ed.), Symbiotic Nitrogen Fixation in Plants, pp. 77–90. Cambridge University Press, London.

Genetello, C., N. Van Larebeke, M. Holsters, A. de Picker, M. Van Montagu, and J. Schell. 1977. Ti plasmids of *Agrobacterium* as conjugative plasmids. Nature 265:561–563.

Hooykaas, P. J. J., P. M. Klapwijk, M. P. Nuti, R. A. Schilperoort, and A. Rörsch. 1977. Transfer of the *Agrobacterium tumefaciens* Ti plasmid to avirulent Agrobacteria and to Rhizobium *ex planta*. J. Gen. Microbiol. 98:477–484.

Hooykaas, P. J. J., C. Roobol, and R. A. Schilperoort. 1979. Regulation of the transfer of Ti plasmids in *Agrobacterium tumefaciens*. J. Gen. Microbiol. 110:99–109.

Julliot, J. S., and P. Boistard. 1979. Use of RP_4^- prime plasmids constructed *in vitro* to promote a polarized transfer of the chromosome in *Escherichia coli* and *Rhizobium melilott*. Mol. Gen. Genet. 173:289–298.

Kerr, A., P. Manigault, and J. Tempé. 1977. Transfer of virulence *in vivo* and *in vitro* in *Agrobacterium*. Nature 265:560–561.

Klein, G. A., P. Jemison, R. A. Haak, and A. G. Mathijsse. 1975. Physical evidence of a plasmid in *Rhizobium japonicum*. Biochim. Biophys. Acta 44:357–361.

Kondorosi, A., E. G. Kiss, T. Forrai, E. Vincze, and Z. Barfalvi. 1977. Circular linkage map of *Rhizobium meliloti* chromosome. Nature 268:525–527.

Ledeboer, A. M. 1978. Large plasmids in Rhizobiaceae. Ph.D. thesis, Leiden University, Leiden, The Netherlands.

Ledeboer, A. M., A. J. M. Krol, J. J. M. Dons, R. Spier, R. A. Schilperoort, I. Zaenen, N. van Larebeke, and J. Schell. 1976. On the isolation of Ti plasmid from *Agrobacterium tumefaciens*. Nucleic Acid Res. 3:449–463.

MacGregor, A. N., and M. Alexander. 1971. Formation of tumor-like structures on legume roots by *Rhizobium* J. Bacteriol. 105:728–732.

Meade, H. M., and E. R. Signer. 1977. Genetic mapping of *Rhizobium meliloti*. Proc. Natl. Acad. Sci. USA 74:2076–2078.

Meyers, J. A., D. Sanchez, L. P. Elwell, and S. Falkow. 1976. Simple agarose gel electrophoretic method for the identification and characterization of plasmid deoxyribonucleic acid. J. Bacteriol. 127:1529–1537.

Nuti, M. P., A. M. Ledeboer, A. A. Lepidi, and R. A. Schilperoort. 1977. Large plasmids in different *Rhizobium* species. J. Gen. Microbiol. 100:241–248.

Olivares, J., E. Montoya, and A. Palomares. 1977. Some effects derived from the presence of extrachromosomal DNA in *Rhizobium meliloti*. In: W. E. Newton, J.

R. Postgate, and C. Rodriguez-Barrueco (eds.), Recent Developments in Nitrogen Fixation, pp. 375–385. Academic Press, London.

Petit, A., J. Tempé, A. Kerr, M. Holsters, M. van Montagu, and J. Schell. 1978. Substrate induction of conjugative activity of *Agrobacterium tumefaciens* Ti plasmids. Nature 271:570–572.

Planqué, K., and J. Kijne. 1977. Binding of pea lectins to a glycan type polysaccharide in the cell walls of *Rhizobium leguminosarum*. FEBS Lett. 73:59–65.

Sanders, R. E., R. W. Carlson, and P. Albersheim. 1978. A *Rhizobium* mutant incapable of nodulation and normal polysaccharide excretion. Nature 271:240–242.

Sciaky, D., A. L. Montoya, and M. D. Chilton. 1978. Fingerprints of *Agrobacterium* Ti plasmids. Plasmid 1:238–253.

Skotnicki, M. L., and B. G. Rolfe. 1978. Transfer of nitrogen fixation genes from a bacterium with the characteristics of both *Rhizobium* and *Agrobacterium*. J. Bacteriol. 133:518–526.

Sutton, W. D. 1974. Some features of the DNA of *Rhizobium* bacteroids and bacteria. Biochim. Biophys. Acta 366:1–10.

Tshitenge, G., N. Luyundula, P. F. Lurquin, and L. Ledoux. 1975. Plasmid deoxyribonucleic acid in *Rhizobium vigna* and *Rhizobium trifolii*. Biochim. Biophys. Acta 414:357–361.

Van Larebeke, N., G. Engler, M. Holsters, S. van den Elsacker, I. Zaenen, R. A. Schilperoort, and J. Schell. 1974. Large plasmid in *Agrobacterium tumefaciens* essential for crown-gall inducing ability. Nature 252:169–170.

Watson, B., T. C. Currier, M. P. Gordon, M. D. Chilton, and E. W. Nester. 1975. Plasmid required for virulence of *Agrobacterium tumefaciens*. J. Bacteriol. 123:255–264.

Wolpert, J. S., and P. Albersheim. 1976. Host symbiont interactions. I. The lectins of legumes interact with the O-antigen containing lipopolysaccharides of their symbiont Rhizobia. Biochem. Biophys. Res. Comm. 70:729–737.

Zaenen, I., N. Van Larebeke, H. Teuchy, M. Van Montagu, and J. Schell. 1974. Supercoiled circular DNA in Crown gall inducing *Agrobacterium* strains. J. Mol. Biol. 86:109–127.

Zurkowski, W., and Z. Lorkiewicz. 1976. Plasmid deoxyribonucleic acid in *Rhizobium trifolii*. J. Bacteriol. 128:481–484.

Nitrogen Fixation, Volume II
Edited by W. E. Newton and W. H. Orme-Johnson

Determinants of Host Specificity in the *Rhizobium*-Clover Symbiosis

F. B. Dazzo

Several essential events precede active N_2 fixation in the *Rhizobium*-legume symbiosis. These include mutual host-symbiont recognition of the *Rhizobium* species on the legume rhizoplane, specific rhizobial adherence to root hairs, root hair curling, specific shepherd's crook formation, root hair infection, and root nodulation. The symbiosis is characterized by a high degree of host specificity. For instance, *Rhizobium trifolii* infects and nodulates white clover roots but not alfalfa or soybean roots. Host specificity is expressed at a very early step of infection of the root hair by infective rhizobia prior to the penetration of the root hair cell wall and formation of the infection thread (McCoy, 1932; Li and Hubbell, 1969; Napoli and Hubbell, 1975). The recognition of the symbionts on the root surface constitutes one of the first in a series of critical interactions that must be accomplished for the symbiosis to succeed.

Current efforts to explain the biochemical basis for host specificity in the *Rhizobium*-legume symbiosis emphasize the specific interactions of complementary macromolecules on the surfaces of the legume root hairs and the appropriate rhizobial symbiont. These complementary macromolecules consist of carbohydrate-binding proteins, called lectins, on legume roots and acidic heteropolysaccharides present exclusively on the surface of the homologous rhizobial symbionts. It has been proposed that this protein-carbohydrate interaction constitutes a specific recognition event that

This work was supported by grants from the National Science Foundation awarded to David H. Hubbell and Winston J. Brill, and by the Michigan Agricultural Experiment Station. Article no. 8601 from the Michigan Agricultural Experiment Station.

results in the selective adherence of *Rhizobium* to infection sites on root hairs of the appropriate legume host.

IMPORTANCE OF STUDIES ON HOST SPECIFICITY

An understanding of the mechanism of host recognition should not only advance our knowledge of the developmental events in the symbiosis, but also indicate ways in which *Rhizobium* and the plant species may be manipulated genetically to increase the range of agricultural crops that can enter efficient N_2-fixing symbioses (Behringer, 1978). These studies also relate directly to physiological recognition processes of plant hosts and phytopathogens, with the possibility of reducing the burden of plant pathogenesis on agricultural crops by the induction of disease resistance (Graham, Sequeira, and Huang, 1977).

TESTING THE LECTIN RECOGNITION HYPOTHESIS

Hamblin and Kent (1973) found that a phytohemagglutinin (lectin) from kidney bean could bind to *Rhizobium phaseoli*, the symbiont of red kidney beans. The lectin from kidney beans binds to the blood group A substance on human erythrocytes. Consequently, these authors used human group A erythrocytes to detect the agglutinin on epidermal root surfaces, and in saline extracts of seeds, nodules, and roots below the nodules of bean. They proposed that the lectin binds the bacteria to the root surface.

Soybean—*Rhizobium japonicum* Interactions

Bohlool and Schmidt (1974) suggested the possible role of lectins in host specificity and formalized the lectin recognition hypothesis. They demonstrated that a fluorescein isothiocyanate (FITC)–labeled soybean lectin preparation only bound to strains of *R. japonicum* capable of nodulating soybean. Bhuvaneswari, Pueppke, and Bauer (1977) have confirmed these observations with highly purified FITC- and ^{3}H-labeled soybean lectin and have shown that the binding to *R. japonicum* was reversed by *N*-acetylgalactosamine, a potent specific inhibitor of binding by soybean lectin. Wolpert and Albersheim (1976) reported similar lectin binding specificities with rhizobial lipopolysaccharide (LPS) preparations, although they did not demonstrate that the lectin binding was reversible with the respective sugar inhibitors. Other observations of the soybean lectin–*R. japonicum* interactions include: biphasic lectin binding to bacteria suggestive of multiple carbohydrate receptors (Bhuvaneswari, Pueppke, and Bauer, 1977); transient appearance of lectin receptors on the *R. japonicum* cell surface (Bhuvaneswari, Pueppke, and Bauer, 1977); binding of lectin to polar tips of some *R. japonicum* strains (Bohlool and Schmidt, 1976; Tsien and Schmidt,

1977); rapid decline of lectin levels in developing soybean seedling roots (Pueppke et al., 1978); induction of lectin-binding surface receptors on *R. japonicum* in the soybean rhizosphere (Bhuvaneswari and Bauer, 1978); requirement of encapsulated cells for binding to the lectin and soybean root hairs (Bal, Shantharam, and Ratnam, 1978); and the reports by Albersheim and Wolpert (1976, 1977) that lectins, including one from soybean, are enzymes that specifically degrade the LPS of their symbiont rhizobia.

Anomalous Interactions

There are reports of a number of anomalous interactions of lectins with rhizobia that do not correspond with their nodulation properties (Dazzo and Hubbell, 1975b; Chen and Phillips, 1976; Brethauer and Paxton, 1977; Law and Strijdom, 1977; Rougé and Labroue, 1977). Possibly, legumes may contain several lectins of different carbohydrate specificities (e.g., Kauss and Glaser, 1974) and lectins having functions other than *Rhizobium* recognition may have been examined. Often, only "classical" lectins that agglutinate erythrocytes have been considered, and therefore legume proteins that may bind to and distinguish nodulating and non-nodulating strains of *Rhizobium* but do not agglutinate erythrocytes may have been overlooked. There is no a priori reason why erythrocytes and rhizobia should have the same carbohydrate receptors that bind the lectin. Other explanations for the anomalous binding reactions include: the use of excessively high seed lectin concentrations (up to 13 mg/ml); improper bacterial growth conditions; or use of media containing yeast extract that affects the apparent composition of *Rhizobium* polysaccharides (Humphrey, Edgley, and Vincent, 1974). Furthermore, some studies omitted sugar-inhibited controls to rule out nonspecific binding of lectin.

Rhizobium Specificity with Tropical Legumes

The major exception to the general rule of restricted host range in *Rhizobium*-legume combinations is associated with the tropical legumes. These legumes are considered the primitive ancestors of the temperate legumes (Norris, 1959) and are nodulated by the promiscuous cowpea miscellany *Rhizobium* strains. These rhizobia infect through void spaces created by epithelial desquamation and lateral root emergence instead of infection thread formation in root hairs (Allen and Allen, 1940; Napoli, Dazzo, and Hubbell, 1975b). Recognition processes governing specificity in these more primitive legumes are less specific, as shown by the lectin concanavalin A (con A) from the tropical legume *Canavalia ensiformis* (jackbean, nodulated by the "cowpea rhizobia"; Fred, Baldwin, and McCoy, 1932) binding in a biochemically specific manner (α-methyl-mannoside inhibitable) with surface determinants shared among many rhizobia, regardless of their ability to nodulate jackbean (Dazzo and Hubbell, 1975b).

These receptors apparently reside on neither the O antigen nor exopolysaccharide (Wolpert and Albersheim, 1976). This anomalous interaction may involve the curious hydrophobic cavity in each protomer of con A (Reeke et al., 1975) with nonpolar constituents on the cell envelope surface of *Rhizobium*. It is important to note that con A undergoes conformational changes when associated with saccharide binding (Reeke et al., 1975) and these changes might have affected potentially important nonpolar interactions within the cavity.

Clover–*Rhizobium trifolii* Interactions

My co-workers and I have tested the lectin recognition hypothesis in the *R. trifolii*–clover symbiosis since the infection process with this combination is more fully understood (Napoli and Hubbell, 1975). We looked directly for legume proteins on roots that bind specifically at very low concentrations with unique surface receptors of *Rhizobium*. Antigenic differences were found between infective and related noninfective strains of *R. trifolii* using hyperimmune antisera obtained by a persistent immunization schedule (Dazzo and Hubbell, 1975a). Subsequently, we found that the surface of infective encapsulated *R. trifolii* contains an immunochemically unique polysaccharide that is antigenically cross-reactive with a component on clover epidermal cells (Dazzo and Hubbell, 1975c). This unique cross-reactive antigen (CRA) contains receptors that bind to the recently purified and characterized clover lectin called *trifoliin* (Dazzo, Yanke, and Brill, 1978). Electron microscopy shows that primary bacterial attachment sites on cells in the process of "docking" (Reissig, 1977) consist of the fibrillar capsule of *R. trifolii* in physical contact with electron-dense globular aggregates lying on the outer periphery of the fibrillar clover root hair cell wall (Dazzo and Hubbell, 1975c; Figure 1); *R. japonicum* adherence to soybean root hairs is similar (Bal, Shantharam, and Ratnam, 1978). Dazzo and Hubbell (1975c) proposed a model to explain this specificity (Figure 2) in which trifoliin recognizes these unique surface receptors and cross-bridges them to form the correct molecular structure for selective adherence of the bacteria to the root hair surface. This cross-bridging model is a modification of earlier hypotheses (Hamblin and Kent, 1973; Bohlool and Schmidt, 1974) proposing the role of lectin in adsorption and specific recognition of rhizobia on the root surface. However, the current model proposes two other necessary elements of the specificity-determining complex: the unique CRAs on the rhizobia and on the plant cell that bind the multivalent lectin.

There is much experimental evidence to support this model. Only rhizobia that infect clover have the surface CRA (Dazzo and Hubbell, 1975c; Dazzo and Brill, 1979) and the trifoliin receptor (Dazzo, Yanke, and Brill, 1978). Clover root hairs preferentially adsorb infective *R. trifolii* (Dazzo, Napoli, and Hubbell, 1976; Figure 3) and its CRA (Dazzo and Brill, 1977;

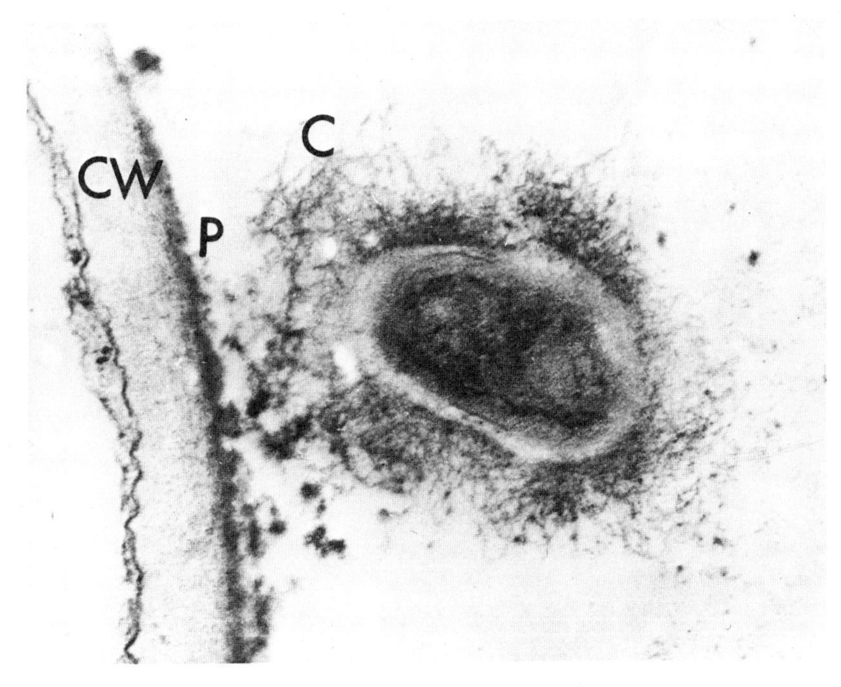

Figure 1. Transmission electron micrograph of infective *Rhizobium trifolii* NA30 in association with a strawberry clover root hair. The fibrillar capsule (C) of the bacterium is in contact with globular particles (P) on the outer periphery of the root hair cell wall (CW). From F. Dazzo and D. Hubbell (1975c), with permission from the American Society for Microbiology.

Figure 4). The receptor sites on clover roots that bind *R. trifolii* (Figure 5) and its CRA match the distribution of immunologically detectable trifoliin on clover root surfaces (Dazzo, Yanke, and Brill, 1978; Figure 6). These receptor sites accumulate at root hair tips and diminish toward the base of the root hair. By contrast, the CRA is uniformly distributed on both root hairs and undifferentiated epidermal cell walls in the root hair region of the clover seedlings (Dazzo and Brill, 1979). Trifoliin is multivalent in binding and specifically agglutinating *R. trifolii* (Dazzo, Yanke, and Brill, 1978). The sugar 2-deoxyglucose specifically inhibits agglutination of *R. trifolii* by trifoliin and the anti–clover root CRA antibody (Dazzo and Hubbell, 1975c; Dazzo and Brill, 1977) and also the specific binding of *R. trifolii* or its capsular polysaccharide to clover root hairs (Dazzo, Napoli, and Hubbell, 1976; Dazzo and Brill, 1977; Table 1), as well as specifically eluting trifoliin from intact clover roots or trifoliin-coated *R. trifolii* (Dazzo and Brill, 1977; Dazzo, Yanke, and Brill, 1978). As a negative control, 2-deoxyglucose did not inhibit specific adsorption of *R. meliloti* or its capsular polysaccharides to alfalfa root hairs (Dazzo, Napoli, and Hubbell, 1976; Dazzo and Brill,

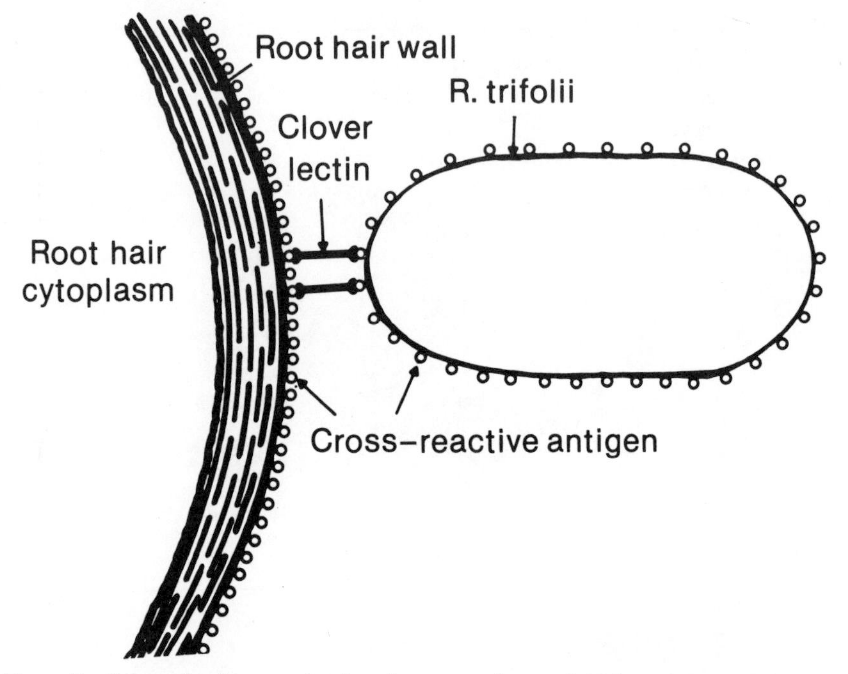

Figure 2. Schematic diagram showing the proposed cross-bridging of the cross-reactive antigen receptors of *Rhizobium trifolii* and clover root hairs with the clover lectin trifoliin. From F. Dazzo and D. Hubbell (1975c), with permission from the American Society for Microbiology.

1977). This *Rhizobium*-host combination constitutes a different cross-inoculation group (Burton and Wilson, 1939). Such inhibition studies involve the sugar combining with the site on the lectin or the antibody that is normally occupied by the polysaccharide and this implies a close, but not necessarily identical, structure of the inhibitor and antigenic determinant. Some lectins undergo conformational changes when associated with saccharide binding (Reeke et al., 1975). *Rhizobium* cells coated with a lectin have increased ability to adsorb to host specific roots (Solheim, 1975; Dazzo, Napoli, and Hubbell, 1976). This observation suggests that clover roots have receptor sites that bind trifoliin and that trifoliin does not occupy all the plant receptor sites for this root lectin. Only clover roots have the *R. trifolii*-specific CRA on their surface (Dazzo and Brill, 1979).

Does the unique CRA on the capsular polysaccharide bind *R. trifolii* to clover root hairs? Trifoliin and anti-clover root CRA bind to the same or similar overlapping determinants on the surface of encapsulated *R. trifolii* and exposure of this polysaccharide antigenic determinant is necessary for attachment of the bacteria to clover root hairs (Dazzo and Brill, 1979). Three experiments provided important clues to the physiological roles of these surface polysaccharides. Two experiments utilized anti-CRA Fab monovalent fragments that bind immunospecifically to cell surface antigens, but cannot cross-bridge antigens on neighboring cells to agglutinate them. In the first experiment, these anti–clover root CRA Fab fragments specifically blocked the agglutination of *R. trifolii* by purified trifoliin. In the second experiment, the anti-CRA Fab fragments specifically blocked the ability of *R. trifolii* to bind to clover root hairs. In both experiments, Fab fragments purified from normal rabbit preimmune IgG were without effect.

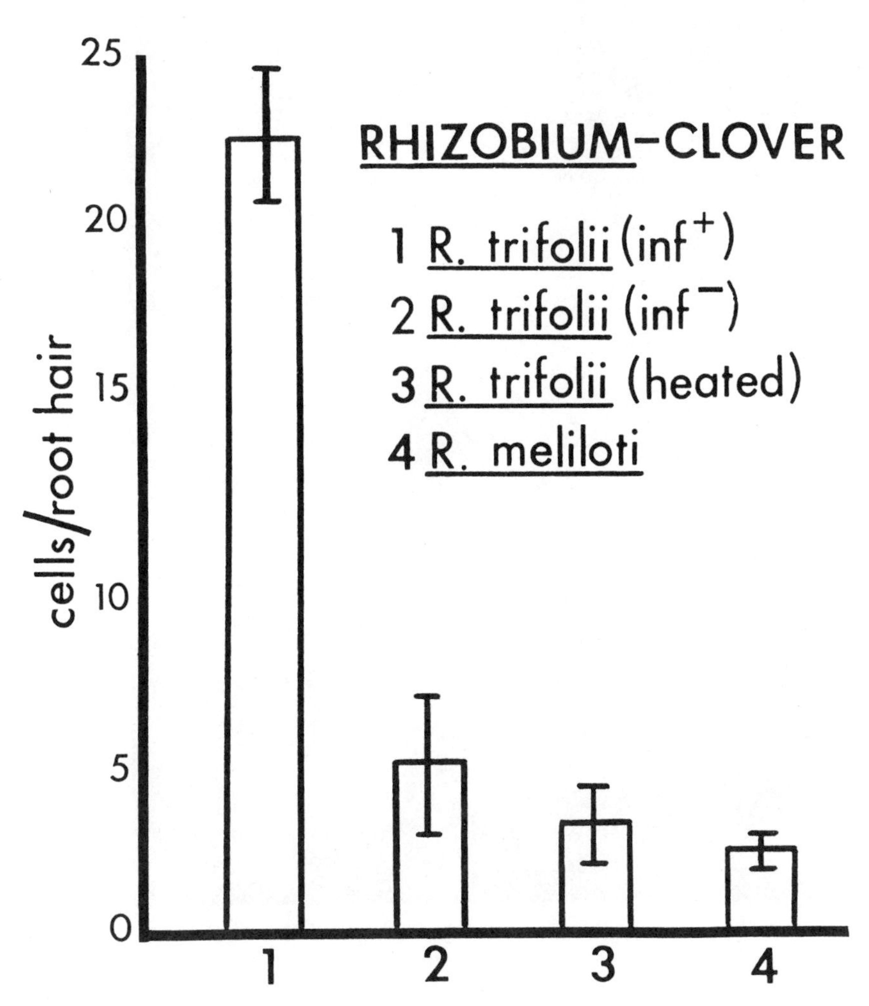

Figure 3. Adsorption of *Rhizobium* to clover root hairs. Bars represent means (± SD) of cells adsorbed to clover root hairs 200 μm in length after 12-hr incubation. Data taken from F. Dazzo, C. Napoli, and D. Hubbell (1976).

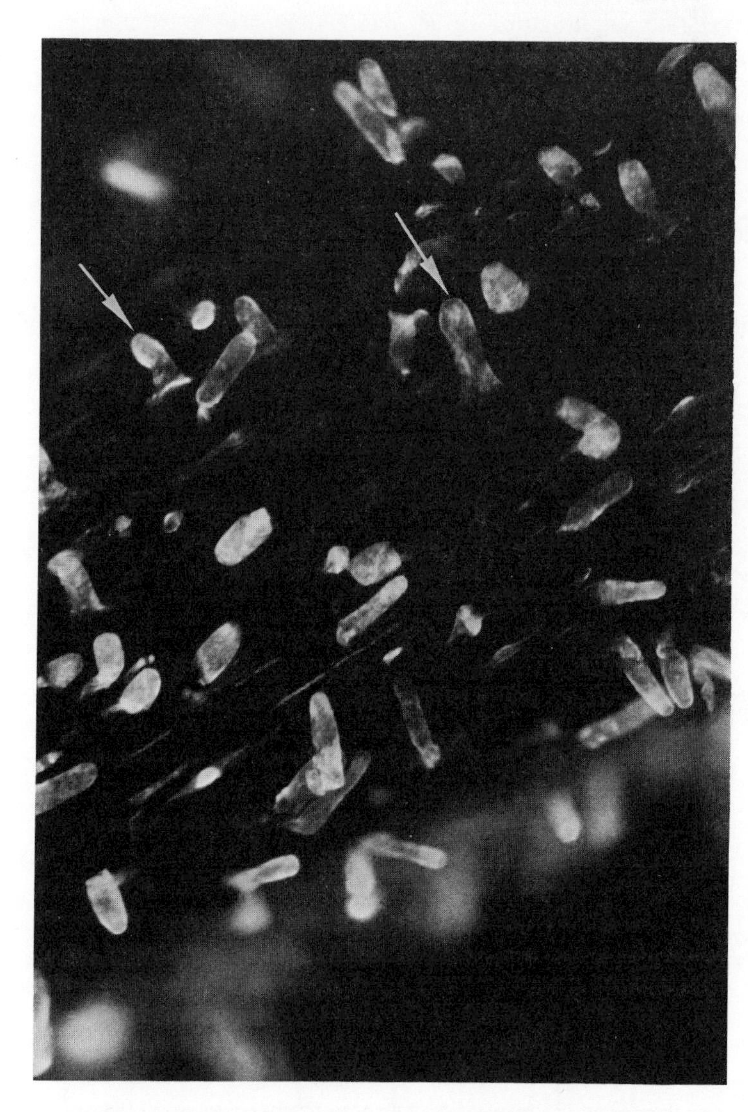

Figure 4. Fluorescence micrograph of a clover root with root hairs that have bound fluorescein isothiocyanate-conjugated capsular polysaccharide from *Rhizobium trifolii*. Root hair tips had a brilliant yellow-green fluorescence. From F. Dazzo and W. Brill (1977), with permission from the American Society for Microbiology.

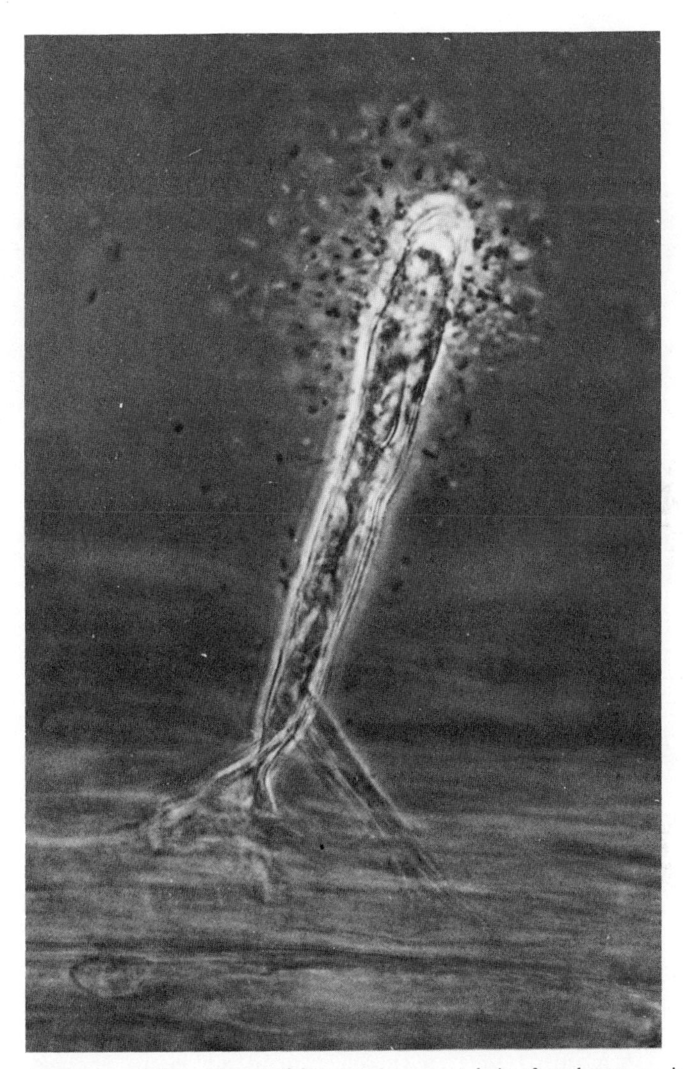

Figure 5. Adherence of *Rhizobium trifolii* to a clover root hair after short term incubation with bacteria in excess. Roots were rinsed after incubation with the bacteria. Note that the bacteria accumulate at the root hair tip and diminish in numbers toward the base of the root hair.

In the third experiment, intergeneric transformation (Page and Sadoff, 1976; Page, 1978) was conducted with *Azotobacter vinelandii nif*$^-$ mutants and *R. trifolii* donor deoxyribonucleic acid (Bishop et al., 1977). The genetic hybrids were selected on nitrogen-free medium for correction of *nif* on a mutant strain UW1O that has a structural defect in the MoFe protein of nitrogenase. The *nif*$^+$ hybrids were examined for *Rhizobium*-specific traits and 13% of them expressed the clover root CRA on their cell surface. Only

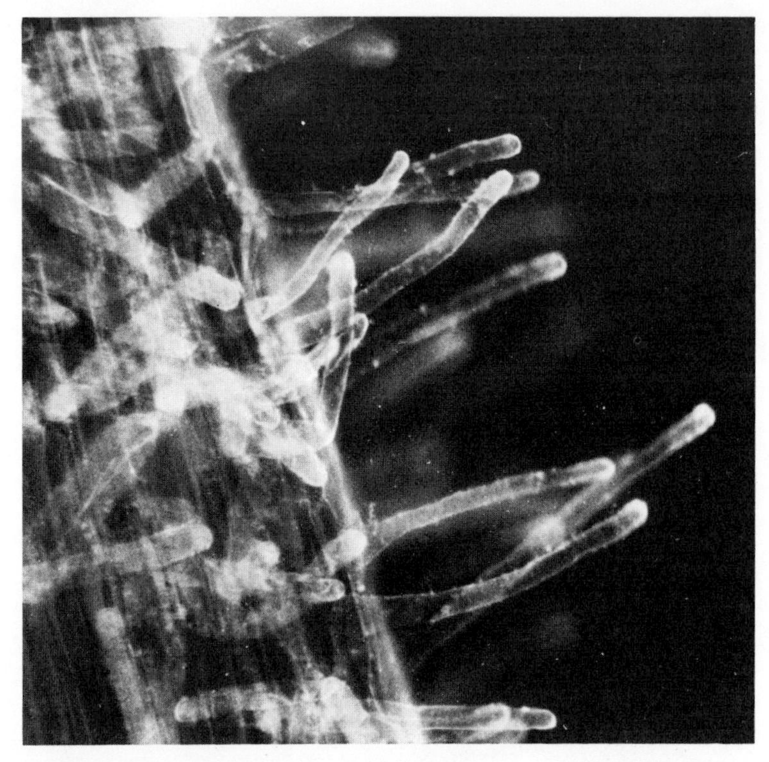

Figure 6. Immunofluorescent detection of trifoliin on clover root hairs with homologous anti-trifoliin antibody.

the same hybrids expressed the trifoliin receptor. These data indicate that the genetic co-transformation frequency of the genes that control the synthesis of the trifoliin receptor and CRA on *R. trifolii* is 100% and suggest that they are indeed the same determinants (Bishop et al., 1977). These hybrids have been stable for over 1 year. Back-cross of a CRA$^+$ transformant DNA into UW1O with selection of nif$^+$ again showed 100% co-transformation of the CRA marker and the trifoliin receptor. Only the *A. vinelandii–R. trifolii* hybrids that carry the trifoliin receptor bound in high numbers to clover root hairs. This binding was inhibited by 2-deoxyglucose (Dazzo and Brill, 1979). Critical negative controls in this experiment included *A. vinelandii* mutant strain UW10, a nif$^+$ revertant of UW10 that had not been transformed, and *A. vinelandii* nif$^+$ transformants (that lack CRA) obtained by transformation with DNA from *R. trifolii* and *R. japonicum*. The latter hybrid carries *R. japonicum*-specific surface antigens (Bishop et al., 1977). Thus, the attachment of *R. trifolii* to clover root hairs involves an interaction of the CRA with the root hair surface, and the

Table 1. Agglutination of *Rhizobium* by protein eluted from intact clover roots[a]

Bacterium	Additions to root washing solution[b]	Specific activity (agglutination units/mg of protein)
R. trifolii	2-Deoxyglucose	1240
R. trifolii	α-D-Glucose	42
R. trifolii	—	59
R. trifolii	2-Deoxyglucose, protease[c]	<5
R. meliloti	2-Deoxyglucose	<5
R. japonicum	2-Deoxyglucose	<5

[a] From Dazzo, Napoli, and Hubbell (1976), courtesy of American Society for Microbiology.

[b] Concentration of sugar was 30 mM in 0.01 M phosphate buffered saline (pH 7.2); sugar was removed by dialysis prior to assay.

[c] Root washings were incubated with 2-deoxyglucose or protese (1 mg/ml) 4 hr prior to assay.

determinant involved in bacterial binding is related immunochemically to a component on the host cell wall. It must be emphasized, however, that the nature of the association of CRA with native polysaccharides on the *A. vinelandii* cell is undefined at this time.

PROPERTIES OF TRIFOLIIN

Trifoliin was extracted from ground seeds in phosphate-buffered saline (PBS) containing sodium ascorbate and insoluble polyvinylpyrrolidone (PVP), and purified by a combination of gel filtration, ultrafiltration, and ion exchange chromatography (Dazzo, Yanke, and Brill, 1978). A 310-fold purification was achieved, as measured by specific agglutinating activity (agglutinating units/mg of protein). Purity of the trifoliin was demonstrated by: the elution of a symmetrical peak of 280 nm absorbance from a Sephadex G200 column; the presence of a single sharp band in 6% and 8% polyacrylamide gels stained with Coomassie Blue R-250 for protein and stained with periodate-Schiff's reagent for carbohydrate; and single immunoprecipitin bands in Ouchterlony immunodiffusion gels with trifoliin and antiserum prepared against crude seed extract or against purified trifoliin. Trifoliin at concentrations as low as 0.1–0.2 μg of protein/ml specifically agglutinated the symbiont. Purified clover protein is composed of subunits with molecular weight in SDS gels of approximately 50,000, has an isoelectric point of 7.3, and contains approximately 6 μmol of reducing sugar/mg of protein. Aggregates of trifoliin at pH 7.0 were 10 nm in diameter, the same dimensions as soybean lectin aggregates (Bauer, 1977). Electron-dense aggregates (Dazzo and Hubbell, 1975c; Figure 1) and subcellular particles (Figure 1; see also Figure 5C of Dart, 1971) have been observed on clover root hair walls at *R. trifolii* adsorption sites. Trifoliin was eluted from intact clover roots in the presence of PBS containing 2-deoxyglucose, sodium ascorbate, and PVP (Dazzo, Yanke, and Brill, 1978). The 2-deoxyglucose facilitates (thirtyfold) the elution of trifoliin from roots, presumably by dissociating the interaction of this lectin with binding sites on the root surface. The protein was then concentrated by ultrafiltration and purified to homogeneity by preparative slab gel electrophoresis. Trifoliin from clover seedling roots shares the following properties with trifoliin from seeds: the same electrophoretic mobility in native and SDS-polyacrylamide gels; immunochemical reaction of identity using antiserum to trifoliin from seeds; and specificity of agglutination toward *R. trifolii* inhibited by 2-deoxyglucose.

RHIZOBIAL ADSORPTION TO ROOT HAIRS

Quantitative microscopic assays (Dazzo, Napoli, and Hubbell, 1976; Dazzo, 1979; Figure 3; Table 2) and numerous electron microscopic studies (Napoli and Dazzo, unpublished data) revealed multiple mechanisms of rhizobial adsorption to clover root hairs. This fact should be borne in mind in adsorption studies related to the expression of host specificity. Nonspecific mechanisms allow rhizobia to attach at low background levels (2–5 cells/200 μm of clover root hairs; Figure 3) whereas specific mechanisms of adsorption bind infective *R. trifolii* cells at 22–27 cells/200 μm of root hair. The differences are significant at the 99.5% confidence level (Dazzo, Napoli, and Hubbell, 1976). The specific attachment mechanism was inhibited by 2-deoxyglucose and therefore may involve trifoliin binding (Dazzo, Napoli, and Hubbell, 1976).

Table 2. Inhibition of adsorption of *Rhizobium* spp. to legume root hairs[a]

Strain	Treatment[b]	Adsorbed bacteria (means ± SD)
R. trifolii strain 0403–*T. repens*	None	24.70 ± 1.80
	2-Deoxyglucose	2.60[c] ± 1.06
	2-Deoxygalactose	23.55 ± 2.34
	α-D-Glucose	24.00 ± 2.75
R. meliloti strain F28–*M. sativa*	None	35.50 ± 4.11
	2-Deoxyglucose	34.25 ± 4.79

[a] From Dazzo, Napoli, and Hubbell (1976), courtesy of American Society for Microbiology.

[b] Final sugar concentration was 30 mM.

[c] Significantly less than untreated control at 99% level.

Soybean root cells in culture selectively attach homologous *R. japonicum* cells, with the bacterial attachment inhibited by pectinase treatment of the root cells (Reporter, Raveed, and Norris, 1975). Only small numbers of *R. japonicum* and *Rhizobium leguminosarum* cells adsorb to roots of non-nodulating soybean and pea plant varieties, respectively, as compared to roots of their normal nodulating host varieties (Elkan, 1962; Degenhardt, LaRue, and Paul, 1976).

Other data also suggest multiple mechanisms of bacterial adherence to roots. Nissen (1971) differentiated specific from nonspecific mechanisms of bacterial adsorption to barley roots by their sensitivity to acid pH and requirements for Ca^{2+} and plant-derived energy. Werner, Wilcoxson, and Zimmermann (1975) found that multiple mechanisms of adsorption to roots could be identified by their different sensitivities to dissociation by high salt concentrations. Chen and Phillips (1976) found that the kinetics of rhizobial adsorption to pea roots progressed first at a rapid rate and later more slowly. They also identified the root hair tips as important sites of rhizobial attachment. However, data from viable plate counts (Peters and Alexander, 1966) and scintillation counts of radiolabeled bacteria on roots are open to question since these techniques count unadsorbed bacteria that are not washed away as well as bacteria adsorbed to undifferentiated epidermal cells. Other complications are the entrapment of flocs of cells and the variability of surface areas among individual plant roots. Furthermore, Kotarski and Savage (1978) have shown that use of radiolabeled cells for adsorption studies is complicated by the narrow range of linearity in counts per min versus cell number, and the transfer of radiolabeled metabolites from the bacterium to the host tissue. Use of FITC-labeled cells for quantitative adsorption studies is also questionable since the bacterial cell surface has been modified. The direct microscopic assay is reproducible (Dazzo, Napoli, and Hubbell, 1976), and has a high signal-to-noise ratio since the only bacterial cells that are scored are those with their native surface in physical contact with root hair cell walls having a standardized length and surface area and with a uniform state of physiological development.

A third mechanism that functions in the firm, irreversible adsorption of rhizobia to clover root hairs may exist, since many bacterial cells cannot be desorbed after 12 hr of incubation. Many *Rhizobium* species, including noninfective mutants, produce extracellular cellulose microfibrils in culture (Napoli, Dazzo, and Hubbell, 1975a). These neutral exopolysaccharides flocculate cells, often as they enter stationary phase of growth. It has been proposed that exopolysaccharide microfibrils may serve to anchor the bacteria firmly to the root surface once recognition has been established (Napoli, Dazzo, and Hubbell, 1975a). This maintenance of an intimate physical contact between the bacterium and the root hair could permit the localized biochemical interactions to trigger specific shepherd's crook formation and successful infection. Cellulose microfibrils clearly cannot account for specific recognition of rhizobia by legumes, since it is a neutral exopolysaccharide produced by many species of rhizobia (Napoli, Dazzo, and Hubbell, 1975a). Napoli (1976) reported that fibrillar appendages appear on *R. trifolii* growing in aseptically collected clover root exudate. Similar appendages have been observed on *Rhizobium meliloti* or *R. trifolii* adsorbed to barrel medic (Dart and Mercer, 1964) or clover root hairs (Dart, 1971; Figure 7), respectively. These appendages are detected on the bacteria only after prolonged incubation with the root.

Solheim (1975) has proposed a model of rhizobial adsorption based initially on studies of the curling factor (Solheim and Raa, 1973), and later on the effect of *Vicia* lectin on rhizobial attachment to *Vicia* roots (Solheim, 1975). His model indicates that the root hair wall has a "stabilizing factor" (presumably a lectin) that is secreted from the root and binds to a curling factor on the surface of *Rhizobium*. The bacteria containing the plant "stabilizing factor" then acquire a strong affinity for receptor sites on the root. The major difference between Solheim's model and that of Dazzo and Hubbell (1975c) (see Figure 2) is whether the lectin is released or remains attached to the roots. Detection of trifoliin in root washings with isotonic

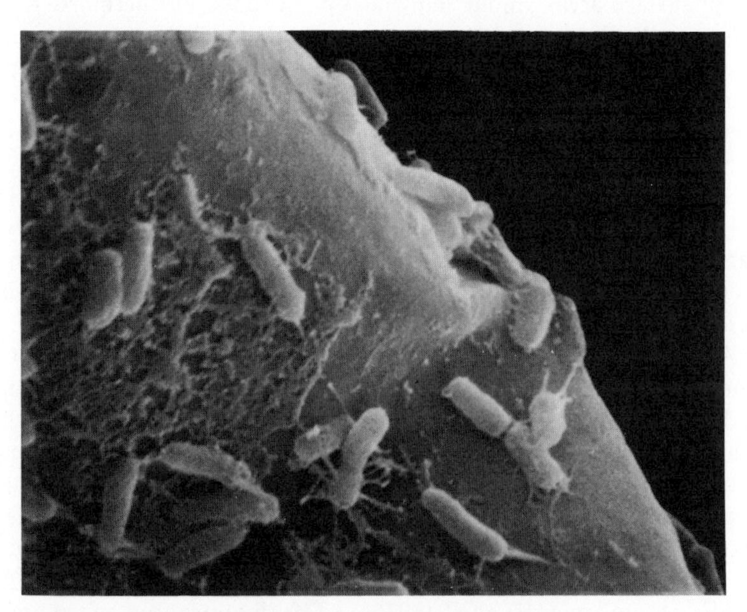

Figure 7. Scanning electron micrograph of *Rhizobium trifolii* 0403 on a clover root hair after 4 days incubation. Note the granular appearance of the root hair surface and the extracellular microfibrils associated with the bacteria.

buffer (Dazzo and Brill, 1977) and agglutination of homologous rhizobial cells in hydroponic culture suggest that some lectin may be released. Both models may be operative, the second one suggesting a salvage mechanism for the eluted multivalent lectin. Preparations of capsular polysaccharides have curling activity (Dazzo and Hubbell, 1975c; Dazzo and Brill, unpublished data) and lectin-coated rhizobia have a mechanism for specifically recognizing binding sites on roots of their homologous legume host (Dazzo, Napoli, and Hubbell, 1976; Solheim, 1975; Solheim and Raa, 1973).

LECTIN RECEPTOR ON *RHIZOBIUM*

A major controversy exists concerning the nature of the carbohydrate receptor for the lectin on the *Rhizobium* cell. The lectin receptors were identified as capsular polysaccharide (Dazzo and Hubbell, 1975c), lipopolysaccharide (Wolpert and Albersheim, 1976), or a glycan (Planqué and Kijne, 1977). They are all characterized as surface, acidic heteropolysaccharides (Dazzo and Hubbell, 1975c; Planqué and Kijne, 1977; Carlson and Albersheim, 1977). Electron microscopic examination of *R. trifolii* revealed a very prominent fibrillar capsular that surrounds the cell and stains positive for acidic polysaccharide with ruthenium red (Dazzo and Brill, 1979; Figure 8). Fluorescence microscopic techniques indicate greater reactivity of anti-CRA and purified trifoliin with encapsulated as compared with unencapsulated *R. trifolii* cells separated by differential centrifugation (Dazzo and Hubbell, 1975c; Dazzo and Brill, 1977; Dazzo, Yanke, and Brill, 1978). The same observation has been made with *R. japonicum* and soybean lectin (Bal, Shantharam, and Ratnam, 1978; Bhuvaneswari and Bauer, 1978). The purified acidic heteropolysaccharide from capsular preparations of *R. trifolii* (Dazzo and Brill, 1979) interacts specifically with purified trifoliin and anti–clover root CRA, although it lacks heptose, 2-keto-3-deoxyoctonic acid, and endotoxic lipid A, which are chemical markers present in lipopolysaccharide from the same strain. As shown with many gram-negative bacteria (Jann and Westphal, 1975; Luderitz, 1977), these acidic polysaccharides can be transferred under certain conditions to the core-lipid A and replace the netural O antigen polysaccharide (Dazzo and Brill, 1979). The invariable presence of a small, neutral polysaccharide in the capsule suggests that it normally associates with the acidic polysaccharide, perhaps to perpetuate the capsular structure on the bacterial cell surface (Dazzo and Brill, 1979). Most important was the finding (Dazzo and Brill, 1979) that, although sugar component differences were found between the corresponding acidic heteropolysaccharides of two distinct *R. trifolii* strains, their determinants that bound trifoliin and the anti–clover root CRA were conserved. These results emphasize the existence of differences in car-

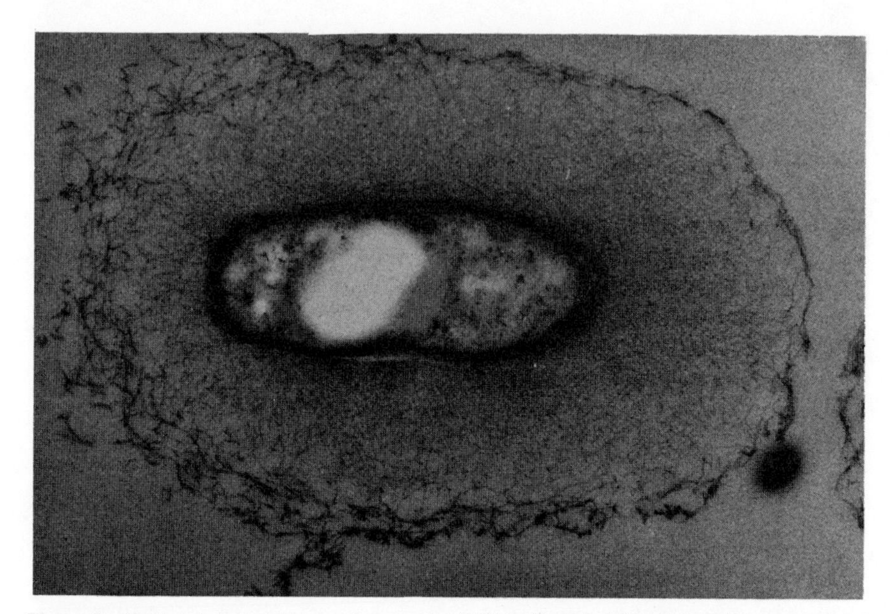

Figure 8. Transmission electron micrograph of *Rhizobium trifolii* 0403. Cells were stained with ruthenium red to show the acidic polysaccharide fibrils of the surrounding capsule.

bohydrate composition that are unimportant to their interaction with complementary lectins and recognition with the host. The structural diversity of LPS and exopolysaccharides of *Rhizobium* is enormous and therefore it is critically important to distinguish important residues that function in recognition from structures unimportant to this process.

Recently, detailed analyses of nodulating and related non-nodulating mutant strains of *Rhizobium* have revealed structural changes in their LPS that may relate to their nodulation capabilities (Saunders, Carlson, and Albersheim, 1978; Maier and Brill, 1978). An *R. leguminosarum* mutant strain does not nodulate peas and 27% of its total LPS mass is anthrone-reactive carbohydrate as compared to 63% of the total LPS mass from the wild-type *R. leguminosarum* (Saunders, Carlson, and Albersheim, 1978). The mutant strain was selected to produce small, nongummy colonies. The glycosyl and antigenic composition of the O antigen of the mutant and wild-type strains does not seem to be different. The original hypothesis was that the mutant strain has reduced its production of extracellular polysaccharide and not of LPS (Saunders, Carlson, and Albersheim, 1978). However, a more plausible hypothesis is that the phenotype of this mutant strain is due to a genetic defect of the synthesis or to control of the enzyme that polymerizes in a block fashion the O-antigenic oligosaccharide chains to form the long repeating O antigen on the LPS molecule. A recent analysis of nodulating and non-nodulating mutant strains of *R. japonicum* showed that the antigenic difference on their cell surface resides in the somatic O antigens (Maier and Brill, 1978). Discrete, silver nitrate-reactive components are missing in acid hydrolysates of LPS and acidic heteropolysaccharides from the *R. japonicum* non-nodulating mutant strains. These results suggest that the mutant strains may be defective in the ability to add the components onto the O antigen chain or to modify them after the intact LPS is assembled. Thus, several different structural modifications of intact LPS may result in the inability of the *Rhizobium* mutant strains to nodulate their legume host.

REGULATION OF RECOGNITION

How is the recognition process regulated? Do both symbionts participate in regulating the components of the recognition process? Answers to these important questions are beginning to emerge. It has been known for many years that fixed nitrogen (e.g., NH_4^+, NO_3^-) is one of the many environmental factors that limit the development and the success of the *Rhizobium*-legume symbiosis in nature (for a review see Fred, Baldwin, and McCoy, 1932). Recently, Dazzo and Brill (1978) found that the immunologically detectable levels of trifoliin and the specific binding of *R. trifolii* to root hairs decreased in a parallel fashion as the concentrations of either NO_3^- or NH_4^+ were increased in the rooting medium. The levels of these fixed ions were well below the levels that stunted seedling growth. Interestingly, both of these parameters were increased at specific low levels of NO_3^-. Thus, low levels of NO_3^- enhanced the recognition process and high levels shut it off. Garcia et al. (1978) have observed that adherence of *Azospirillum brasilense* to root hairs of the grass pearl millet is suppressed when the roots are grown in critical concentrations of NO_3^-. One measureable ultrastructural change was the degranulation in the surface topography of the root hair cell. These observations have opened new avenues to investigate how symbiotic development is regulated by this key nutrient in soil.

The lectin receptor on the *R. japonicum* cell surface may also be subject to regulation since its appearance is transient during cell growth (Bhuvaneswari, Pueppke, and Bauer, 1977). The appearance of the trifoliin receptor on *R. trifolii* is also transient and correlates with the distinct appearance of the capsule surrounding the cell (Dazzo, Urbano, and Brill, 1979). *Rhizobium trifolii* retains the ability to interact with trifoliin when grown in sterile soil extract or clover root exudate (Dazzo and Hubbell, 1975c). Interestingly, some *R. japonicum* strains that lacked soybean lectin receptors when grown in bacterial culture media would acquire this property when grown in root exudate (Bhuvaneswari and Bauer, 1978). Some of the intergeneric hybrids of *A. vinelandii*–*R. trifolii* deposited the CRA in a patchy distribution on their cell surface (Bishop et al., 1977). These hybrids offer potentially useful intermediates for studies of the regulation of the synthesis and compartmentation of this polysaccharide that functions in symbiotic recognition.

POLARITY

Some *R. japonicum* strains growing exponentially have cell surface polarity (Bohlool and Schmidt, 1976; Tsien and Schmidt, 1977) as illustrated by binding of antibody or soybean lectin. Electron microscopy shows accumulation of reserve polymer and exopolysaccharide at opposite cell poles. Bohlool and Schmidt (1976) speculated that the polar attachment of rhizobia to root surfaces may occur on the legume rhizoplane as a prelude to nodulation. The polar antigens common to many rhizobia (Bohlool and Schmidt, 1976) could reflect Braun lipoproteins, which are exposed at poles just after cell division and separation during exponential growth. These lipoproteins are covalently attached to peptidoglycan through peptide linkages. They extend through the periplasmic space and associate with the outer membrane through nonpolar interactions with the covalently attached long-chain fatty acids. The lipoproteins of many gram-negative bacteria are antigenically related. Ballou (1977) has argued that the polar binding of soybean lectin to *R. japonicum* may reflect only an enhanced exposure of surface receptors owing to a disorganization of wall structure in the area of the rapidly growing cell extension. Whether these polar receptors of rhizobia occur on the rhizoplane remains to be determined. The low background adsorption of heterologous rhizobia to legume root hairs is polar, and thus the role of this striking polarity in specific recognition and in the infection process is open to question.

SPECULATION ON THE FUNCTION OF CRA ON CLOVER ROOTS

The occurrence of antigenic constituents common to host and parasite is well documented (DeVay and Adler, 1976). Quantitative adsorption studies with seedlings from several legumes indicate that the CRA related to *R. trifolii* is unique to the host clover (Dazzo and Brill, 1979). At this time, the evidence strongly indicates that the CRA on the clover root is the receptor for trifoliin. Perhaps a normal physiological function of trifoliin is to cross-link polysaccharide structures during tip growth of the root hair cell wall (see Kauss and Glaser, 1974). The CRA would then serve as the "address" for trifoliin during its mobilization through the plasma membrane of the root hair to the cell wall. The advantage of N_2 fixation would favor the high accumulation of *Rhizobium* necessary for successful infection and symbiotic development. Designing the same complementary sites in the polypeptide chains that would accommodate both the plant cell wall structures and the rhizobial surface polysaccharide would be considered a genetically conservative venture. Functional multivalency of trifoliin, as has been shown with concanavalin A (Reeke et al., 1975), could then be accomplished by symmetrical aggregation of identical subunits as products of one gene. Selective pressures imposed by invasive pathogens may have favored polysaccharide and lectin binding diversities of the host. It has also been proposed that common antigenic constituents between pathogen and host may foil host defense mechanisms associated with hypersensitivity (Chen and Phillips, 1976; DeVay and Adler, 1976). Perhaps the CRA could serve as a scavenger to salvage trifoliin that has eluted from clover roots and would provide a specific receptor for the adherence of trifoliin-coated *Rhizobium*, as proposed in the model of Solheim (1975). The presence of the plant CRA, as a component of the "specificity-determining complex," may provide an additional host target subject to regulation and control.

CONCLUSIONS

In summary, it is proposed that selective adherence of *Rhizobium trifolii* to clover root hairs is an early expression of host specificity, which may be initiated by cross-bridging of similar polysaccharide determinants on the surfaces of both symbionts by a multivalent, host-coded lectin. The most fundamental unanswered question is how this recognition process of mechanical maneuvering of *Rhizobium* at the root hair surface *interfaces with the informational exchange* that then triggers the very specific events of shepherd's crook formation (Yao and Vincent, 1976) and invagination (Napoli and Hubbell, 1975) of the root hair cell wall to result in successful infection. Therein lies the molecular basis for host specificity in the *Rhizobium*-clover symbiosis.

ACKNOWLEDGMENTS

I would like to thank David H. Hubbell, of the University of Florida, Gainesville, and Winston J. Brill, of the University of Wisconsin, Madison, for their encouragement and support.

REFERENCES

Albersheim, P., and J. S. Wolpert. 1976. The lectins of legumes are enzymes which degrade the lipopolysaccharides of their symbiont rhizobia. Plant Physiol. Suppl. 58:79.

Albersheim, P., and J. Wolpert. 1977. Molecular determinants of symbiont-host selectivity between nitrogen-fixing bacteria and plants. In: B. Solheim and J. Raa (eds.), Cell Wall Biochemistry Related to Specificity in Host-Plant Pathogen Interactions, pp. 373–376. Universitetsforlaget, Oslo.

Allen, O. N., and E. K. Allen. 1940. Response of the peanut to inoculation with rhizobia, with special reference to morphological development of the nodules. Bot. Gaz. 102:121–142.

Bal, A. K., S. Shantharam, and S. Ratnam. 1978. Ultrastructure of *Rhizobium japonicum* in relation to its attachment to root hairs. J. Bacteriol. 133:1393–1400.

Ballou, C. 1977. Cell wall structure and recognition in yeasts. In: B. Solheim and J. Raa (eds), Cell Wall Biochemistry Related to Specificity in Host-Plant Pathogen Interactions, pp. 364–371. Universitetsforlaget, Oslo.

Bauer, W. D. 1977. Lectins as determinants of host specificity. Basic Life Sci. 9:283–297.

Behringer, J. E. 1978. *Rhizobium* recognition. Nature 271:205–206.

Bhuvaneswari, T. V., and W. D. Bauer. 1978. The role of lectins in plant-microorganism interactions. III. Influence of rhizosphere/rhizoplane culture conditions on the soybean lectin-binding properties of rhizobia. Plant Physiol. 62:71–74.

Bhuvaneswari, T. V., S. G. Pueppke, and W. D. Bauer. 1977. Role of lectins in plant-microorganism interactions. I. Binding of soybean lectin to rhizobia. Plant Physiol. 60:486–491.

Bishop, P. E., F. B. Dazzo, E. R. Appelbaum, R. J. Maier, and W. J. Brill. 1977. Intergeneric transfer of genes involved in the *Rhizobium*-legume symbiosis. Science 198:938–940.

Bohlool, B. B., and E. L. Schmidt. 1974. Lectins: A possible basis for specificity in the *Rhizobium*-legume root nodule symbiosis. Science 185:269–271.

Bohlool, B. B., and E. L. Schmidt. 1976. Immunofluorescent polar tips of *Rhizobium japonicum:* Possible site of attachment or lectin binding. J. Bacteriol. 125:1188–1194.

Brethauer, T. S., and J. D. Paxton. 1977. The role of lectin in soybean-*Rhizobium japonicum* interactions. In: B. Solheim and J. Raa (eds.), Cell Wall Biochemistry Related to Specificity in Host-Plant Pathogen Interactions, pp. 381–387. Universitetsforlaget, Oslo.

Burton, J. C., and P. W. Wilson. 1939. Host plant specificity among the *Medicago* in association with root nodule bacteria. Soil Sci. 47:293–303.

Carlson, R., and P. Albersheim. 1977. Host-symbiont interaction: The structural diversity of *Rhizobium* lipopolysaccharides. Plant Physiol. 59:51.

Chen, A. T., and A. Phillips,. 1976. Attachment of *Rhizobium* to legume roots as the basis for specific interactions. Physiol. Plant. 38:83–88.

Dart, P. J. 1971. Scanning electron microscopy of plant roots. J. Exp. Bot. 22:163–168.

Dart, P. J., and E. V. Mercer. 1964. The legume rhizosphere. Arch. Mikrobiol. 47:344–378.

Dazzo, F. B. 1979. Adsorption of microorganisms to roots and other plant surfaces. In: G. Bitton and K. Marshall (eds.), Adsorption of Microorganisms to Surfaces. John Wiley & Sons, Inc., New York.

Dazzo, F. B., and W. J. Brill. 1977. Receptor site on clover and alfalfa roots for *Rhizobium*. Appl. Environ. Microbiol. 33:132–136.

Dazzo, F. B., and W. J. Brill. 1978. Regulation by fixed nitrogen of host-symbiont recognition in the *Rhizobium*-clover symbiosis. Plant Physiol. 62:18–21.

Dazzo, F. B., and W. J. Brill. 1979. Bacterial polysaccharide which binds *Rhizobium trifolii* to clover root hairs. J. Bacteriol. 137:1367–1373.

Dazzo, F. B., and D. H. Hubbell. 1975a. Antigenic differences between infective and noninfective strains of *Rhizobium trifolii*. Appl. Microbiol. 30:172–177.

Dazzo, F. B., and D. H. Hubbell. 1975b. Concanavalin A: Lack of correlation between binding to *Rhizobium* and specificity in the *Rhizobium*-legume symbiosis. Plant Soil 43:713–717.

Dazzo, F. B., and D. H. Hubbell. 1975c. Cross-reactive antigens and lectin as determinants of symbiotic specificity in the *Rhizobium*-clover association. Appl. Microbiol. 30:1017–1033.

Dazzo, F. B., C. A. Napoli, and D. H. Hubbell. 1976. Adsorption of bacteria to roots as related to host-specificity in the *Rhizobium*-clover symbiosis. Appl. Environ. Microbiol. 32:168–171.

Dazzo, F. B., M. R. Urbano, and W. J. Brill. 1979. Transient appearance of lectin receptors on *Rhizobium trifolii*. Curr. Microbiol. 2:15–20.

Dazzo, F. B., W. E. Yanke, and W. J. Brill. 1978. Trifoliin: A *Rhizobium* recognition protein from white clover. Biochim. Biophys. Acta 539:276–286.

Degenhardt, T. L., T. A. LaRue, and E. A. Paul. 1976. Investigation of a non-nodulating cultivar of *Pisum sativum*. Can. J. Bot. 54:1633–1636.

DeVay, J. E., and H. E. Adler. 1976. Antigens common to hosts and parasites. Annu. Rev. Microbiol. 30:147–168.

Elkan, G. H. 1962. Comparison of rhizosphere microorganisms of genetically related nodulating and non-nodulating soybean lines. Can. J. Microbiol. 8:79–87.

Fred, E. B., I. L. Baldwin, and E. McCoy. 1932. Root Nodule Bacteria and Leguminous Plants, p. 343. University of Wisconsin Press, Madison, Wisconsin.

Garcia, M. U., D. H. Hubbell, M. H. Gaskins, and F. B. Dazzo. 1978. Adsorption and infection of grass roots by *Azospirillum brasilense* SP7. Paper presented at the 3rd Kettering-Steenbock International Symposium on Nitrogen Fixation, June 12–16, Madison, Wisconsin.

Graham, T. L., L. Sequeira, and T. R. Huang. 1977. Bacterial lipopolysaccharides as inducers of disease resistance in tobacco. Appl. Environ. Microbiol. 34:424–432.

Hamblin, J., and S. P. Kent. 1973. Possible role of phytohaemagglutin in *Phaseolus vulgaris* L. Nature New Biol. 245:28–30.

Humphrey, B., M. Edgley, and J. M. Vincent. 1974. Absence of mannose in extracellular polysaccharides of fast-growing rhizobia. J. Gen. Microbiol. 81:267–270.

Jann, K., and O. Westphal. 1975. Microbial polysaccharides. In: M. Sela (ed.), The Antigens, Vol. 3, pp. 1–125. Academic Press, Inc., New York.

Kauss, H., and C. Glaser. 1974. Carbohydrate-binding proteins from plant cell walls and their possible involvement in extension growth. FEBS Lett. 45:304–307.

Kotarski, S. F., and D. C. Savage. 1978. Radioisotopic assay of lactobacilli adhering to murine stomach epithelia. Proceedings of the American Society for Microbiology, Abstracts of the Annual Meeting, p. 15 (B11). American Society for Microbiology, Washington, D.C.

Law, I. J., and B. W. Strijdom. 1977. Some observations on plant lectins and *Rhizobium* specificity. Soil Biol. Biochem. 9:79–84.

Li, D., and D. H. Hubbell. 1969. Infection thread formation as a basis of nodulation specificity in *Rhizobium*-strawberry clover associations. Can. J. Microbiol. 15:1133–1136.

Luderitz, O. 1977. Endotoxins and other cell wall components of Gram-negative bacteria and their biological activities. In: D. Schlessinger (ed.), Microbiology—1977, pp. 239–246. American Society for Microbiology, Washington, D.C.

Maier, R. J., and W. J. Brill. 1978. Involvement of *Rhizobium japonicum* O-antigen in soybean nodulation. J. Bacteriol. 133:1295–1299.

McCoy, E. F. 1932. Infection by *Bact radicicola* in relation to the microchemistry of the host cell walls. Proc. R. Soc. Lond. [Biol.] 110:514–533.

Napoli, C. A. 1976. Physiological and ultrastructural aspects of the infection of clover (*Trifolium fragiferum*) by *Rhizobium trifolii* NA30. Ph.D. dissertation, University of Florida, Gainesville, p. 103.

Napoli, C. A., F. B. Dazzo, and D. H. Hubbell. 1975a. Production of cellulose microfibrils by *Rhizobium*. Appl. Microbiol. 30:123–131.

Napoli, C., F. Dazzo, and D. Hubbell. 1975b. Ultrastructure of infection and "common antigen" relationships in *Aeschynomene*. Proceedings of the 5th Australian Legume Nodulation Conference. C.S.I.R.O., Brisbane, Australia. Rhizobium Newsletter Suppl. 20:35–37.

Napoli, C. A., and D. H. Hubbell. 1975. Ultrastructure of *Rhizobium*-induced infection threads in clover root hairs. Appl. Microbiol. 30:1003–1009.

Nissen, P. 1971. Choline sulfate permease: Transfer of information from bacteria to higher plants? II. Induction processes. In: L. G. Ledoux (ed.), Informative Molecules in Biological Systems, pp. 201–212. North Holland Publishing Company, Amsterdam.

Norris, D. O. 1959. Legume bacteriology in the tropics. J. Aust. Inst. Agric. Sci. 25:202.

Page, W. J. 1978. Transformation of *Azotobacter vinelandii* strains unable to fix nitrogen with *Rhizobium* spp. DNA. Can. J. Microbiol. 24:209–214.

Page, W. J., and H. L. Sadoff. 1976. Physiological factors affecting transformation of *Azotobacter vinelandii*. J. Bacteriol. 125:1080–1087.

Peters, R. J., and M. Alexander. 1966. Effect of legume root exudates on the root nodule bacteria. Soil Science 102:380–387.

Planqué K., and J. W. Kijne. 1977. Binding of pea lectins to a glycan type polysaccharide in the cell walls of *Rhizobium leguminosarum*. FEBS Lett. 73:64–66.

Pueppke, S. G., K. Keegstra, A. L. Ferguson, and W. D. Bauer. 1978. The role of lectins in plant-microorganism interactions. II. Distribution of soybean lectin in tissues of *Glycine max* (L). Merr. Plant Physiol. 61:779–784.

Reeke, J. N., J. W. Becker, B. A. Cunningham, J. L. Wang, I. Yahara, and G. M. Edelman. 1975. Structure and function of Concanavalin A. Adv. Exp. Med. Biol. 55:13–33.

Reissig, J. L. 1977. An overview. In: J. L. Reissig (ed.), Microbial Interactions. Receptors and Recognition, Series B., Vol. 3, pp. 399–415. Chapman and Hall, London.

Reporter, M., D. Raveed, and G. Norris, 1975. Binding of *Rhizobium japonicum* to cultured soybean root cells: Morphological evidence. Plant Sci. Lett. 5:73–76.

Rougé, P., and L. Labroue. 1977. On the role of phytohemagglutinin in the specificity of binding to *Pisum* of *Rhizobium leguminosarum* compatible strains. C. R. Acad. Sci. [D] (Paris) 284:2423–2427.

Saunders, R. E., R. W. Carlson, and P. Albersheim. 1978. A *Rhizobium* mutant incapable of nodulation and normal polysaccharide secretion. Nature 271:240–242.

Solheim, B. 1975. A model of the recognition-reaction between *Rhizobium trifolii* and *Trifolium repens*, and possible role of lectins in the infection of legumes by rhizobia. Paper presented at the North Atlantic Treaty Organization Conference on Specificity in Plant Diseases, Advanced Study Institute, May 4–16, 1975, Sardinia.

Solheim, B., and J. Raa. 1973. Characterization of the substances causing deformation of root hairs of *Trifolium repens* when inoculated with *Rhizobium trifolii*. J. Gen. Microbiol. 77:241–247.

Tsien, H. C., and E. L. Schmidt. 1977. Polarity in the exponential-phase *Rhizobium japonicum* cell. Can. J. Microbiol. 23:1274–1284.

Werner, D., J. Wilcockson, and E. Zimmermann. 1975. Adsorption and selection of rhizobia with ion exchange papers. Arch. Microbiol. 105:27–32.

Wolpert, J. S., and P. Albersheim. 1976. Host-symbiont interactions. I. The lectins of legumes interact with the O-antigen-containing lipopolysaccharides of their symbiont *Rhizobia*. Biochem. Biophys. Res. Commun. 70:729–737.

Yao, P. Y., and J. M. Vincent. 1976. Factors responsible for the curling and branching of clover root hairs by *Rhizobium*. Plant Sci. 45:1–16.

Nitrogen Fixation, Volume II
Edited by W. E. Newton and W. H. Orme-Johnson

Host-Symbiont Interactions: Recognizing *Rhizobium*

C. Napoli, R. Sanders, R. Carlson, and P. Albersheim

Phylogenetic relationships among all organisms are most accurately determined by comparisons of DNA base sequences. However, less definitive but more easily analyzed features of organisms are generally used to determine taxonomic relationships. Many bacteria have been categorized on the basis of their surface properties (Lüderitz, Jann, and Wheat, 1968). We propose to extend to *Rhizobium* the use of surface characteristics as criteria for the taxonomic classification and identification of gram-negative bacteria.

The present criterion for inclusion in the genus *Rhizobium* is the ability to induce nodules on the roots of leguminous plants. Nodulation is a relatively specific interaction since the bacteria, in most instances, are limited to a restricted number of botanically related legumes with which they can enter into a nitrogen-fixing symbiosis. These limitations are the basis for the concept of cross-inoculation groups of nodulating rhizobia and host legumes. Legumes belong to the same cross-inoculation group if they are infected by the same species of *Rhizobium*. Six cross-inoculation groups are currently recognized and the rhizobia associated with each are placed in a single species (Table 1). A seventh nodulation group is comprised of rhizobia designated as the cowpea miscellany that nodulate the great majority of legumes that have been investigated, but not the major crop legumes. These bacteria do not have specific ranking and are considered to be symbiotically promiscuous because they cross-nodulate a wide variety of legumes. Unlike

Supported by grants from the Rockefeller Foundation (GA-AS-7707), the National Science Foundation (#PCM75-13897), and the Department of Energy (EY-76-S-02-1426).

Table 1. Cross-inoculation groups

Rhizobium species	Representative preferred hosts
R. leguminosarum	*Pisum, Vicia, Lens, Cicer, Lathyrus*
R. trifolii	*Trifolium*
R. phaseoli	*Phaseolus vulgaris*
R. meliloti	*Medicago, Melilotus, Trigonella*
R. lupini	*Lupinus, Ornithopus*
R. japonicum	*Glycine Max*
R. species	Cowpea group

the legumes in the other cross-inoculation groups, those nodulated by the cowpea rhizobia belong to many different genera. The wisdom of subdividing the genus *Rhizobium* into species on the basis of host range is problematical and any classification based on a single characteristic, other than DNA base sequence, will result in categories that contain genetically diverse members. The species of a *Rhizobium* is easy to ascertain if the bacteria are isolated directly from host plant nodules. However, this identification may be difficult and somewhat ambiguous if the bacteria are isolated from the soil or if the isolate is a noninfective variant. In this case, serological data using antisera raised against whole bacterial cells have been employed as an additional criterion for species and strain identification (Dudman, 1977).

Each *Rhizobium* species is further divided into strains. Strains are numbered according to where they were isolated. The various strains of a *Rhizobium* species may have a varying ability to nodulate and fix nitrogen in the legumes constituting a single cross-inoculation group. Ideally, strains of bacteria should be defined as having unique DNA base sequences. The standard analytical procedures for identifying a *Rhizobium* are clearly insufficient to define a strain, because two strains of a single *Rhizobium* species could have similar antigenic patterns yet differ in other important characteristics.

This paper describes four useful methods, based on bacterial surface properties, for *Rhizobium* strain identification. The four methods are: fingerprint analysis of the sugar composition of *Rhizobium* lipopolysaccharides; reaction of *Rhizobium* with lipopolysaccharide-specific antibodies; reaction of *Rhizobium* with flagella-specific antibodies; and reaction with *Rhizobium* bacteriophage. In addition to these identification techniques, the nodulation assay for preferred host range will continue to be used for determining the species of an isolate.

NODULATION ASSAY

The cross-inoculation groups have served to rationalize the problem of providing suitable strains of *Rhizobium* for commercial leguminous crops.

Most of the *Rhizobium* species used in agricultural production in temperate climate regions fall into clearly defined nodulation groups such as pea, clover, alfalfa, or beans. This situation has led to the conclusion that host range in nodulation assays could serve as the taxonomic basis of species identification in the *Rhizobium* (Norris, 1956). However, it has been discovered that the nodulation groups are not mutually exclusive (Dixon, 1969). This promiscuity is most common in the lupin, cowpea, and soybean groups, but also occurs between the clover and pea groups (Kleczkowski, Nutman, and Bond, 1944). An additional complication with the nodulation groups is the lack of reciprocal nodulation within a group. This problem is illustrated within the alfalfa group, which is composed of three genera of legumes. Two of the genera in this group are nodulated by different *Rhizobium meliloti* strains that do not cross-nodulate between the two genera. The third genus is nodulated by almost all strains of *R. meliloti* (Vincent, 1974). Much dissatisfaction has been expressed with regard to the present taxonomic classification of *Rhizobium* based on the cross-inoculation groups (Dixon, 1969; Wilson, 1944) and alternative methods have been proposed (Graham, 1964; DeLey and Rassel, 1965; Norris, 1965). However, the cross-inoculation grouping is presently the only accepted classification scheme (Jordon and Allen, 1974).

The nodulation assay remains a useful test for identifying the most likely *Rhizobium* species of an isolate. A nodulation study requires the testing of a number of legumes in order to determine the cross-inoculation group of a rhizobial isolate. Methods for nodulation tests are summarized in Vincent (1970). A method not covered by Vincent involves plastic growth pouches developed by Northrup, King and Company (Weaver and Frederick, 1972) and distributed commercially by Scientific Products. The growth pouches have the advantage that they require little space and thus many legume hosts can be tested while utilizing a small amount of space. Although this method is more applicable for small seeded legumes, it can be used for large seeds, such as soybean and pea.

FINGERPRINT ANALYSIS OF THE SUGAR COMPOSITION OF *RHIZOBIUM* LIPOPOLYSACCHARIDES

The cell wall of gram-negative bacteria, which include the genus *Rhizobium*, is a complex structure containing a layer of peptidoglycan and an outer membrane that contains the dominant cell surface antigen, the lipopolysaccharide (Lüderitz, Staub, and Westphal, 1966). Cell surface lipopolysaccharides (LPS) are complex macromolecules organized into three distinct structural regions: the O antigen–repeating oligosaccharide chain; the core polysaccharide; and Lipid A (Osborn, 1969). The O antigens for the genus *Salmonella* are used to define species (Lüderitz, Staub, and

Westphal, 1966; Lüderitz, Jann, and Wheat, 1968), whereas in the genus *Escherichia*, which contains only one species, the O antigens are used to define strains (Lüderitz, Jann, and Wheat, 1968; Ørskov et al., 1977).

The lipopolysaccharides of *Salmonella*, *Escherichia*, *Shigella*, and related genera in the Enterobacteriaceae can be classified, according to sugar compositions, into chemotypes (Osborn, 1969). The simplest chemotypes contain five different sugars, and the more complex contain eight or nine different sugars. These sugars may include common hexoses, hexosamines, 6-deoxyhexoses, and pentoses. The O antigen chains frequently contain rare sugars that are characteristic of lipopolysaccharides, including 3,6-dideoxyhexoses, 2-amino-2,6-dideoxyhexoses, and 3-amino-3,6-dideoxyhexoses. Certain sugars are usually present in all chemotypes. These are glucose, galactose, glucosamine, and two sugars unique to LPS, L-glycero-D-mannoheptose and 3-deoxy-D-mannooctulosonic acid, commonly called 2-keto-3-deoxyoctonate (KDO).

The lipopolysaccharides of *Rhizobium* have not been well characterized. The partial compositions of the lipopolysaccharide preparations from several strains of *Rhizobium trifolii* and *Rhizobium leguminosarum* have been reported (Humphrey and Vincent, 1969; Lorkiewicz and Russa, 1971; Planqué and Kijne, 1977; Russa and Lorkiewicz, 1974; Zajac, Russa, and Lorkiewicz, 1975). Lipopolysaccharide preparations have been extracted by the hot phenol method from two strains of *R. trifolii* (Humphrey and Vincent, 1969). Both LPS preparations contained lipid, KDO, glucose, mannose, and fucose, and one lacked heptose and had a higher carbohydrate content than the other. The *R. trifolii* LPS differed from the LPS of other Enterobacteriaceae in that the phosphorus content was low and glucuronic acid was present. LPS preparations from several other strains of *R. trifolii* contain glucuronic acid, glucosamine, galactose, glucose, mannose, xylose or fucose, rhamnose, ribose, and KDO (Lorkiewicz and Russa, 1971). Rough mutants (lacking the O antigen chain) of the *R. trifolii* strains have been found to be deficient in rhamnose (Lorkiewicz and Russa, 1971). Other studies (Planqué and Kijne, 1977; Zajac, Russa, and Lorkiewicz, 1975) have reported that lipopolysaccharide preparations from strains of *R. trifolii* and *R. leguminosarum* contain rhamnose, fucose, ribose, arabinose, xylose, mannose, galactose, glucose, heptose, KDO, and an unidentified sugar. Evidence obtained by our laboratory (Carlson et al., 1978) suggests that many of the LPS preparations used in the above studies were contaminated with other polysaccharides.

The lipopolysaccharides of three strains each of *R. leguminosarum*, *Rhizobium phaseoli*, and *R. trifolii* have been purified and partially characterized in our laboratory (Carlson et al., 1978). Their sugar compositions are given in Table 2. The neutral and amino sugars were analyzed as their volatile alditol acetates, whereas KDO and uronic acids were analyzed

Table 2. Composition of *Rhizobium* lipopolysaccharides[a]

	R. leguminosarum strain					*R. phaseoli* strain				*R. trifolii* strain			
	128C53					127K17						2S	
LPS constituents	(1)[b]	(2)	(3)	3HOQ1	128C63	(1)	(2)	127K24	127K14	0403	162S7	(1)	(2)
Mannose	6.9	5.3	10.7	11.0	2.8	1.9	1.8	1.7	+	2.4	1.9	1.7	2.3
Galactose	2.9	1.5	2.7	1.1	2.7	2.1	3.1	1.2	+	2.5	1.1	2.2	2.5
Glucose	1.5	1.6	1.1	13.4	2.3	1.3	2.1	2.4	+	3.1	0.9	1.0	0.4
Glucosamine	1.6	1.0	1.9	2.2	2.6	2.1	—	Tr	—	2.7	1.5	1.6	2.2
2-keto-3-deoxyoctanoate	9.2	—	6.1	3.2	2.9	3.4	5.6	2.9	+	2.4	7.2	2.7	2.2
Uronic acids	6.2	—	7.4	4.7	7.3	6.2	10.8	+	+	8.8	2.2	7.0	7.6
Fucose	9.1	8.3	17.8	10.4	4.1	2.4	1.7	3.7	+	2.5	4.0	7.5	8.2
Rhamnose	5.9	5.3	12.1	7.4	0.0	0.0	0.0	0.7	0	0.0	0.6	7.5	8.9
3-*O*-methylhexose	0.0	0.0	0.0	0.0	0.0	0.0	0.0	0	0	0.0	0.0	1.1	1.0
2-*O*-methyl-6-deoxyhexose	0.0	0.0	0.0	0.0	0.0	0.0	0.0	4.2	0	2.9(1)	4.5(2)	0.0	0.0
3-*O*-methyl-6-deoxyhexose	0.0	0.0	0.0	0.0	0.0	0.0	0.0	0	0	0.0	0.0	3.6	3.7
2,3-di-*O*-methyl-6-deoxyhexose	0.0	0.0	0.0	0.0	1.3	0.0	0.0	0	+	0.0	0.0	0.0	0.0
3,4-di-*O*-methyl-6-deoxyhexose	0.0	0.0	0.0	0.0	0.0	0.0	0.0	0	0	0.0	0.7	0.0	0.0
N-methyl-3-amino-3,6-dideoxyhexose	0.0	0.0	0.0	0.0	2.8	1.5	1.9	0	0	3.7	0.0	0.0	0.0
2-amino-2,6-dideoxyhexose	0.0	0.0	0.0	0.0	0.0	0.5	1.3	0	+	0.0	0.0	0.0	0.0
Heptose	0.0	0.0	0.0	0.0	2.4	0.0	0.0	0	0	3.2	0.0	0.0	0.0
Lipid A	—	—	42.0	—	48.0	—	—	—	—	32.0	24.0	25.0	—

[a] Numerical values are percent of dry weight of LPS. + = component present but not quantitated; 0 = component not present; – = not determined; Tr = trace amounts detected.

[b] No special hydrolysis procedures were used to determine glucosamine. Therefore, the amounts of glucosamine reported here are probably low.

colorimetrically. LPS from all these *Rhizobium* strains contain fucose, galactose, glucose, mannose, glucosamine, uronic acids, and KDO. Most possess one or more methylated sugar. Each strain of the three different species contains sugars that are unique to that strain, with only one exception; *R. leguminosarum* strains 3HOQ1 and 128C53 have lipopolysaccharides with the same sugar compositions. With this one exception, all 16 *Rhizobium* strains examined in our laboratory constitute different chemotypes, that is, possess different sugar compositions.

We propose that lipopolysaccharide sugar compositions analysis be performed as an aid in strain identification. The methods involved can be carried out by a trained technician. The composition of neutral and amino sugars is determined effectively by gas chromatographic analysis of the alditol acetate derivitives (Albersheim et al., 1967) by visual comparison of recorder tracings without the use of electronic integration. Such recorder tracings for five *Rhizobium* strains are compared in Figure 1. The effectiveness of this analysis could be enhanced by examination of the peaks eluting from the gas chromatograph by quantitative electronic integration and by mass spectrometric analysis, but these procedures are too sophisticated and expensive to be routinely available. Therefore, it is fortunate that visual comparison of the gas chromatograph outputs will yield a great deal of information toward the identification of *Rhizobium* strains.

It is not true that every strain of a *Rhizobium* species has a unique lipopolysaccharide sugar composition. However, it is likely that most strains will have structurally unique LPS, as in the *Salmonella* (Lüderitz, Staub, and Westphal, 1966; Lüderitz, Jann, and Wheat, 1968). Some LPS probably have identical sugar compositions but vary either in the anomeric configurations of the glycosidic linkages or by the specific carbon atoms acting as recipients of the glycosidic bonds. These differences in glycosidic linkages would not be detected by simple sugar analysis of the lipopolysaccharides, but can usually be distinguished by antibodies directed against the purified lipopolysaccharide (Lüderitz, Staub, and Westphal, 1966). Thus, immunochemical analysis of the lipopolysaccharides is a method that is complementary to fingerprint analysis of the lipopolysaccharide sugar compositions.

LIPOPOLYSACCHARIDE-SPECIFIC ANTIBODIES

Serological heterogeneity, as determined by immunization with whole bacterial cells, exists within a species of *Rhizobium*. Three broad serological groups have been defined in the genus: 1) *R. trifolii*, *R. leguminosarum*, and *R. phaseoli*; 2) *R. meliloti*; and 3) *R. japonicum* and *R. lupini* (Graham, 1969). Strains of *R. trifolii* and *R. leguminosarum* have been found to contain no single antigen or set of antigens that are common to the genus or

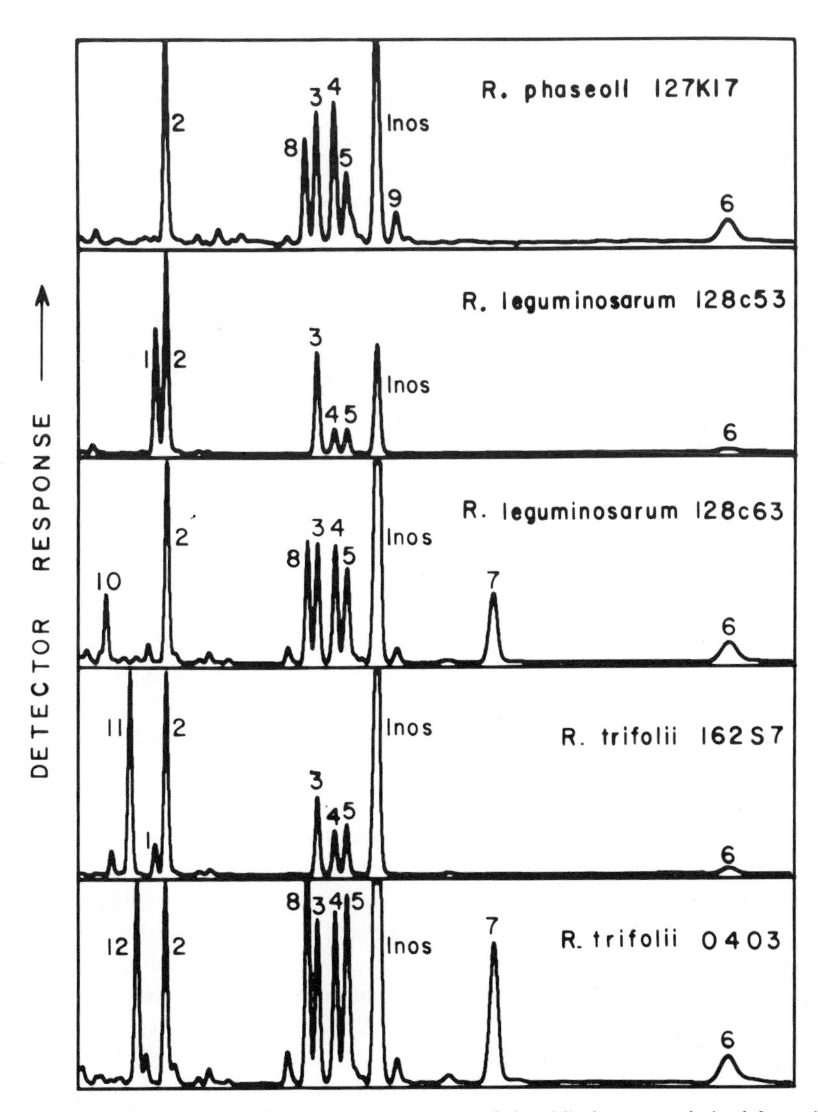

Figure 1. A comparison of the gas chromatograms of the alditol acetates derived from the sugars of five different LPS. The sugars are numbered as follows: 1 = rhamnose, 2 = fucose, 3 = mannose, 4 = galactose, 5 = glucose, 6 = 2-glucosamine, 7 = heptose, 8 = *N*-methyl-3-amino-3,6-dideoxyhexose, 9 = 2-amino-2,6-dideoxyhexose, 10 = 2,3-di-*O*-methyl-6-deoxyhexose, 11 = 2-*O*-methyl-6-deoxyhexose, and 12 = 2-*O*-methyl-6-deoxyhexose, INOS = myoinositol (added as an internal standard).

even to a single species (Kleczkowska and Thornton, 1944). However, cross-agglutination by a single antiserum of strains of *R. trifolii*, *R. leguminosarum*, and *R. phaseoli* has been observed (Graham, 1963). At least nine distinct somatic antigen groups in *R. trifolii* and 15 such groups in *R. meliloti* have been recognized (Purchase, Vincent, and Ward, 1951). The somatic antigens of 28 strains of *R. japonicum* have been reported to reside in at least 17 serological groups containing at least 24 distinct antigens (Date and Decker, 1965). Only four serological groups were detected among 40 strains of *R. japonicum* studied (Skrdleta, 1965).

Widely cross-reactive group antigens have been reported to be liberated from broken or leaking (calcium-deficient) rhizobia (Humphrey and Vincent, 1965). The number of such internal antigens shared among certain *Rhizobium* species may reflect taxonomic relatedness among these species. Patterns of internal antigens delineate *R. trifolii*, *R. leguminosarum*, and *R. phaseoli* as one serological group, whereas *R. meliloti* has been grouped with fast-growing rhizobia isolated from *Lotus* and *Leucaena* (Humphrey and Vincent, 1965; Vincent and Humphrey, 1970). *Rhizobium japonicum* has been grouped with slow-growing *R. lupini* and the cowpea rhizobia (Vincent and Humphrey, 1973). However, these serological groupings of *Rhizobium* species do not assist in identification of *Rhizobium* isolates.

Partially purified LPS preparations from two strains of *R. trifolii* have been found to possess distinctive agglutination patterns (Humphrey and Vincent, 1969). This observation, in conjunction with the fact that these preparations differ in chemical composition, suggests that antisera raised against purified LPS would aid in the identification of *Rhizobium* isolates. This is not surprising since the sereological specificity of lipopolysaccharides was the basis of the highly successful classification of *Salmonella* by Kauffmann and White (Lüderitz, Staub, and Westphal, 1966).

We have raised antisera (Humphrey and Vincent, 1965) against the purified lipopolysaccharides of one strain each of *R. phaseoli*, *R. leguminosarum*, and *R. trifolii* (Carlson et al., 1978). We have used the Ouchterlony immunodiffusion technique, as well as whole cell agglutination studies, to determine the antigenic specificity of the antisera. Both techniques demonstrate that, with the exception of *R. leguminosarum* strains 3HOQ1 and 128C53, the antisera can specifically differentiate the lipopolysaccharides of each *Rhizobium* strain. The antiserum formed against *R. phaseoli* strain 127K17 LPS reacts strongly with purified lipopolysaccharide from this strain to yield a precipitin band on double diffusion gel analysis (Carlson et al., 1978). Figure 2 illustrates a double diffusion assay in which *R. phaseoli* strain 127K17 LPS antisera reacts with the cells of that strain, but not with the cells of six other *R. phaseoli* strains. Moreover, this same antiserum reacts neither with the purified LPS of six other strains of *R. phaseoli* nor with lyophilized cells and the purified lipopolysaccharides of strains of *R.*

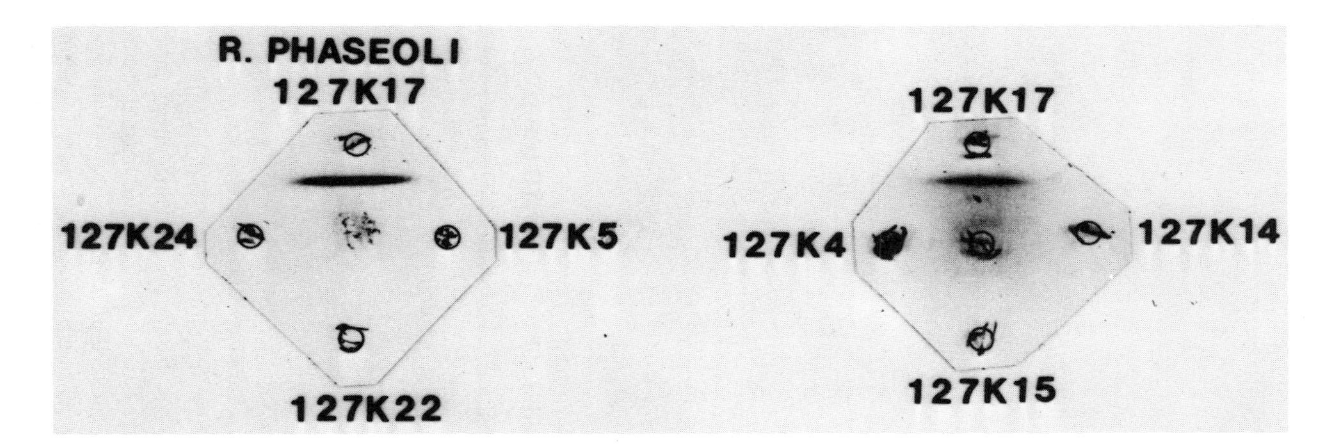

Figure 2. A microdouble diffusion gel. Undiluted rabbit antiserum (center wells), formed against the LPS purified from *R. phaseoli* strain 127K17, was diffused against lyophilized cells of each of seven *R. phaseoli* strains (outer wells). Suspensions of lyophilized cells, 100 mg/ml in physiological buffered saline, of *R. phaseoli* strains 127K17, 127K22, and 127K4 and 50 mg/ml of *R. phaseoli* strains 127K24, 127K5, 127K15, and 128K14 were prepared. Each of the outer wells contains about 20 μl of the cell suspension of the indicated *R. phaseoli* strain. The center well contains 20 μl of antiserum.

leguminosarum and *R. japonicum*. This immunological strain specificity is further demonstrated by the antiserum formed against the purified *R. japonicum* strain 61A23 LPS, forming a precipitate with the LPS of that strain but not with the LPS of two other *R. japonicum* strains, one strain of *R. leguminosarum*, and two strains of *R. phaseoli* (including strain 127K17). The specificity of these antisera is also demonstrated in agglutination studies. For example, antiserum raised against the lipopolysaccharide of *R. japonicum* strain 61A23 can be diluted 800 times and still agglutinate the cells of 61A23, whereas it fails to agglutinate the cells of two other *R. japonicum* strains and four other *Rhizobium* strains distributed among three other *Rhizobium* species even when diluted only 50 times, the most concentrated antiserum tested.

We have found that the compositional and immunochemical characteristics of *Rhizobium* lipopolysaccharides are stable features of these organisms. For example, the antibodies against *R. phaseoli* strain 127K17 obtained in 1976 (from Dr. Joe Burton of The Nitragin Company) also agglutinated and formed a precipitin band with a second isolate of that strain obtained from the same source in 1977. In addition, the sugar composition and serology of the purified *R. phaseoli* strain 127K17 LPS cultured in a yeast extract–mannitol medium were identical to the sugar composition of the lipopolysaccharide purified from the same *R. phaseoli* strain cultured 1 year later in a defined medium. These results lead us to believe that antibodies formed against purified lipopolysaccharides will greatly assist in the identification of *Rhizobium* strains.

REACTION WITH FLAGELLIN-SPECIFIC ANTIBODIES

A second major class of cell surface antigens of gram-negative bacteria is found on the flagella (Smith and Koffler, 1971; Silverman and Simon, 1977) and is referred to as the H antigen. The bacterial flagellum is composed of three structurally distinct parts; the filament, the hook, and the basal body (DePamphilis and Adler, 1971). The filament and hook may be immunologically distinct (Lawn, 1967; Dimmitt and Simon, 1970). The filament is generally made up of a single protein component flagellin (Smith and Koffler, 1971; Silverman and Simon, 1977), although there are reports that the filament may be composed of more than one protein (Smith and Koffler, 1971) and that some flagellins are glycoproteins (see Smith and Koffler, 1971). Antiserum can be prepared using either intact flagella or flagellin subunits (Smith and Koffler, 1971). The limited studies on flagella antigens of *Rhizobium* suggest that there is not strain specificity with respect to flagella. Three flagellar antigen groups were found in 16 strains of *R. meliloti* and two groups were detected among 12 strains of *R. trifolii*

(Purchase, Vincent, and Ward, 1951). Our laboratory is currently investigating the distribution of flagellar antigen groups within species of *Rhizobium* using purified flagella to prepare immune antiserum. Previous studies did not use highly purified flagella preparations and cross-reactions could have been due to contaminating antigens common to more than one strain. Thus, it remains to be seen whether the members of a single nodulation group have common flagellar structures.

REACTION WITH BACTERIOPHAGES

Bacteriophages have been demonstrated for all the main groups of *Rhizobium* (Vincent, 1974). Most of these phages will lyse more than one strain of *Rhizobium* (Conn, Bottcher, and Randall, 1945; Carlson et al., 1978). Phages prepared against strains of *R. meliloti* and *R. japonicum* seem to be species specific, that is, these phages do not form plaques on those bacteria tested that nodulate other legume species. Phages isolated using strains of *R. trifolii*, *R. leguminosarum*, and *R. phaseoli* will, in general, lyse bacteria of some strains of all three of these species. To a first approximation, these phages will not lyse strains of *R. meliloti* and *R. japonicum* (see Table 3).

A simple screening procedure can be developed for strain identification using the above generalities. An unknown bacterium can be relegated to the *R. meliloti*, the *R. japonicum*, or the *R. trifolii–R. leguminosarum–R. phaseoli* group on the basis of its ability to be lysed by phages prepared against bacteria from each of these groupings. In the cases of *R. japonicum* and *R. meliloti*, this preliminary phage lysis pattern is likely to be adequate for species identification. Additional phages are necessary to distinguish among *R. phaseoli*, *R. trifolii*, and *R. leguminosarum*. In all cases, additional phages or other methods are needed to assign a strain designation to a bacterium. Phage lysis pattern is a technically simple adjunct to the procedures described above and can reduce the number of immunochemical tests needed to identify a bacterium by assigning the bacterium to the soybean, the alfalfa, or the pea-bean-clover group. Moreover, in rare cases, phage lysis can definitively identify a bacterium that does not possess a unique lipopolysaccharide chemotype or serotype. For example, although *R. leguminosarum* strains 128C53 and 3HOQ1 have the same LPS sugar compositions and antiserum prepared against *R. leguminosarum* 128C53 LPS reacts with lipopolysaccharide from both *R. leguminosarum* strains 128C53 and 3HGQ1, five phages isolated against strain 128C53 fail to lyse strain 3HOQ1 (Table 3), thus defining these bacteria as nonidentical.

Phages with known host ranges could be maintained in *Rhizobium* collection centers and used for strain identification. In addition, phage sus-

Table 3. Phage sensitivity of *Rhizobium* strains

Rhizobium strain used to isolate phage	*R. leguminosarum*			*R. phaseoli*			*R. trifolii*	*R. japonicum*	*R. meliloti*				
	128c53	3HOQ1	128c63	127K17	127K14	127K24	162S7	61A23	RM2	102F58	102F51	2001	2009
R. leguminosarum 128c53													
Phage 1	+	–	–	+	+	–	+	–	–	–	–	–	–
Phage 2	+	–	–	+	+	–	+	–	–	–	–	–	–
Phage 3	+	–	+	+	–	+	+	–	–	–	–	–	–
Phage 4	+	–	–	+	–	+	+	–	–	–	–	–	–
Phage 5	+	–	–	+	+	+	+	–	–	–	–	–	–
R. phaseoli 127K17													
Phage 1	–	–	–	+	–	–	+	–	–	–	–	–	–
Phage 2	+	–	+	+	–	–	+	–	–	–	–	–	–
Phage 3	+	–	–	+	+	–	+	–	–	–	–	–	–
Phage 4	+	–	+	+	–	–	+	–	–	–	–	–	–
Phage 5	+	–	–	+	–	–	+	–	–	–	–	–	–
R. trifolii 162S7													
Phage 1	+	–	+	+	–	+	+	–	–	–	–	–	–
Phage 2	+	–	–	+	+	+	+	–	–	–	–	–	–
Phage 3	+	–	+	+	–	–	+	–	–	–	–	–	–
R. japonicum 61A23													
Phage 1	–	–	–	–	–	–	–	+	–	–	–	–	–
Phage 2	–	–	–	–	–	–	–	+	–	–	–	–	–
Phage 3	–	–	–	–	–	–	–	+	–	–	–	–	–
R. meliloti Rm²													
Phage 1	–	–	–	–	–	–	–	–	+	–	–	–	–
Phage 2	–	–	–	–	–	–	–	–	+	–	–	–	–
Phage 3	–	–	–	–	–	–	–	–	+	–	–	–	–
Phage 4	–	–	–	–	–	–	–	–	+	–	–	–	–
Phage 5	–	–	–	–	–	–	–	–	+	–	–	–	–
Phage 6	–	–	–	–	–	–	–	–	+	–	–	–	–

ceptibility can be used as one easy-to-measure parameter of whether a *Rhizobium* introduced into a legume field is present in the nodules of the legume.

CONCLUSIONS

Basic and applied research in nitrogen-fixing legume-*Rhizobium* symbioses depend on the proper identification of the *Rhizobium* strains under investigation. The current methods for such identification rely on an analysis of the ability of *Rhizobium* strains to nodulate a restricted number of legume hosts and on the ability of the strains to interact with antisera formed against whole *Rhizobium* cells. The use of nodulation patterns and whole cell immunology has led to the organization of *Rhizobium* into biologically related groups. These criteria are not adequate to differentiate between strains of *Rhizobium* within a species. This paper has summarized methods and procedures for the identification of *Rhizobium* species and strains. These procedures are a direct application of the results of basic studies to correlate nodulation (host range) characteristics with specific cell surface characteristics. We are hopeful that the procedures developed will be useful to national and international centers that maintain *Rhizobium* collections as well as to experiment stations and individual research laboratories.

REFERENCES

Albersheim, P., D. J. Nevins, P. D. English, and A. Karr. 1967. A method for the analysis of sugars in plant cell-wall polysaccharides by gas-liquid chromatography. Carbohydr. Res. 5:340–345.

Carlson, R. W., R. E. Sanders, C. Napoli, and P. Albersheim. 1978. Host-symbiont interactions III. The purification and partial characterization of *Rhizobium* lipopolysaccharides. Plant Physiol. 62:912–917.

Conn, H. J., E. J. Bottcher, and C. Randall. 1945. The value of bacteriophage in classifying certain soil bacteria. J. Bacteriol. 49:359–373.

Date, R. A., and A. M. Decker. 1965. Minimal antigenic constitution of 28 strains of *Rhizobium japonicum*. Can. J. Microbiol. 11:1–8.

DeLey, J., and J. Rassel. 1965. DNA base composition, flagellation and taxonomy of the genus *Rhizobium*. J. Gen. Microbiol. 41:85–91.

DePamphilis, M. L., and J. Adler. 1971. Fine structure and isolation of the hook-basal body complex of flagella from *Escherichia coli* and *Bacillus subtilis*. J. Bacteriol. 105:384–395.

Dimmitt, K., and M. Simon. 1970. Antigenic nature of bacterial flagellar hook structures. Infect. Immun. 1:212–213.

Dixon, R. O. D. 1969. Rhizobia (with particular reference to relationships with host plants). Annu. Rev. Microbiol. 23:137–157.

Dudman, W. F. 1977. Serological methods and their application to dinitrogen-fixing organisms. In: R. W. F. Hardy and A. H. Gibson (eds.), A Treatise on Dinitrogen Fixation, Section IV, Agronomy and Ecology, pp. 487–508. John Wiley & Sons, Inc, New York.

Graham, P. H. 1963. Antigenic affinities of root-nodule bacteria of legumes. Antonie van Leuwenhock 29:281–291.

Graham, P. H. 1964. The application of computer techniques to the taxonomy of the root-nodule bacteria of legumes. J. Gen. Microbiol. 35:511–517.

Graham, P. H. 1969. Analytical serology of the Rhizobiaceae. In: F. Kwapinski (ed.), Analytical Serology of Microorganisms, Vol. 2, pp. 353–378. Wiley-Interscience, New York.

Humphrey, B. A., and J. M. Vincent. 1965. The effect of calcium nutrition on the production of diffusible antigens by *Rhizobium trifolii*. J. Gen. Microbiol. 41:109–118.

Humphrey, B., and J. M. Vincent. 1969. The somatic antigens of two strains of *Rhizobium trifolii*. J. Gen. Microbiol. 59:411–425.

Jordon, D. C., and O. N. Allen. 1974. Family III Rhizobiaceae. In: R. E. Berchanan and N. E. Gibbons (eds.), Bergey's Manual of Determinative Bacteriology, pp. 261–264. The Williams & Wilkins Co., Baltimore.

Kleczkowska, J., P. S. Nutman, and G. Bond. 1944. Note on the ability of certain strains of rhizobia from peas and clover to infect each other's host plants. J. Bacteriol. 48:673–675.

Kleczkowski, A., and H. G. Thornton. 1944. A serological study of the root nodule bacteria from the pea and clover inoculation groups. J. Bacteriol. 48:661–673.

Lawn, A. M. 1967. Simple immunological labelling method for electron microscopy and its application to the study of filamentous appendages of bacteria. Nature 214:1151–1152.

Lorkiewicz, Z., and R. Russa. 1971. Immunochemical studies of *Rhizobium* mutants. Plant Soil, Special Vol., pp. 105–109.

Lüderitz, O., K. Jann, and R. Wheat. 1968. Somatic and capsular antigens of Gram-negative bacteria. Comp. Biochem. 26A:105–228.

Lüderitz, O., A. M. Staub, and O. Westphal. 1966. Immunochemistry of O and R antigens of *Salmonella* and related Enterobacteriaceae. Bacteriol. Rev. 30:192–255.

Norris, D. O. 1956. Legumes and the *Rhizobium* symbiosis. Empire J. Exp. Agric. 24:247–270.

Norris, D. O. 1965. Acid production by *Rhizobium;* a unifying concept. Plant Soil 22:143–166.

Ørskov, I., B. Ørskov, B. Jann, and K. Jann. 1977. Serology, chemistry and genetics of O and K antigens of *Escherichia coli*. Bacteriol. Rev. 41:667–710.

Osborn, M. G. 1969. Structure and biosynthesis of the bacterial cell wall. Annu. Rev. Biochem. 38:501–538.

Planqué, K., and J. W. Kijne. 1977. Binding of pea lectins to a glycan type polysaccharide in the cell walls of *Rhizobium leguminosarum*. FEBS Lett. 73:64–66.

Purchase, H. F., J. M. Vincent, and L. M. Ward. 1951. Serological studies of the root nodule bacteria. IV. Further analysis of isolates from *Trifolium* and *Medicago*. Proc. Linn. Soc. New South Wales 76:1–6.

Russa, R., and Z. Lorkiewicz. 1974. Fatty acids present in the lipopolysaccharide of *Rhizobium trifolii*. J. Bacteriol. 119:771–775.

Silverman, M., and M. I. Simon. 1977. Bacterial flagella. Annu. Rev. Microbiol. 31:397–419.

Skrdleta, V. 1965. Somatic serogroups of *Rhizobium japonicum*. Plant Soil 23:43–48.

Smith, R. W., and H. Koffler. 1971. Bacterial flagella. Adv. Microbiol. Physiol. 6:219–330.

Vincent, J. M. 1970. A Manual for the Practical Study of Root-Nodule Bacteria. I.B.P. Handbook No. 15, p. 164. Blackwell Scientific Publications, Oxford.
Vincent, J. M. 1974. Root-nodule symbiosis with *Rhizobium*. In: A. Quispel (ed.), The Biology of Nitrogen Fixation, pp. 265–341. American Elsevier Publishing Co., New York.
Vincent, J. M., and B. Humphrey. 1970. Taxonomic significant group antigens in *Rhizobium*. J. Gen. Microbiol. 63:379–382.
Vincent, J. M., and B. Humphrey. 1973. Group antigens in slow-growing rhizobia. Arch. Microbiol. 89:79–82.
Weaver, R. W., and L. R. Frederick. 1972. A new technique for most-probable-number counts of rhizobia. Plant Soil 36:219–222.
Wilson, J. K. 1944. Over five hundred reasons for abandoning the cross-inoculation groups of legumes. Soil Sci. 58:61–69.
Zajac, E., R. Russa, and Z. Lorkiewicz. 1975. Lipopolysaccharide as receptor for *Rhizobium* phage 1P. J. Gen. Microbiol. 90:365–367.

Nitrogen Fixation, Volume II
Edited by W. E. Newton and W. H. Orme-Johnson

Role of Soybean Lectin in the Soybean–*Rhizobium japonicum* Symbiosis

W. D. Bauer

The idea that host plant lectins might be involved in the recognition of symbiotic and pathogenic microorganisms has recently received a great deal of attention (Albersheim and Anderson-Prouty, 1975; Bauer and Bhuvaneswari, 1978; Beringer, 1978; Sequeira, 1978). As yet, however, there does not seem to be any compelling evidence that lectins are or are not involved in the ability of plants to recognize and respond to particular microorganisms. Thus, the work and ideas presented in this brief review should not be taken dogmatically.

Studies of the role of soybean lectin have focused on three basic questions:

1. Are there correlations between the ability of soybean lectin to bind to various rhizobia and the ability of the rhizobia to infect and nodulate soybean?
2. Is a *R. japonicum*-binding lectin present in the roots of the host plant at times and locations that permit involvement of the lectin in determining host specific infection?
3. To what component of *R. japonicum* does soybean lectin bind, and are the structure, location, and synthesis of the receptor component consistent with its postulated role as a determinant of host specificity?

CORRELATIONAL STUDIES

There have been several studies of the correlation between soybean lectin binding and host specificity of rhizobia (Bohlool and Schmidt, 1974;

Wolpert and Albersheim, 1976; Bhuvaneswari, Pueppke, and Bauer 1977; Brethauer, 1977; Bhuvaneswari and Bauer, 1978). Without exception, these correlational studies have used soybean seed lectin (SBL) rather than lectin isolated from the roots. Unless the seed lectin can be shown to be present in the roots at appropriate times and locations, the evidence obtained with seed lectin binding studies must be regarded as indirect. At best, correlative evidence obtained with seed lectin shows only that the host plant *can* synthesize a carbohydrate-binding protein capable of distinguishing between homologous and heterologous rhizobia.

Of the 48 *Rhizobium* strains tested in the first experiments by Bohlool and Schmidt (1974), all 23 of the heterologous strains failed to bind fluorescein isothiocyanate–labeled soybean lectin (FITC-SBL), whereas 22 of the 25 homologous strains did bind the labeled lectin. Subsequently, however, Chen and Phillips (1976) and Law and Strijdom (1977) failed to find a correlation between the ability of various *Rhizobium* strains to nodulate soybean and to bind FITC-SBL. They also found that some heterologous strains of rhizobia did bind FITC-SBL, although FITC-SBL did not bind to all of the homologous *R. japonicum* strains (in agreement with the pattern observed by Bohlool and Schmidt).

The studies by Brethauer (1977) introduced some important improvements in methodology. By using affinity chromatography-purified SBL, radiolabeled with ^{125}I, and by using a sugar hapten control for each binding experiment to eliminate artifacts due to nonspecific binding, Brethauer was able to overcome some of the experimental uncertainties in the earlier report. Brethauer examined the binding of ^{125}I-SBL to 28 strains of rhizobia belonging to the cowpea miscellany. Thirteen of these strains nodulated both soybean (*Glycine max* L. Merr.) and cowpea (*Vigna unguiculata* L. Walp.), whereas the other 15 strains nodulated cowpea but not soybean. Some of the results obtained by Brethauer (1977) are shown in Tables 1 and 2. It can be seen that 6 of 12 soybean-cowpea nodulating strains consistently bind SBL in a hapten-reversible manner, as do 3 of the 15 cowpea nodulating strains. There was no strong correlation between ability to bind SBL and ability to nodulate soybean.

Our studies (Bhuvaneswari, Pueppke, and Bauer, 1977; Bhuvaneswari and Bauer, 1978) generally confirmed the observations made by Bohlool and Schmidt (1974) and Brethauer (1977). When cultured in a defined liquid medium, most of the soybean-nodulating strains of *Rhizobium* bind SBL in a hapten-reversible (i.e., biochemically specific) manner. None of the various heterologous strains tested are able to bind the lectin except for one of the cowpea strains (3G4b4) obtained from Brethauer. Our initial studies indicated, however, that the lectin binding properties of *R. japonicum* cultures vary considerably with culture age. As shown in Figure 1, *R. japonicum* strains such as 3I1b 138 do not have any SBL-binding cells in stationary phase cultures. SBL-binding cells appear after one or two generations of

Table 1. SBL binding to soybean-cowpea–nodulating rhizobia

Strain (*R. japonicum*)	Days of growth in liquid media[a]		
	7	14	21
126	2.70	0.80	0.40
61A24	0.40	1.20	0.50
61A89	0.02	0.00	0.00
3I1b 110	0.90	1.60	2.50
5063	0.10	0.01	0.00
5104	0.05	0.02	0.03
10140	1.10	2.10	1.30
61A72	0.00	0.10	0.00
3I1b 83	0.00	0.00	0.00
3I1b 85	0.00	0.00	0.00
61A92	0.00	0.00	0.01
61A93	0.00	0.00	0.00

From Brethauer (1977).

[a] μg SBL/10^9 bacteria.

growth in fresh medium. By mid-log phase, most of the cells are capable of binding the lectin, but the proportion of lectin-binding cells declines rapidly as the cultures approach stationary phase. Most of the *R. japonicum* strains we have examined behave in this way, but some strains have no SBL-binding cells at any stage of growth in our synthetic media and one strain (3I1b 123) has a pattern that is the inverse of the one in Figure 1.

These results suggest that the receptors for SBL on the *R. japonicum* cell surface are transient, rather than constitutive, and that the synthesis of the receptors might be under some sort of specific regulation. We therefore cultured various *Rhizobium* strains in root exudate medium or in association with the roots of soybean seedlings to see if such culture conditions affected the lectin binding properties of the bacteria. Under these culture

Table 2. SBL binding to cowpea-nodulating rhizobia

Strain[a]	Days of growth in liquid media[b]		
	7	14	21
Crotalara 32Z1	0.6	0.4	0.7
Cowpea 64lle	0.5	0.2	0.2
Peanut 3G4b4	1.3	1.1	1.1

[a] Twelve other cowpea-nodulating strains tested failed to bind more than 0.01 μg SBL/10^9 cells. From Brethauer (1977).

[b] μg SBL/10^9 bacteria.

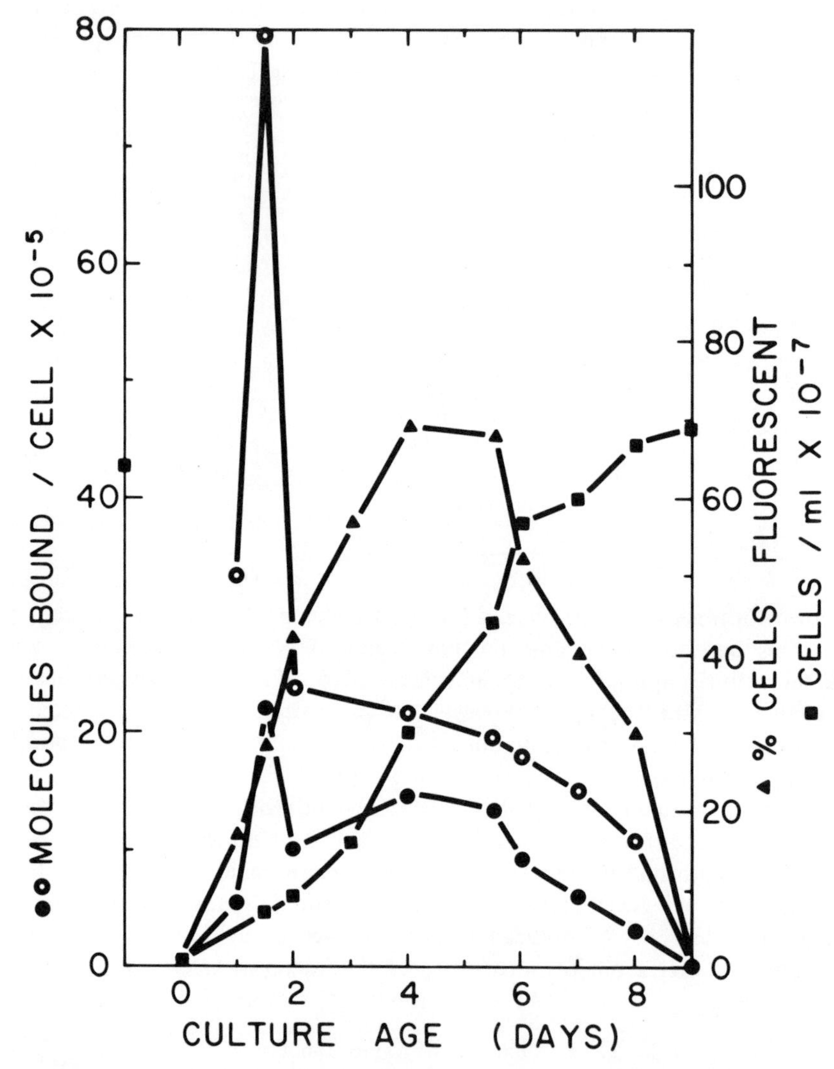

Figure 1. Changes in lectin binding properties of *R. japonicum* with culture age. Stationary phase cells of strain 3I1b 138 were transferred to fresh defined medium at a dilution of 100-fold. The new cultures were harvested at intervals and the cells washed and assayed for binding of ^{3}H-SBL and FITC-SBL. The number of cells/ml (■), the percentage of cells exhibiting fluorescence due to FITC-SBL binding (▲), and the average number of SBL molecules bound/cell at saturating lectin concentration (●) are indicated. By dividing the average number of SBL molecules bound/cell by the fraction of cells capable of binding SBL, the average number of SBL molecules bound/lectin-positive cell (○) is obtained. The binding shown is that which could be inhibited by the addition of 5 mM *N*-acetylgalactosamine (94%–99% of the total SBL bound). Reprinted from Bhuvaneswari, Pueppke, and Bauer (1977) with permission of the publishers.

conditions, all of the *R. japonicum* strains tested developed receptors for SBL (Bhuvaneswari and Bauer, 1978). Six of these same *R. japonicum* strains had no SBL-binding cells when cultured in defined media. Thus, soybean roots produce substance(s) that induce these homologous strains of rhizobia to synthesize receptors for the lectin. This effect is analogous to the synthesis of new cell surface components by bacterial pathogens of animals when they are cultured in vivo (Smith, 1977; Costerton, Geesey, and Cheng, 1978). In contrast, soybean root culture conditions did not induce strains of *R. phaseoli*, *R. trifolii*, *R. lupini*, or *R. meliloti* to synthesize receptors capable of binding SBL in a hapten-reversible manner (Bhuvaneswari and Bauer, 1978). In this respect, the binding of SBL to root-cultured rhizobia correlates perfectly with the ability of the rhizobia to nodulate soybean.

Two of the cowpea-nodulating strains examined by Brethauer were also grown in soybean rhizosphere/rhizoplane cultures. SBL-binding cells were detected in both the defined medium and the rhizosphere/rhizoplane cultures of one of these strains (3G4b4). SBL-binding cells were detected exclusively in the rhizoplane cultures of the other strain (6411e) (Bhuvaneswari and Bauer, 1978). The SBL binding ability of these two cowpea strains is thus not correlated with ability to nodulate soybean. One can tentatively conclude that either SBL binding has nothing to do with the process of infection and nodulation or SBL binding is only one early step in this process and that these cowpea strains fail to complete some other step in the process.

Wolpert and Albersheim (1976) have used a rather different experimental approach to investigate correlations between lectin binding and the host specificity of nodulation. They obtained lipopolysaccharide (LPS) preparations from four strains of rhizobia, each of which nodulated a different legume host. Affinity columns were prepared with purified seed lectin from each of the host legumes by covalently attaching the lectins to agarose beads. The rhizobial LPS preparations were then passed through each of the columns. A portion of the carbohydrate material from each of the rhizobial LPS preparations was found to interact with the column containing the host legume lectin, but not with the columns containing lectin from nonhost legumes. The specificity of the *Rhizobium* carbohydrate–legume lectin interactions thus correlated in each case with the host specificity of nodulation. *R. japonicum* was one of the *Rhizobium* species examined in these experiments and SBL was one of the lectins used.

LECTIN IN SOYBEAN ROOTS

As yet, little effort has been focused on the problem of determining whether or not a *R. japonicum*-binding lectin is present in soybean roots at times and places appropriate for involvement in the infection process. Our study of the distribution of soybean seed lectin (SBL) in various tissues of soybean seedlings during the first few weeks of growth has not provided a very satisfactory answer to this question (Pueppke et al., 1978). Tissue extracts from seed parts and young seedlings were assayed by hemagglutination, isotope dilution of radiolabeled SBL, and a radioimmunoassay technique. High SBL concentrations were present in the cotyledons, seed coats, and embryo axes of the seeds. Low SBL concentrations were found in the leaves, stems, and roots. The SBL concentrations in all tissues were found to decline rather rapidly during the first 2 weeks of growth and dropped below the limits of detection by the time of cotyledon abscission (14–16 days after planting). Similar patterns of lectin tissue distribution and changes with age have been observed in several other legumes (Howard, Sage, and Horton, 1972; Rougé, 1974; Rougé, 1975).

These results, however, do not tell us whether the disappearance of SBL with age from the roots and other tissues means that the lectin is no longer there or whether the assay and extraction methods are inadequate to solubilize and detect the SBL. Nor do we know whether a *Rhizobium*-binding lectin other than SBL is present in the roots. There is some evidence that soybean roots contain at least one other lectin besides SBL (Pueppke et al., 1978), but the other lectin(s), extractable with Triton X-100 buffer, seem to be of the "all β lectin" type (Keegstra, personal communication). One interesting problem raised by these tissue distribution studies is the origin of the SBL in leaves, stems, and roots. Since the embryo axes do not appear to contain enough SBL to account for the quantities detected in these other tissues, the SBL must either be transported from the cotyledons or synthesized in the roots, stems, and leaves. Talbot and Etzler (1978) have recently discovered a protein that may be a prolectin species of the *Dolichos biflorus* seed lectin in stems and leaves of that plant.

Another approach to studying the role of SBL in *Rhizobium* infection is to determine whether any soybean varieties exist that lack SBL. Pull, Pueppke, and Hymowitz (1978) have reported that SBL is not present in the seeds of five of the 102 cultivars of soybean they examined. Genetic studies by Orf et al. (1978) indicate that the presence of SBL in the seeds is controlled by a single dominant gene. Because the soybean cultivars lacking SBL are nodulated by *R. japonicum*, Pull, Pueppke, and Hymowitz (1978) concluded that SBL is probably not required for the initiation of the nodulation process. The presence or absence of SBL or other *Rhizobium*-binding lectins in the roots of these five cultivars has not yet been determined.

Attempts in our laboratory to detect SBL on the surface of soybean roots by the immunofluorescent and hapten washing techniques used for clover lectin (Dazzo and Brill, 1977) have not been successful. High background fluorescence due to nonspecific binding of FITC-labeled immunoglobulins has thus far prevented us from drawing any conclusions from the immunofluorescence experiments. We are currently attempting to prepare labeled *R. japonicum* lectin receptor material as a means of detecting and localizing *R. japonicum*-binding lectins in situ in intact roots.

RHIZOBIUM JAPONICUM LECTIN RECEPTORS

Most of the available evidence indicates that the receptors for SBL on the *R. japonicum* cell surface are polysaccharides of the bacterial capsule. Light microscope observations using FITC-SBL and the India ink technique for visualizing bacterial capsules indicate that all of the cells of *R. japonicum* that bind FITC-SBL have capsules and that the position of the fluorescent label on the cells appears to correspond to the shape of the capsular envelope (Bhuvaneswari and Bauer, unpublished data). Recent ultrastructural studies with ferritin-labeled lectin (FN-SBL) have demonstrated quite convincingly that SBL binds to the capsular material surrounding the bacteria and does not bind to either the flagella or the outer membrane surfaces (Bal, Shantharam, and Ratnam, 1978; Calvert et al., 1978). Figure 2 illustrates the binding of FN-SBL to the capsular material of *R. japonicum* strain 3I1b 138 cells. Freeze-etch and whole mount techniques reveal a similar pattern of lectin binding (Calvert et al., 1978). Wolpert and Albersheim (1976) suggested that the *Rhizobium* lectin receptors are lipopolysaccharide in nature. This suggestion is not substantiated by the ultrastructural studies. LPS is found primarily on the surface of the outer membrane and not in the capsule. Moreover, recent chemical evidence indicates that the sugar compositions of LPS from different strains and species of rhizobia are different and unique for each *strain*, rather than for each species (Carlson et al., 1978).

Mutants are also being examined to determine which cell surface components are required for successful, specific nodulation. Sanders, Carlson, and Albersheim (1978) selected several mutants of *R. leguminosarum* for diminished ability to produce exopolysaccharide. Cells of the mutants that they obtained were unable to nodulate the host plant, whereas revertants of each of the mutants regained both exopolysaccharide production and nodulating ability. Maier and Brill (1978) reported that *R. japonicum* mutants that were selected for inability to nodulate soybean synthesized altered LPS. We are currently attempting to select for *R. japonicum* mutants that synthesize altered SBL receptors so that they are unable to bind the lectin. The nodulation ability of such mutants will be compared with that of the wild type to test the requirement for active SBL receptors in the infection process.

The isolation, purification, and chemical characterization of the SBL receptors from *R. japonicum* are proving somewhat more difficult than anticipated. It appears that receptor material can be obtained from both the

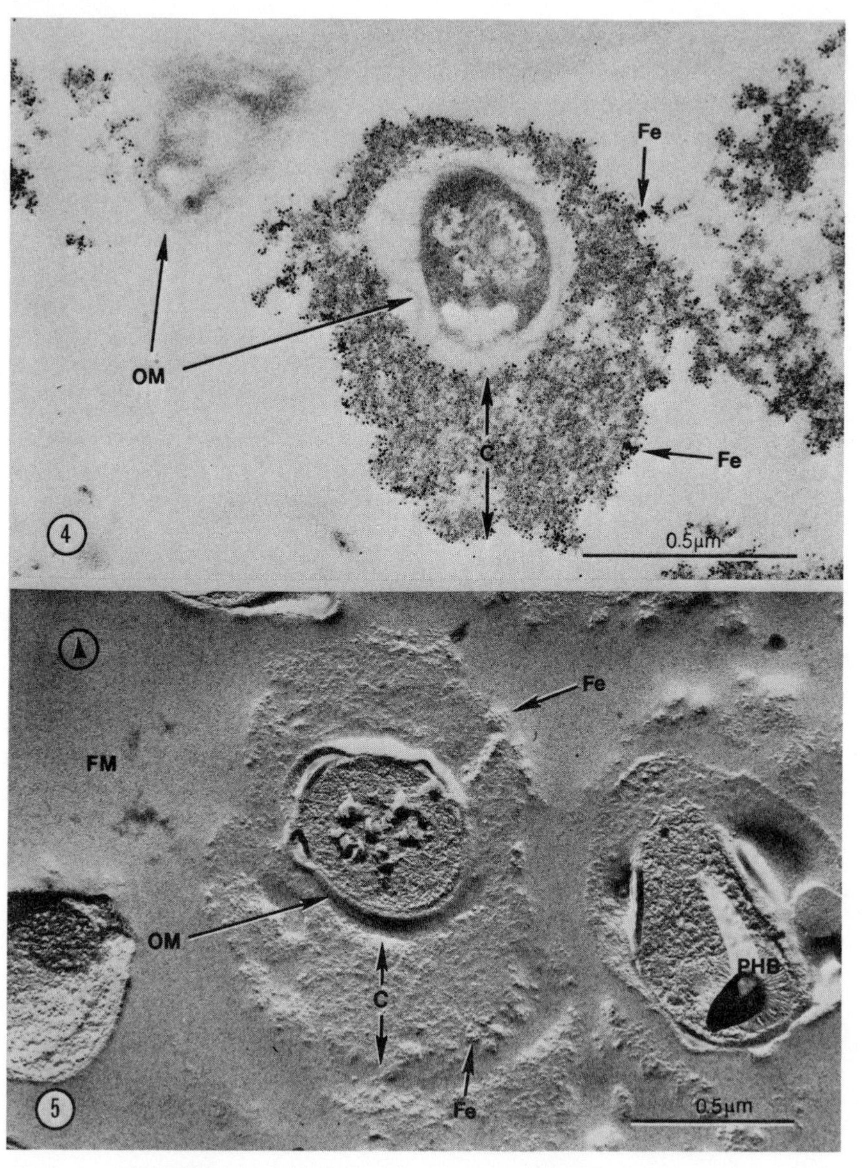

Figure 2. Binding of ferritin-labeled SBL to cells of *Rhizobium japonicum*. The upper micrograph shows bacterial cells in cross-section. The capsule (C) is stained with ruthenium red. Ferritin-labeled SBL (Fe) can be seen throughout the capsular material. The lectin does not bind to the outer membranes (OM) of the cells. The lower micrograph shows a similar pattern of lectin binding with freeze-etch techniques. PHB is polyhydroxybutyrate, FM is frozen medium. The cryoprotectant is DMSO. From Calvert et al. (1978) with permission of the National Research Council of Canada.

capsules of the bacteria and from cell-free culture filtrates. The sugar compositions of polysaccharides from these two sources are rather similar, although there are differences in the relative amounts of certain sugars both between the two sources and at different culture ages (Mort and Bauer, 1978). The capsular and exopolysaccharide preparations from both *R. japonicum* 3I1b 138 and 3I1b 110 contain mannose, glucose, galactose, galacturonic acid, acetyl groups, and a sugar tentatively identified as 4-*O*-methyl-galactose. The galactose content of the capsular polysaccharide material decreases with culture age, as does the ability of the cells to bind SBL (Mort and Bauer, 1978). We are currently attempting to purify the receptor molecules by affinity chromatography.

CONCLUSIONS AND PROSPECTS

It seems reasonably well established from the lectin binding studies that *R. japonicum* cells have carbohydrate components on their cell surfaces that distinguish them from most heterologous rhizobia. Because the host plant produces a protein (SBL) that can interact with these components in a biochemically specific fashion, it would be easy to conclude that the lectin and the bacterial receptors are the major or initial determinants of host specificity. However, we are a considerable distance from showing that the SBL-receptor interaction occurs in vivo at the sites of infection and we are even further from understanding how this interaction, if it occurs, is translated into specific cell-cell recognition.

REFERENCES

Albersheim, P., and A. J. Anderson-Prouty. 1975. Carbohydrates, proteins, cell surfaces and the biochemistry of pathogenesis. Annu. Rev. Plant Physiol. 26:31–52.

Bal, A. K., S. Shantharam, and S. Ratnam. 1978. Ultrastructure of *R. japonicum* in relation to its attachment to root hairs. J. Bacteriol. 133:1393–1400.

Bauer, W. D., and T. V. Bhuvaneswari. 1978. Possible role of lectins in plant-microorganism interactions, specifically legume/*Rhizobium* symbiosis. In: N. S. Subba Rao (ed.), Recent Advances in Biological Nitrogen Fixation.

Beringer, J. E. 1978. *Rhizobium* recognition. Nature 271:206–207.

Bhuvaneswari, T. V., and W. D. Bauer. 1978. Role of lectins in plant-microorganism interactions. III. Binding of soybean lectin to root-cultured rhizobia. Plant Physiol. 62:71–74.

Bhuvaneswari, T. V., S. G. Pueppke, and W. D. Bauer. 1977. Role of lectins in plant-microorganism interactions. I. Binding of soybean lectin to rhizobia. Plant Physiol. 60:486–491.

Bohlool, B. B., and E. L. Schmidt. 1974. Lectins: A possible basis for specificity in *Rhizobium*. Science 185:269–271.

Brethauer, T. S. 1977. Soybean lectin binds to rhizobia unable to nodulate soybean. Lectins may not determine host specificity. Master's thesis, University of Illinois.

Calvert, H. E., M. Lalonde, T. V. Bhuvaneswari, and W. D. Bauer. 1978. Role of lectins in plant-microorganism interactions. IV. Ultrastructural localization of soybean lectin binding sites on *Rhizobium japonicum*. Can. J. Microbiol. 24:785–793.

Carlson, R., and P. Albersheim. 1978. Host-symbiont interactions. II. The purification and partial characterization of *Rhizobium* lipopolysaccharides. Plant Physiol. 62:912–917.

Chen, A. T., and D. A. Phillips. 1976. Attachment of *Rhizobium* to legume roots as the basis for specific interactions. Physiol. Plant. 38:83–88.

Costerton, J. W., G. G. Geesey, and K.-J. Cheng. 1978. How bacteria stick. Sci. Am. 238:86–92.

Dazzo, F. B., and W. J. Brill. 1977. Receptor site on clover and alfalfa for *Rhizobium*. Appl. Env. Microbiol. 33:132–136.

Howard, I. K., H. J. Sage, and C. B. Horton. 1972. Studies on the appearance and location of hemagglutinins from a common lentil during the life cycle of the plant. Arch. Biochem. Biophys. 194:323–327.

Law, I. J., and B. W. Strijdom. 1977. Some observations on plant lectins and *Rhizobium* specificity. Soil Biol. Biochem. 9:79–84.

Maier, R. J., and W. J. Brill. 1978. Involvement of *Rhizobium japonicum* O antigen in soybean nodulation. J. Bacteriol. 133:1295–1299.

Mort, A. J., and W. D. Bauer. 1978. The chemical basis of lectin binding to *Rhizobium japonicum*. Plant Physiol. Annual Meetings Abstracts.

Orf, J. H., T. Hymowitz, S. P. Pull, and S. G. Pueppke. 1978. Inheritance of soybean seed lectin. Crop Sci. In press.

Pueppke, S. G., K. Keegstra, A. L. Ferguson, and W. D. Bauer. 1978. Role of lectins in plant-microorganism interactions. II. Distribution of soybean lectin in tissues of *Glycine max* (L.) Merr. Plant Physiol. 61:779–784.

Pull, S. P., S. G. Pueppke, and T. Hymowitz. 1978. Soybean lines lacking the 120,000 dalton seed lectin. Science. In press.

Rougé, P. 1974. Étude de la phytohémagglutinine des graines leutille au cours de la germination et des premiers stadis du development de la plante. Évolution dans les cotyledons. C. R. Acad. Sci. [D] (Paris) 278:449–452.

Rougé, P. 1975. Devenir des phytohémagglutinines provenant des diverses parties de la granine dans les jeunes germinations du Pois. C. R. Acad. Sci [D] (Paris) 280:2105–2118.

Sanders, R. E., R. W. Carlson, and P. Albersheim. 1978. A *Rhizobium* mutant incapable of nodulation and normal polysaccharide secretion. Nature 271:240–242.

Sequeira, L. 1978. Lectins and their role in host-pathogen specificity. Annu. Rev. Phytopathol. 16:453–481.

Smith, H. 1977. Microbial surfaces in relation to pathogenicity. Bacteriol. Rev. 41:475–500.

Talbot, C. F., and M. E. Etzler. 1978. Isolation and characterization of a protein from leaves and stems of *Dolichos biflorus* that cross reacts with antibodies to the seed lectin. Biochemistry 17:1474–1479.

Wolpert, J. S., and P. Albersheim. 1976. Host-symbiont interactions. I. The lectins of legumes interact with O-antigen containing lipopolysaccharides of their symbiont *Rhizobia*. Biochem. Biophys. Res. Commun. 70:729–737.

Section II
Nonleguminous Associations

Nitrogen Fixation, Volume II
Edited by W. E. Newton and W. H. Orme-Johnson

On The Nature of the Endophyte Causing Root Nodulation in *Comptonia*

J. G. Torrey, D. Baker, D. Callaham,
P. Del Tredici, W. Newcomb, R. L. Peterson,
and J. D. Tjepkema

We have in pure culture an actinomycete that, when applied to seedling roots of the woody dicot *Comptonia peregrina*, causes the formation of numerous root nodules that are capable of substantial rates of fixation of atmospheric nitrogen. The organism was isolated by microdissection and enzyme treatment of surface-sterilized nodules and cultured in a series of nutrient media. Details of the isolation and culture procedures have been published (Callaham, Del Tredici, and Torrey, 1978).

COMPTONIA ISOLATE IN CULTURE

In liquid culture, the actinomycete grows as a filamentous, septate, and branched mycelial mat. The filaments vary from about 0.2–1.2 μm in diameter and branching occurs randomly at right or oblique angles. The organism may be cultured in agar plates, where the filaments grow totally beneath the agar, or streaked out on agar plates, where colonies develop. Young filaments are gram-negative; older hyphae are gram-negative or variable. At the ultrastructural level, the filament is characterized by a thin

---

This research has been supported in part by the Maria Moors Cabot Foundation for Botanical Research of Harvard University, and in part by Research Grant DEB 77-02249 from the U.S. National Science Foundation and by an operating grant from the National Research Council of Canada to R. L. P.

wall, diffuse nucleoid regions with fibrils, numerous ribosomes, and vesicular structures with reticulate or fibrillar contents.

After a period of hyphal growth in culture, terminal and/or intrahyphal sporangia are formed. At the tip of a hyphal branch, repeated septations occur, first at right angles to the length of the filament and then parallel to it until a swollen compartmentalized sac is formed. The spores within become rounded and walled off from each other. The sporangium itself seems to be defined only by the former hyphal membrane. Intrahyphal sporangia form in similar fashion by septation and further subdivision but in the middle of a filament that has swollen in diameter. Sporangia may reach 30–60 μm in length and contain dozens to perhaps a hundred spores. The spores tend to be rounded or oval in shape and about 1.5–3.5 μm in diameter or length. When first formed, the spore walls are thin and the spores show little vesiculation with small nucleoid areas. On maturing, the spores have a thick, two-layered cell wall, more elaborate vesicular structures with fibrillar substructure, and evident nucleoid area. The details of fine structure have been described by Newcomb et al. (1979) and by Lalonde (1978).

At the present time, we do not know much about spore release from the sporangia nor about spore germination. From preliminary experiments with preparations of spores and of broken fragments of filaments plated into nutrient agar, we believe the organism can be propagated from either structure. We have no information on the nature of the free-living actinomycete in the soil.

ACTINOMYCETE IN ROOT NODULES OF *COMPTONIA*

The organism that causes nodule formation on field-grown plants of *C. peregrina* is remarkably similar in structural detail to the *Comptonia* isolate, lending credence to the belief they are the same organism. Lalonde (1978), using immunofluorescence methods, has demonstrated the identity of the endophyte in *Comptonia* nodules with the cultured isolate. The actinomycete enters young seedling roots in the region of elongating root hairs (Callaham et al., 1978) and causes marked root hair deformation. The actinomycete penetrates the root hair wall by a process not fully understood and within the root hair it elicits a remarkable cytoplasmic activity involving endoplasmic reticulum and Golgi bodies, and leading to polysaccharide formation and encapsulation of the intruder. Encapsulation is an ongoing process that accompanies the endophyte throughout its entire life within the host, so that every structure of the actinomycete within the host is enclosed by a polysaccharide, and probably pectic, capsule. The entering filament branches and grows down the root hair into the enlarged basal part of the root hair cell, penetrates by local dissolution of the cell wall, and enters the outer cortical cells of the root. The cells in the vicinity are stimulated to divide and enlarge, forming a localized hypertrophy termed the prenodule (Callaham and Torrey, 1977). The endophyte branches repeatedly and successively invades the cortical cells of the root, forming a beachhead for further invasion. Hyphal branches occupy the central areas of these cortical cells.

At this stage of nodule initiation, modified lateral rootlike primordia are initiated in the pericycle at or near the site of the initial root hair infection. In *Comptonia*, the number of primordia initiated varies from a few up to a dozen; in *Myrica gale*, the number is usually two or three (Torrey and Callaham, 1979). Each of these primordia develops through the root cortex, incorporating divided cortical cells into the body of the primordial meristem. In the *Myrica*-type nodule, seen in *Myrica* and *Comptonia*, these primary nodule primordia, after passing through an arrested stage during which the cortex is invaded by endophyte coming from the prenodule, elongate vertically upward as nodule roots. The nodule roots are determinate structures, reaching 2–3 cm in length (Torrey and Callaham, 1978). The endophyte invades only the swollen basal cortical tissues and does not enter the root tissues themselves.

The invasive stage of nodule development is characterized by the growth of encapsulated filaments into the cortical tissues of the primary nodule lobes and the ramifying of the filaments, nearly filling the infected cells. Not all the cells of the cortex are invaded, and these uninfected cells are usually characterized by the presence of starch grains or tannin deposits or both. The invasive stage is followed by vesicle formation in most actinomycetal nodules studied. A detailed study of the actinomycete in *Comptonia* nodules has been published (Newcomb et al., 1979). The terminal portions of the hyphae become swollen into globose or flask-shaped structures that have characteristic septations. They are, like the rest of the endophyte, surrounded by a pectic capsule. These vesicles may occupy the outer periphery of the cortical cells or may be distributed throughout the cell. According to some authors (Akkermans, 1971; Becking, 1977), the vesicles are the site of nitrogenase formation and activity. Isolated vesicles maintained under anaerobic conditions in vitro have been shown to possess low nitrogenase activity (Van Straten, Akkermans, and Roelofsen, 1977). However, nodules of *Casuarina*, which effectively fix nitrogen, have not been shown to form vesicles (Tyson and Silver, 1979). It may be that nitrogenase has more than one site in the infected nodule cells.

Vesicle formation may be a morphological state of the actinomycete that occurs only in the endophytic condition within the host. No regular formation of vesicles has been observed in the *Comptonia* isolate grown in a range of cultural conditions. There is to date no evidence of nitrogenase activity in the cultured organism grown in a range of complex or synthetic media. Vesicle form depends on the host in which the endophyte resides. The *Comptonia* isolate invades *Comptonia* and *M. gale* in essentially the same manner and the vesicles that form in the mature nodule are peripheral, flask-shaped, and septate. Lalonde (1979) has reported that the *Comptonia* isolate, which forms nodules on *Alnus glutinosa* by cross-inoculation, develops globose vesicles more typical of the form usually seen in *Alnus* with its own endophyte. Therefore, the form and position of the vesicles must result from the interacting genomes of the host and the endophyte.

A final stage in the development of the actinomycete within the host nodule is the formation of sporangia, a sequence of events well described in the *Comptonia* isolate grown in culture but only poorly understood in the nodule. Knowledge of the developmental sequence in the cultured organism helps in our interpretation and understanding of the developmental events within the nodule. Van Dijk and Merkus (1976) described "spindle" formation in nodules of *A. glutinosa* and recognized it as an event in the development of the "bacteroid" stage. As they properly pointed out, these were related events in the initiation of a sporangium that, at maturation, formed thick-walled spores. Their light microscopy and electron microscopy photographs of developing sporangia from actinomycete filaments within cortical cells of *Alnus* can be matched by these stages in the cultured *Comptonia* that has been isolated (Lalonde, 1978; Newcomb et al., 1978). Others have reported sporangia formation in nodules of *Alnus* (Lalonde, 1978) or of *M. gale* (Callaham, unpublished data) infected with *Comptonia* isolate. Thus, the interpretation of Van Dijk and Merkus (1976) has been confirmed by the development of essentially identical structures in the *Comptonia* isolate grown in vitro. Conversely, the presence of endophyte sporangia and spores within nodules that resemble in structural detail the actinomycete grown in pure culture further confirms the identity of the organism in the two stages.

METHODS OF ISOLATION AND CULTURE OF THE ACTINOMYCETOUS ENDOPHYTE

Two quite different methods have been used successfully in our laboratory for the isolation of the *Comptonia* actinomycete from root nodules. One method, used at least twice with success on *Comptonia*, involves microdissection, enzyme treatment, micromanipulation, dilution, and culture of the filamentous organism from mature nodules. This method was described in detail by Callaham et al. (1978) and is similar to that used earlier by Quispel (1954), Pommer (1959), Uemura (1964), Lalonde, Knowles, and Fortin (1975) and many others. We have tried to apply the method to nodules of *M. gale*, *Ceanothus americanus*, *Alnus rugosa*, and other species, thus far without success. Clearly, we have more to learn about the isolation process.

The second method (Baker, Kidd, and Torrey, 1979) involves the use of Sephadex columns to separate particles based on size (a gravity-flow system). Certain column fractions of nodule extracts from *Elaeagnus umbellata* and from *M. gale* show an enrichment of the endophyte judged from effectiveness in nodulating seedling plants. The enriched fractions can be further purified on a sucrose–zonal gradient column. When all of these steps are performed aseptically, pure cultures of actinomycetes can be recovered from specific eluent fractions. The method has been tested using homogenates of the *Comptonia* isolate itself. Colonies of the *Comptonia* isolate have been recovered and grown in culture.

Samples of the eluent fractions can be examined by scanning electron microscopy to determine the nature of the infective particles. The method may also be developed as a means of making pure preparations either of spores or of hyphal fragments. Thus far, we have not achieved the isolation of endophytes from other symbiotic actinomycetal nodules. The column method of isolation may prove especially useful with nodules that possess no discrete, outer limiting, suberized cell layers as in *Comptonia* or field-grown *Alnus*. *Elaeagnus* plants grown in water culture develop nodules with highly aerenchymatous surfaces in which surface sterilization by such reagents as mercuric chloride or even sodium hypochlorite is quite impossible. Bacterial populations introduced into the column at the outset with the nodule extract are banded in the Sephadex column and diluted out in the sucrose gradient so that sterile fractions can be recovered. The method shows great promise both for isolation and for biochemical studies of nodules.

METHODS OF TESTING THE ACTINOMYCETE ISOLATE

Many research workers have attempted isolations of the actinomycete organism and many have reported success in isolation but failure to achieve nodulation of seedlings inoculated with the endophyte. Some workers have reported success in this step as well, but their success has not been strongly documented by their test procedures and thus their achievements have not been generally accepted. The question must be asked: What is an acceptable test for effective nodulation by a cultured organism?

The parallel situation exists for testing *Rhizobium*-legume symbioses. In the *Rhizobium* case, the testing is straightforward (Vincent, 1970). Pure cultures of *Rhizobium*, readily isolated from surface-sterilized nodules, are easily grown in defined nutrient media. A suspension of bacterial cells is added to the substrate in which the seedling to be inoculated is growing. The substrate may be agar in tubes, sand in tubes, sand in Leonard jars, moistened filter paper in plastic pouches, water culture, or aeroponics. The root medium may be prepared as a sterile substrate for rigorous proof or with reasonable care for routine tests. Within 5–10 days, morphological structures identifiable as nodules are evident, proving the *infective* capacity of the organism and, within a week or two, one can readily test the whole seedling or the excised root for acetylene reduction—tantamount to demonstrating N_2 fixation and the *effectiveness* of the *Rhizobium*. Rigorous demonstration would involve $[^{15}N]N_2$ incorporation into plant protein.

Comparable tests can be performed for the infectiveness and effectiveness of actinomycetes in inducing nitrogen-fixing root nodules. In the absence of isolated cultured bacteria, the common practice has been to grind up mature nodules in distilled water and to dip the seedling root in the suspension or brush the suspension onto the root. Sometimes only soil extracts taken from around nodulated plants were effective. Seedlings were grown routinely in washed sand, in water culture or on agar slants, in plastic pouches, or aeroponically. Usually, in these procedures aseptic techniques were not followed rigorously. Infectiveness of the nodule suspension could be demonstrated under favorable conditions within about 2 weeks and almost certainly within 4 weeks. Nodulation appearing after longer periods could not be as certainly attributed to the applied suspension, especially if the tests were performed under open greenhouse conditions.

The uncertainty of earlier claims and the criticism of the test methods or test results for actinomycete nodulation by isolated cultured organisms were discussed at length (Bond, 1963, 1967; Fletcher and Gardner, 1974; Quispel, 1974; Lalonde, Knowles, and Fortin, 1975). Because of the frequency and questionable status of claims for successful isolation of actinomycetes effective in causing root nodulation, Bond (1967) suggested a procedure for handling claims for success in demonstrating actinomycete isolation and reinoculation. With the isolate from *Comptonia*, our tests followed accepted procedures. Our first trials were with seedlings of *Comptonia* germinated in washed sand and grown in artificial light in a growth chamber. Subsequent tests involved large numbers of seedling plants grown in aeroponics tanks in the greenhouse. The *Comptonia* isolate was prepared by homogenization of a liquid culture, and inoculation involved either pouring the suspension on the sand or into the nitrogen-free medium of the aeroponics tank. Large numbers of nodules were formed within a week or two and they proved to be active in acetylene reduction (Callaham et al., 1978). Confidence in the identity of the *Comptonia* isolate as the infective organism came also from its activity in cross-inoculation studies with related *Myrica* species.

In the spirit of Bond's 1967 recommendation, we sent cultures of the *Comptonia* isolate, seeds of *Comptonia*, detailed instructions for germination conditions, and young nodulated seedlings of *Comptonia* to Dr. M. Lalonde at the C. F. Kettering Research Laboratory, requesting him to attempt to confirm our experiments. We felt Dr. Lalonde would be both interested in and capable of making impartial tests, since he had extensive experience both with actinomycetes in culture and with actinomycetorhizal plants, especially *Alnus*. This he has done (Lalonde, 1978) and has added new information on cross-inoculations and structure.

In our own laboratory, we have continued with what we felt was a necessary rigorous test of the identity of the infective-effective microorganism—that is, to perform the inoculation of roots of seedling plants grown under axenic conditions in which only the actinomycete added from pure culture was present with the host plant. To grow whole plants under sterile conditions is difficult and there exists an extensive literature describing efforts to grow plants as seedlings or to maturity under sterile conditions (e.g., Hewitt, 1966). In the case of woody dicots, the problems are compounded because native species usually have a seed dormancy that must be overcome. Frequently, microorganisms inhabit the layers between fruit coat and seed coat or the interstices of the seed coat, so that to obtain sterile seedlings may involve seed or fruit scarification, seed treatment (chemical or environmental) for dormancy, and favorable conditions for root and shoot growth in an enclosed environment. To date, our efforts have allowed us to cultivate aseptically grown seedlings of *C. peregrina*, *M. gale*, *Alnus rubra*, *E. umbellata*, and *Elaeagnus angustifolia*. In the first three, we know that the *Comptonia* isolate produces abundant nodulation in sand culture or in aeroponics. In none of these species have we achieved nodule formation in axenic culture in the presence of the *Comptonia* isolate alone.

HOST SPECIFICITY AND CROSS-INOCULATIONS WITH THE *COMPTONIA* ISOLATE

Our experience confirms that of many others concerning the broad specificity of the actinomycete with respect to the host plants it will invade. The subject of cross-inoculability has been reviewed in recent years (Bond, 1963; Rodriguez-Barrueco and Bond, 1976; Hall et al., 1979), yet not until a number of isolations of effective actinomycete endophytes has been made and these isolates tested will we have any idea of the specificity involved in the actinomycetorhizal systems. Thus, for example, our ideas of specificity and of the possible taxonomic classification at the species level based on that specificity (Becking, 1970; Lalonde, 1978) have already begun to break down with tests made with the *Comptonia* isolate.

The *Comptonia* isolate readily infects and produces effective nodules on the following plants tested in aeroponics or water culture: *C. peregrina*, *M. gale*, *Myrica cerifera* (Callaham et al., 1978), *A. glutinosa*, *Alnus crispa* (Lalonde, 1979), *A. rubra*, and *Alnus rugosa*. After careful testing, we have concluded that the *Comptonia* isolate does not nodulate *Casuarina littoralis* (or *Casuarina equisetifolia*), *Casuarina cunninghamiana*, *E. umbellata*, *E. angustifolia*, and *Ceanothus americanus*. Each of these produced effective nodules under our growing conditions when nodule suspensions containing its own endophyte were used. Some peculiarities in the infection process have been observed that caution against careless generalization. *Elaeagnus* seedlings do not form nodules in aeroponics culture with nodule suspensions from field-collected nodules of the corresponding *Elaeagnus* species, although nodules are usually formed at the same age in water culture. We do not understand the differences in physiology that cause this difference in behavior. Strand and Laetsch (1977) reported that *Ceanothus integerrimus* nodulated in water culture when inoculated with soil suspensions but not with nodule suspensions.

The problem of cross-inoculation, as in the case of the *Rhizobium*-legume system, is a fascinating one because, through an understanding of the basis of infection and barriers to infection, we are most likely to be able to broaden the specificity to include other host plants whose roots are not now invaded by actinomycetes. This subject is accessible to study at a number of levels: cross-inoculation using seedlings and nodule suspensions, tests for the susceptibility of plants to a cultured isolate, structural studies using the transmission and scanning electron microscope of the root hair–microorganism–soil interface, or chemical studies of possible chemical interactions involving surface polysaccharides, proteins, or possibly other recognition systems.

OTHER SUCCESSFUL ISOLATIONS OF ACTINOMYCETES FROM ROOT NODULES

Among the several reports in the literature of claims of successful isolation and culture of an actinomycete effective in causing the formation of N_2-fixing root nodules, one of the most interesting was the report of Pommer (1959). Using relatively straightforward methods described earlier by others, Pommer reported that he isolated and grew in pure culture on a quite simple nutrient medium (glucose-asparagine agar) the actinomycete from root nodules of *A. glutinosa*. His drawings of the filamentous structures and of the sporogenous bodies are remarkably similar to the *Comptonia* isolate we have described. Using five test plants and four controls, he applied the culture to *A. glutinosa* seedlings and obtained nodulation in 3 weeks. His work was repeated by Quispel (1960) and Uemura (1964) without success. Unfortunately, Pommer's cultures were not available to others nor tested by them. It seems very likely that Pommer had obtained the effective endophyte from *Alnus* nodules. The only difference between his organism and ours was the presence of small vesicles or "bläschen" illustrated as short side branches on some of the filamentous structures he observed. We have not seen such structures as a regular feature of healthy cultures of the *Comptonia* isolate.

Other successful claims may also have been ignored. Reputed isolations have been carefully and critically reviewed over the years by Bond (1963, 1967), Uemura (1964), Quispel (1974), and Fletcher and Gardner (1974) among others. The major criticism and skepticism leveled at earlier reports were directed at the inadequacy of the methods used in testing the cultured organism for effectiveness in producing nitrogen-fixing nodules on whole plants. Quispel's criticism (1960) of Pommer's claim to successful culture, based on the simplicity of the medium used by the latter, is clearly invalid. The *Comptonia* isolate can be shown to grow (albeit slowly) on very simple nutrient media, including even tap water agar. For some who have isolated actinomycetes from root nodules and failed to demonstrate reinoculation, failure may have been in establishing appropriate growing conditions for seedlings used in testing the effectiveness of the organism. We continue to find testing under rigorously axenic conditions very difficult.

CONCLUSION

The isolation and culture and partial characterization of an actinomycete from the root nodules of a woody dicot capable of nitrogen fixation symbiotically have been achieved and confirmed. Clear demonstration of the isolate as the organism causing nodulation rests on structural grounds, on accepted microbiological techniques, and on physiological and biochemical criteria. Knowing the nature of this organism and some of its structural and biochemical characteristics should make the isolation of the whole range of endophytes in the actinomycetorhizal plants easier. Success in this endeavor will open up, in turn, whole new areas in the study of the biology of these important symbiotic systems.

ACKNOWLEDGMENTS

Appreciation is expressed to G. Kidd for advice on biochemical procedures. The authors are indebted to the following for technical assistance: S. LaPointe, A. Berry, P. Bowne, S. Knowlton, J. McElroy, and R. Van Alstyne.

REFERENCES

Akkermans, A. D. L. 1971. Nitrogen fixation and nodulation of *Alnus* and *Hippophaë* under natural conditions. Ph.D. thesis, University of Leiden, Netherlands.

Baker, D., G. H. Kidd, and J. G. Torrey. 1979. Separation of actinomycete nodule endophytes from crushed nodule suspensions by Sephadex and sucrose-density fractionation. Bot. Gaz. 140 (Suppl.):S49–S51.

Becking, J. H. 1970. Frankiaceae fam. nov. (Actinomycetales) with one new combination and six new species of the genus *Frankia* Brunchorst 1886. Int. J. Syst. Bacteriol. 20:201–220.

Becking, J. H. 1977. Endophyte and association establishment in nonleguminous nitrogen-fixing plants. In: W. Newton, J. R. Postgate, and C. Rodriguez-Barrueco (eds.), Recent Developments in Nitrogen Fixation, pp. 551–567. Academic Press, London.

Bond, G. 1963. The root nodules of non-leguminous angiosperms. Symp. Soc. Gen. Microbiol. 13:72–91.

Bond, G. 1967. Fixation of nitrogen by higher plants other than legumes. Annu. Rev. Plant Physiol. 18:107–126.

Bowes, B., D. Callaham, and J. G. Torrey. 1977. Time-lapse photographic observations of morphogenesis in root nodules of *Comptonia peregrina* (Myricaceae). Am. J. Bot. 64:516–525.

Callaham, D., P. Del Tredici, and J. G. Torrey. 1978. Isolation and cultivation *in vitro* of the actinomycete causing root nodulation in *Comptonia*. Science 199:899–902.

Callaham, D., W. Newcomb, J. G. Torrey, and R. L. Peterson. 1979. Root hair infection in actinomycete-induced root nodule initiation in *Casuarina*, *Myrica* and *Comptonia*. Bot. Gaz. 140 (Suppl.):S1–S9.

Callaham, D., and J. G. Torrey. 1977. Prenodule formation and primary nodule development in roots of *Comptonia* (Myricaceae). Can. J. Bot. 55:2306–2318.

Fletcher, W. W., and I. C. Gardner. 1974. The endophyte of *Myrica gale* nodules. Ann. Microbiol. (Milan) 24:159–172.

Hall, R. B., H. S. McNabb, Jr., C. A. Maynard, and T. L. Green. 1979. Toward development of optimal *Alnus glutinosa* symbioses. Bot. Gaz. 140 (Suppl.): S120–S126.

Hewitt, E. J. 1966. Sand and Water Culture Methods Used in the Study of Plant Nutrition, 2nd Ed. Commonwealth Agricultural Bureau, England.

Lalonde, M. 1978. Confirmation of the infectivity of a free-living actinomycete isolated from *Comptonia peregrina* (L.) Coult. root nodules by immunological and ultrastructural studies. Can. J. Bot. 56:2621–2635.

Lalonde, M. 1979. Immunological and ultrastructural demonstration of nodulation of the European *Alnus glutinosa* (L.) Gaertn. host plant by an actinomycetal isolate from the North-American *Comptonia peregrina* (L.) Coult. root nodule. Bot. Gaz. 140 (Suppl):S35–S43.

Lalonde, M., R. Knowles, and J. Andre Fortin. 1975. Demonstration of the isolation of non-infective *Alnus crispa* var. *mollis* Fern. nodule endophyte by morphological immunolabelling and whole cell composition studies. Can. J. Microbiol. 21:1901–1920.

Newcomb, W., D. Callaham, J. G. Torrey, and R. L. Peterson. 1979. Morphogenesis and fine structure of the actinomycetous endophyte of nitrogen-fixing root nodules of *Comptonia peregrina*. Bot. Gaz. 140 (Suppl.):S22–S34.

Newcomb, W., R. L. Peterson, D. Callaham, and J. G. Torrey. 1978. Structure and host-actinomycete interactions in developing root nodules of *Comptonia peregrina*. Can. J. Bot. 56:502–531.

Pommer, E. H. 1959. Über die Isolierung des Endophyten aus den Wurzelknöllchen *Alnus glutinosa* Gaertn. und über erfolgreiche Re-Infektionsversuche. Ber. Deutsch. Bot. Ges. 72:138–150.

Quispel, A. 1954. Symbiotic nitrogen-fixation in non-leguminous plants. I. Preliminary experiments on the root-nodule symbiosis of *Alnus glutinosa*. Acta Bot. Neerl. 3:495–511.

Quispel, A. 1960. Symbiotic nitrogen fixation in non-leguminous plants. V. The growth requirements of the endophyte of *Alnus glutinosa*. Acta Bot. Neerl. 9:380–396.

Quispel, A. 1974. The endophytes of the root nodules in non-leguminous plants. In: A. Quispel (ed.), The Biology of Nitrogen Fixation, pp. 499–520. North-Holland Publishing Co., Amsterdam.

Rodriguez-Barrueco, C., and G. Bond. 1976. A discussion of the results of cross-inoculation trials between *Alnus glutinosa* and *Myrica gale*. In P. S. Nutman (ed.), Symbiotic Nitrogen Fixation in Plants, IBP 7, pp. 561–565. Cambridge University Press, Cambridge, England.

Strand, R., and W. M. Laetsch. 1977. Cell and endophyte structure of the nitrogen-fixing root nodules of *Ceanothus integerrimus* endosymbiont. Protoplasma 93:165–178.

Torrey, J. G., and D. Callaham. 1978. Determinate development of nodule roots in actinomycete-induced root nodules of *Myrica gale*. Can. J. Bot. 56:1357–1364.

Torrey, J. G., and D. Callaham. 1979. Early nodule development in *Myrica gale*. Bot. Gaz. 140 (Suppl.):S10–S14.

Tyson, J. H., and W. Silver. 1979. Relationship of ultrastructure to acetylene reduction (N_2-fixation) in root nodules of *Casuarina*. Bot. Gaz. 140 (Suppl.):S44–S48.

Uemura, S. 1964. Isolation and properties of microorganisms from root nodules of non-leguminous plants. A review with extensive bibliography. Bull. Govt. For. Exp. Sta. (Tokyo). 167:59–91.

van Dijk, C., and E. Merkus. 1976. A microscopical study of the development of a spore-like stage in the life cycle of the root-nodule endophyte of *Alnus glutinosa* (L.) Gaertn. New Phytol. 77:73–91.

Van Straten, J., A. D. L. Akkermans, and W. Roelofsen. 1977. Nitrogenase activity of endophyte suspensions derived from root nodules of *Alnus*, *Hippophaë*, *Shepherdia* and *Myrica* spp. Nature 266:257–258.

Vincent, J. M. 1970. A Manual for the Practical Study of Root-Nodule Bacteria. I.B.P. Handbook No. 15. Blackwell Scientific Publications. Oxford, England.

Nitrogen Fixation, Volume II
Edited by W. E. Newton and W. H. Orme-Johnson

Analysis of Factors Limiting Nitrogenase (C_2H_2) Activity in the Field

J. Balandreau and P. Ducerf

The advent of the acetylene assay, even with its calibration difficulties, produced a fantastic increase in sensitivity when measuring nitrogen fixation; it made the demonstration of fixation in short term experiments possible even with inefficient systems. This led to a better understanding of the influence of various factors on nitrogenase activity of different systems. In the case of symbiotic systems, many elegant ecological studies appeared at the beginning of the seventies, most of them presented at the Edinburgh IBP meeting in 1973 (Ham, Lawn, and Brun, 1975; Hardy and Havelka, 1975; Sloger et al., 1975; Masterson and Murphy, 1976; Pate, 1976; Sprent, 1976) or even before (Wheeler, 1969). At the same time, for nonsymbiotic systems, incubations of entire intact soil-plant systems gave some clues about the role of factors such as light, age of plants, cultivars, soil moisture, and oxygen pressure in diurnal and seasonal variations (Balandreau et al., 1975; Day et al., 1975; Döbereiner and Day, 1975). Many of these studies were made using simplified growth conditions, either in greenhouses or in growth chambers where only one factor was allowed to vary. Other studies were conducted in the field, especially with legumes, but also simply where one factor was manipulated, e.g., CO_2 (Hardy and Havelka, 1975) or leaf or pod numbers (Ham, Lawn, and Brun, 1975), or in situations where a small number of factors were actually limiting, e.g., nodule numbers and air temperature (Sloger

This research has been partly supported by a grant from the Délégation Générale à la Recherche Scientifique et Technique (Action Concertée Equilibre et Lutte Biologigue).

et al., 1975) or soil and air temperature and hours of sunshine (Masterson and Murphy, 1976).

After the Edinburgh meeting, ecological studies became very rare. The reason was not a lack of interest or the solution of main questions but, rather, methodological limitations.

For nonsymbiotic systems, the absolute need for incubation of entire undisturbed plant-soil systems became evident only recently. Excised nodules of legumes may retain most of their activity for a rather long time, but this is not the case for excised roots of grasses, where the activity disappears when they are extracted from the soil. Even incubation of cores is poorly reliable. Moreover, for both symbiotic and nonsymbiotic systems, there is a limitation when treating data obtained in natural conditions because many factors can vary and influence nitrogenase activity; multiple correlation calculations require large numbers of data points and yield

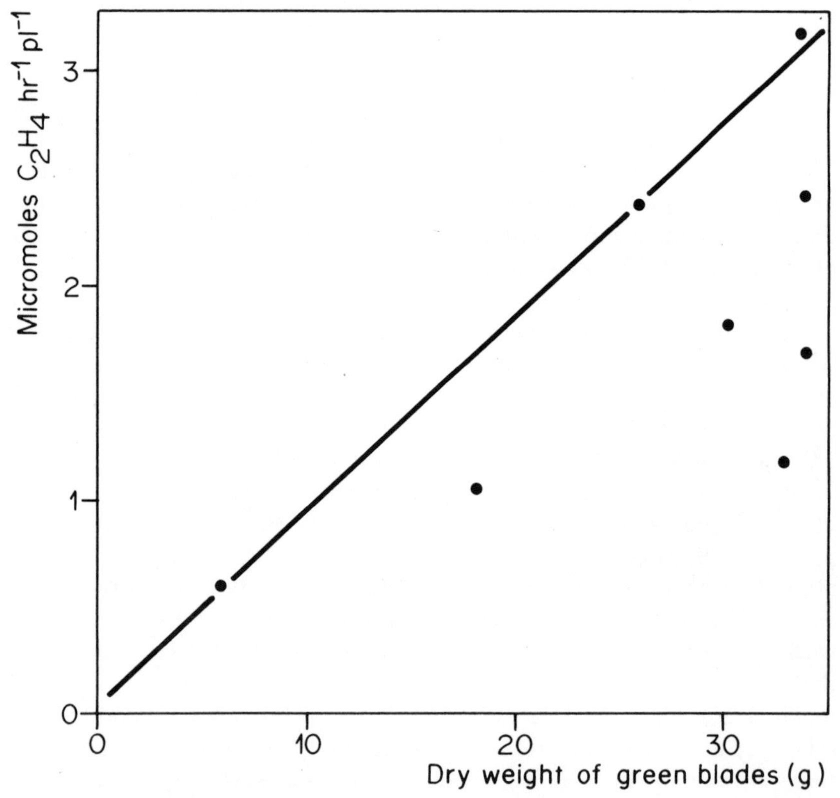

Figure 1. Maize (1975), daily maxima of acetylene-reducing activity (ARA_{max}) versus dry weight of green blades; each point is the maximum recorded value among 40 measurements throughout day and night; measurements were performed 1 day a week during July and August.

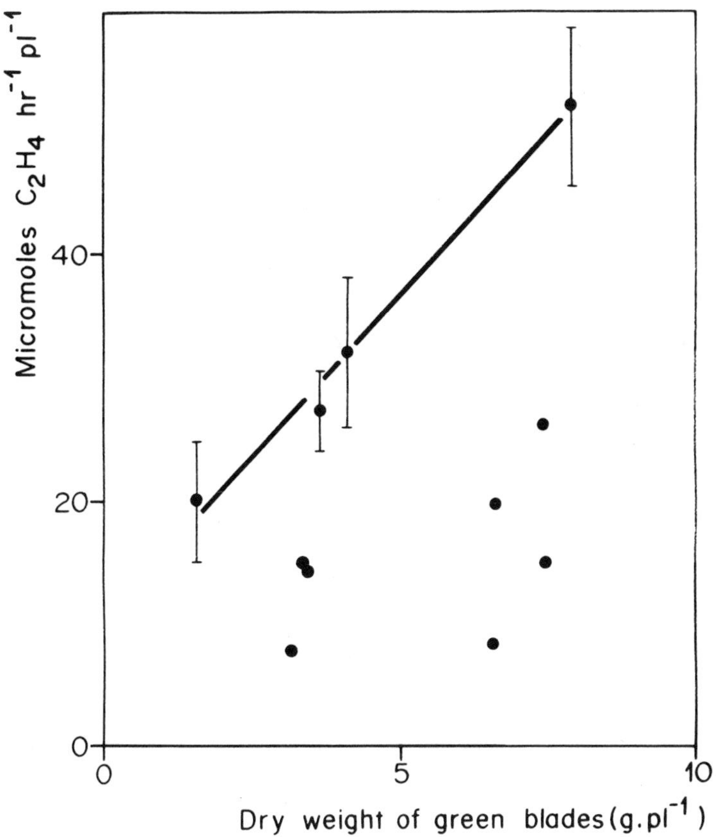

Figure 2. Peanuts (1976), vegetative period of growth, maxima of acetylene-reducing activity (ARA_{max}) versus dry weight of green blades.

imprecise explanations that are often unable to explain individual results. In other words, the true obstacle is the lack of a statistical method of data treatment that will describe natural situations where more than three or four factors can influence nitrogenase activity and estimate the influence of one particular factor, correlate a measured nitrogenase activity with one particular factor, and eventually predict the nitrogenase activity in a particular situation.

We present here data obtained with grass systems (maize, *Dactylis*, rice) in France (Centre National de la Recherche Scientifique, Centre de Pédologie Biologique, Vandoeuvre-les-Nancy; experimental farm of École Nationale Supérieure d'Agronomie et des Industries Alimentaires, Nancy) and with peanuts in Senegal (Centre National de la Recherche Agronomique, Bambey) for which we have tried to study limiting factors in

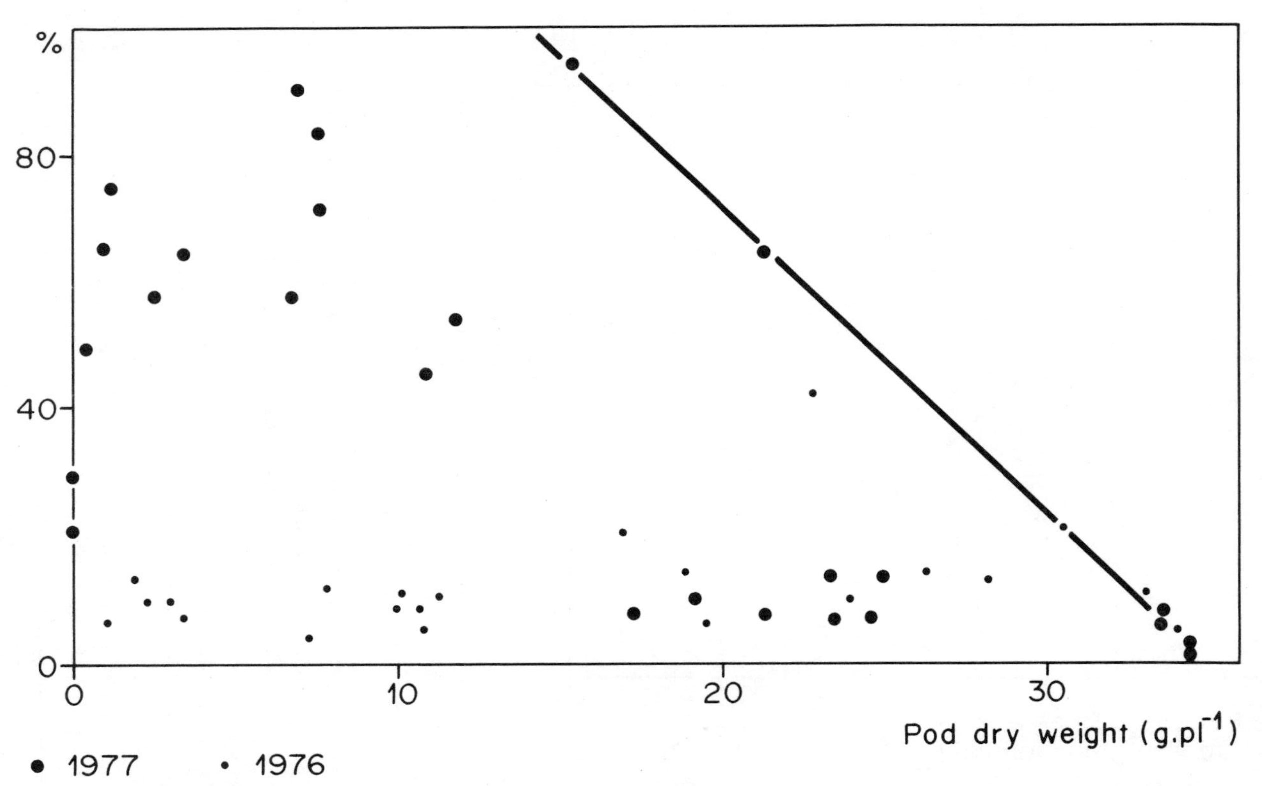

Figure 3. Peanuts (1976 and 1977), daily maxima of acetylene-reducing activity in percent of ARA_{max} (Figure 2) versus pod dry weight.

a rather unconventional way. This data treatment method seems able to fulfill most of the requirements listed above, but still needs to be developed from its present experimental state to one of routine use.

MATERIAL AND METHODS

Plants and Soils

Maize varieties used in these experiments were three new French varieties, INRA 260 (1973), $F_7 \times F_2$ (1975), and LG_{11} (1977); the experimental plots were in a neutral, clay soil of the École Nationale Supérieure d'Agronomie et des Industries Alimentaires experimental farm near Nancy (France). Peanut varieties 7330 (1976) and 55422 (1977) were planted in a very sandy soil of the Centre National de la Recherche of Bambey (Senegal); this last experiment was part of the soil biochemistry program directed by F. Ganry.

Incubations

In situ acetylene-reducing activity (ARA) measurements were made using the apparatus described previously (Balandreau and Dommergues, 1973) except for 1977 peanut measurements, where entire root systems were

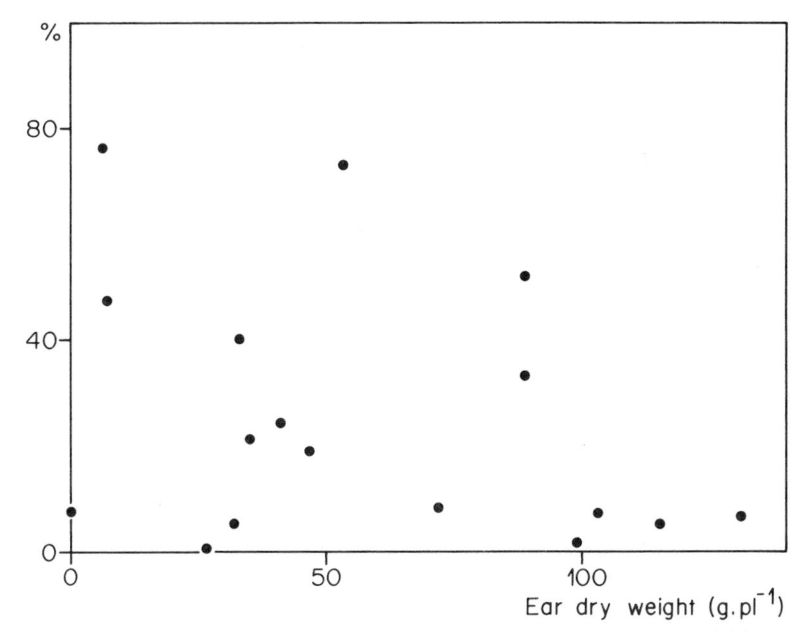

Figure 4. Maize (1977), reproductive period of growth, acetylene-reducing activities versus ear dry weight.

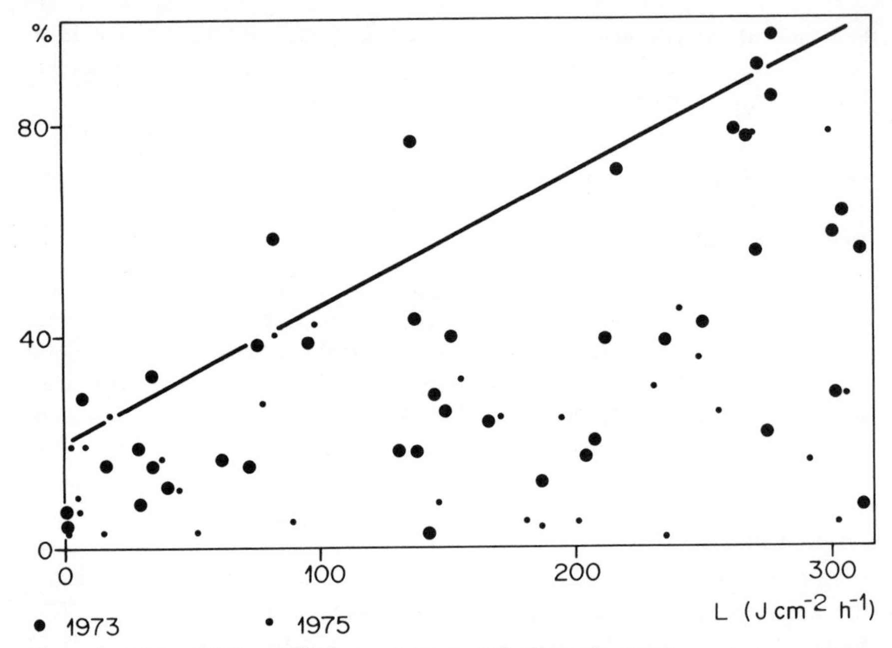

Figure 5. Maize (1973 and 1975), vegetative period of growth, acetylene-reducing activities (as percentage of $ARA_{d\ max}$; cf. Figure 6) versus light energy (L) received for 1 hr.

exposed to 10% acetylene in closed glass vials (Dart, Day, and Harris, 1972) for 30 min.

Environmental Factors

Light intensity was measured with a linear thermopyranometer, which was inexpensive, simple, and reliable and was designed by the Institut National de la Recherche Agronomique station of Versailles, France. Other factors were measured with standard pieces of meteorological apparatus.

Limiting Factor Analysis

The principle is simple. If we plot a measured process *P* against a factor *F*, there are three possibilities:

1. If *F* does not influence *P*, the points will be scattered randomly in the *P*,*F* sector.
2. If *P* depends only on *F*, the points will lie on a curve describing the action of *F* on *P*.
3. If *P* depends also on factors other than *F*, the points for cases where *F* is limiting will define the curve describing the action of *F* on *P*. Points for situations where a factor other than *F* is limiting will lie below this curve.

Frequency of Measurements

In 1973, maize was measured for 5 consecutive days in July (8 measurements a day). In 1975, maize was measured 1 day (5 measurements) every week from the beginning of July to the end of August. In 1976, peanuts were measured 1 day (5 measurements) every week throughout the growth cycle. In 1977, peanuts were measured once at 11 A.M., every 2 or 3 days. For each measurement, five or six replicate plants have been used. The results presented here are related only to daytime measurements.

ANALYSIS OF FACTORS ACTING ON ENERGY AVAILABILITY

The Energy Collector: The Leaves

In both systems, the source of energy for nitrogenase activity is the supply of carbohydrates, which are part of the newly synthesized photosynthates. The amount of this energy source depends on the size of the photosynthetic apparatus, which can be characterized simply by the green blade dry weight, as shown in Figures 1 and 2 for maize and peanuts, respectively. For both

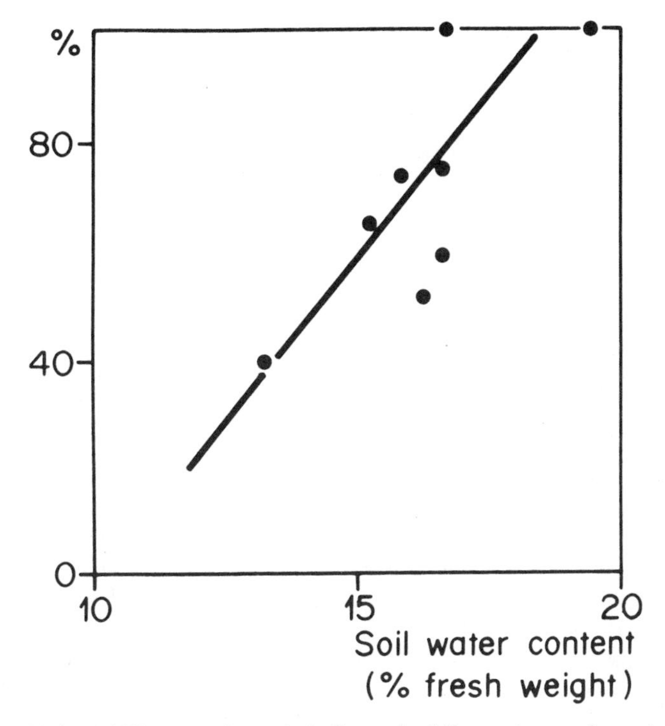

Figure 6. Maize (1975), vegetative period of growth, daily maximum of acetylene-reducing activities (as percentage of ARA_{max}; cf. Figure 1) versus soil water content.

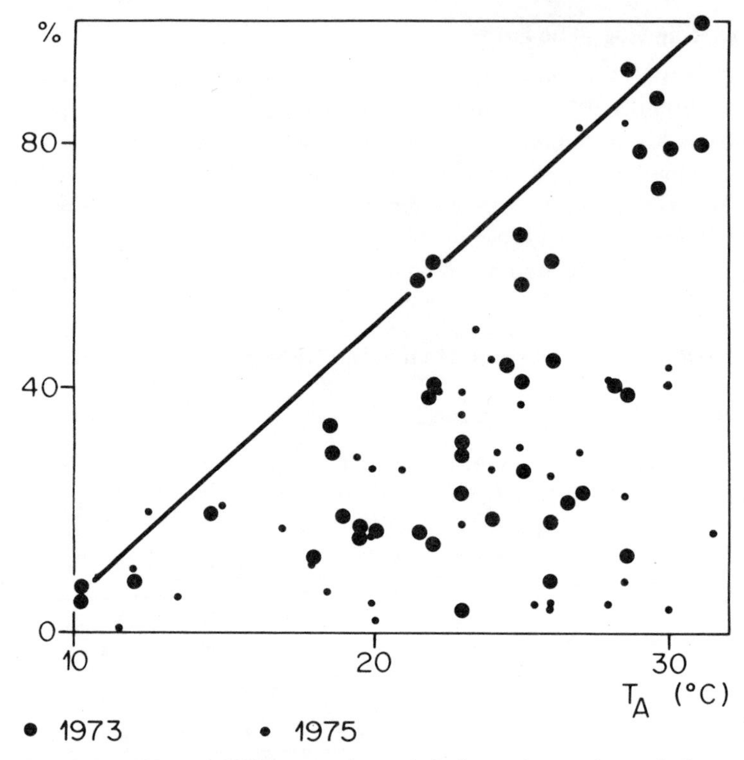

Figure 7. Maize (1973 and 1975), vegetative period of growth, acetylene-reducing activities (as percentage of $ARA_{d\ max}$; cf. Figure 6) versus air temperature (T_A).

plants, the relationship between ARA and leaf dry weight (LDW) is linear. We regard the slope of this straight line (*K*) as a very important characteristic of a given soil-plant system. *K* depends not only on photosynthesis efficiency and partitioning of photosynthates toward exudation but also on soil and microbial factors. *K* varies slightly from cultivar to cultivar (Balandreau et al., 1978) and is probably responsible for the large variations observed between mutants of rice (Dommergues, 1978). *K* varies significantly between species. For maize, it is 95 nmol of C_2H_4 hr^{-1} g of dry blades^{-1} (Figure 1); for *Dactylis* (Figure 1), about 140 nmol (Balandreau, unpublished data); for peanuts, (Figure 2), 5340 nmol; and for rice, possibly still higher (Balandreau et al., 1978), even when measurements were made at limiting low light levels in a phytotron.

This relationship between maximum values of ARA (ARA_{max}) and leaf dry weight allows a normalized expression of results of ARA measurements as a percentage of ARA_{max}. Those units (% ARA/ARA_{max}) are usable throughout the growth cycle, independent of plant size.

A Competing Sink: The Fruits

At initiation of fruiting, a dramatic change occurs—the reproductive organs become another sink for photosynthate. In peanuts this sink competes very actively with the nodule sink, as shown on Figure 3. The pod dry weight (PDW) appears as a limiting factor of nitrogenase activity; the relationship is % $ARA/ARA_{max} = -4.8$ PDW + 168. This result is not new (Brun, 1972; Ham, Lawn, and Brun, 1976), but here, for peanuts, it has been obtained without manipulating the source:sink ratio in a situation where all the factors were allowed to vary.

The competitive interaction between pod filling and nitrogen fixation precludes any improvement of yield by changing the partition of photosynthate. An increase in fixation would be paid for by a decrease in yield. Thus, studies aimed at extending growth over a longer period of time to increase ARA_{max} are desirable.

For maize (Figure 4), the data are limited and the points are too scattered to allow very conclusive evaluation; but the data are compatible with

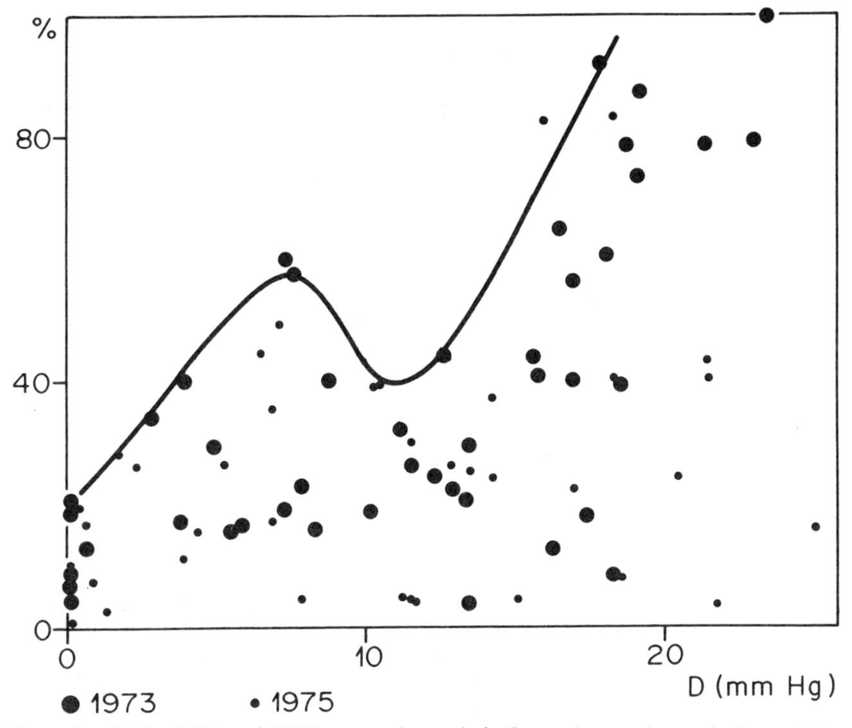

Figure 8. Maize (1973 and 1975), vegetative period of growth, acetylene-reducing activities (as percentage of $ARA_{d\ max}$; cf. Figure 6) versus air water content deficit (D).

ear growth limiting nitrogenase activity. Such a role has already been suggested by Döbereiner and Day (1975) for *Pennisetum purpureum* and by Lee, Castro, and Yoshida (1977) for rice.

The Source of Energy: Light

Figure 5 shows the influence of light on nitrogenase activity in maize plots. The equation of the straight line is $\% \text{ARA}/\text{ARA}_{max} = 0.26\text{ L} + 20$, where L (light energy received) is expressed as J cm^{-2} h^{-1}. For *Dactylis glomerata*, $\% \text{ARA}/\text{ARA}_{max} = 0.29\text{ L} + 11$.

I. Fares-Hamad (Balandreau et al., 1978) studied the action of light on nitrogenase activity in the rhizosphere of rice grown in a phytotron. Assuming that 1 W m^{-2} corresponds to 118 Lux, the equation describing the influence of light on ARA (expressed as % of ARA at 100,00 Lux, extrapolated from her data) is $\% \text{ARA}/\text{ARA}_{max} = 0.29\text{ L} + 10$.

The similarity of those three equations is impressive. The three plants seem to respond to light in the same way irrespective of their C-3 or C-4 metabolic pathway. Counting the points on or near the straight line of Figure 5 gives the rough estimate that light is limiting between 20% and 30% of the time. Surprisingly, for peanuts, light measured 1 or 2 hr before ARA does not appear to be limiting; the points are randomly scattered. This could be an effect of a low translocation rate of photosynthate (Subrahmanyam and Prabhakar, 1975) and/or of a transient storage as starch.

FACTORS INFLUENCING ENERGY USE

Soil Water Content

Figure 6 shows for maize that observed daily maxima of ARA ($\text{ARA}_{d\ max}$) are often lower than the calculated ARA_{max} value. There is a good correlation of $\text{ARA}_{d\ max}$ with soil water content (SWC) and this could be an indirect effect through redox potential and O_2 diffusion, which change the carbohydrate use efficiency (Knowles and O'Toole, 1975). Thus, $\text{ARA}_{d\ max} = \text{ARA}_{max} = 95\text{ LDW}$ if $\text{SWC} > 18\%$, whereas $\text{ARA}_{d\ max} = 95\text{ LDW}\ (0.12\ \text{SWC} - 1.24)$, if $10\% < \text{SWC} < 18\%$.

Ordinates of Figures 5, 7, 8, and 9 are expressed as percentage of $\text{ARA}_{d\ max}$.

Air Temperature

Air temperature appears as a limiting factor for 15% to 25% of measurements. In the case of maize, the effect is on photosynthesis and translocation rates and the decrease in air temperature during the afternoon could

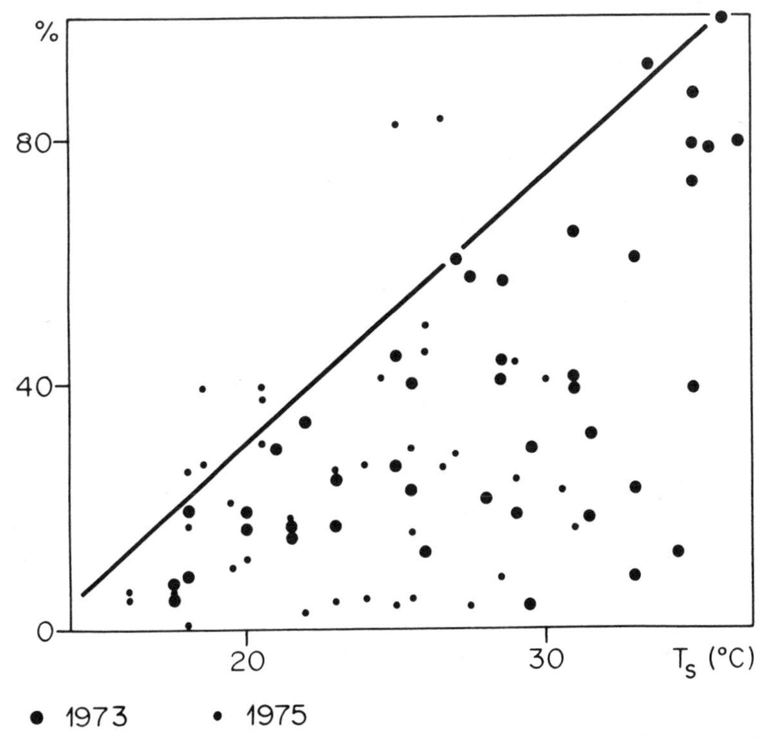

Figure 9. Maize (1973 and 1975), vegetative period of growth acetylene-reducing activities (as percentage of $ARA_{d\ max}$; cf. Figure 6) versus soil temperature (5 cm under the surface) (T_S).

produce a temporary shortage of carbohydrates, explaining the generally very low values of nitrogenase activity at the end of the afternoon.

Air Water Content Deficit

For maize, the influence of this factor seems to be multiphasic, with a first optimum at 5 mm Hg and another higher than 18 mm Hg. The first optimum has also been found for rice in the phytotron (Balandreau et al., 1978). No satisfactory explanation has been found.

Soil Temperature

About 30% of the measurements made on maize could be limited by soil temperature. For 1973 data, the root system morphology remained approximately constant during the 5 days of measurements and the limiting influence of soil temperature (5 cm under the surface) is very clear (large circles on Figure 9). Data for 1975 have been obtained throughout a growth season and the soil temperature at 5 cm is not determinative over such a long period of time because the size of the root system is increasing.

CONCLUSIONS

The proposed methodology of data treatment is usable in field situations where any factor can vary freely. It allows to some extent:

1. Interpretation of individual values of nitrogenase activity.
2. Recognition of the role of individual limiting factors and their modeling (Balandreau et al., 1978).
3. Some predictive power.
4. Establishment of some hierarchy of factors based on the frequency with which they become limiting (Balandreau and Ducerf, 1978).

The *K* value plays a major role in this type of analysis. This value is defined as acetylene-reducing activity of a given soil-plant system without fruit when light is maximum and the single environmental limiting factor, *K*, is expressed on a green blade dry weight basis. This value is likely to become a very important criterion in future plant breeding programs for higher rates of nitrogen fixation since it varies with cultivars and is affected by mutation. However, the *K* value still remains largely a black box. Further experiments are needed to investigate its dependence on soil type, organic matter, clay content, soil combined nitrogen, microorganisms, and, eventually, new, still unknown factors.

ACKNOWLEDGMENTS

We are very grateful to the Institut Sénégalais de la Recherche Agronomique for permission to publish the peanut data.

REFERENCES

Balandreau, J., and Y. Dommergues. 1973. Assaying nitrogenase (C_2H_2) activity in the field. Bull. Ecol. Res. Comm. 17:247–254.

Balandreau, J., and P. Ducerf. 1978. Activité nitrogénasique (C_2H_2) *in situ:* Mesure, analyse des facteurs limitants, comparaison de systèmes fixateurs d'azote. In: Isotopes in Biological Dinitrogen Fixation. IAEA, Vienna, Austria.

Balandreau, J., P. Ducerf, I. Hamad-Fares, P. Weinhard, G. Rinaudo, C. Millier, and Y. Dommergues. 1978. Limiting factors in grass nitrogen fixation. In: J. Döbereiner, R. H. Burris, A Hollaender, A. A. Franco, C. A. Neyra, and D. B. Scott (eds.), Limitations and Potentials for Biological Nitrogen Fixation in the Tropics, pp. 275–302. Plenum Publishing Corp., New York.

Balandreau, J., G. Rinaudo, I. Fares-Hamad, and Y. Dommergues. 1975. Nitrogen fixation in the rhizosphere of rice plants. In: W. D. P. Stewart (ed.), Nitrogen Fixation by Free-living Microorganisms, pp. 57–70. Cambridge University Press, London.

Brun, W. A. 1972. Nodule Activity of soybeans as influenced by photosynthetic source-sink manipulations. Agronomy Abstracts, p. 31. Quoted by: R. W. F. Hardy and U. D. Havelka. 1976. Photosynthate as a major factor limiting nitrogen fixation by field-grown legumes with emphasis on soybeans. In: P. S. Nutman (ed.), Symbiotic Nitrogen Fixation in Plants, p. 430, Cambridge University Press, London.

Dart, P. J., J. M. Day, and D. Harris. 1972. Assay of nitrogenase activity by acetylene reduction. In: Use of isotopes for study of fertilizer utilisation by legume crops. International Atomic Energy Agency Technical Report No. 149, pp. 85–100. IAEA, Vienna, Austria.

Day, J., D. Harris, P. J. Dart, and P. Van Berkum. 1975. The Broadbalk experiment. An investigation of nitrogen gains from non-symbiotic nitrogen fixation. In: W. D. P. Stewart (ed.), Nitrogen Fixation by Free-living Microorganisms, pp. 71–85. Cambridge University Press, London.

Döbereiner, J., and J. M. Day. 1975. Nitrogen fixation in the rhizosphere of tropical grasses. In: W. D. P. Stewart (ed.), Nitrogen Fixation by Free-living Microorganisms, pp. 39–56. Cambridge University Press, London.

Dommergues, Y. 1978. Impact on soil management and plant growth. In: Y. R. Dommergues and S. V. Krupa (eds.), Interactions between Non-pathenogenic Soil Microorganisms and Plants, pp. 443–458. Elsevier Scientific Publishing Company, Amsterdam.

Ham, G. E., R. J. Lawn, and W. A. Brun. 1975. Influence of inoculation, nitrogen fertilizers and photosynthetic source–sink manipulations on field-grown soybeans. In: P. S. Nutman (ed.), Symbiotic Nitrogen Fixation in Plants, pp. 239–253. Cambridge University Press, London.

Hardy, R. W. F., and U. D. Havelka. 1975. Photosynthate as a major factor limiting nitrogen fixation by field-grown legumes with emphasis on soybeans. In: P. S. Nutman (ed.), Symbiotic Nitrogen Fixation in Plants, pp. 421–439. Cambridge University Press, London.

Knowles, R., and P. O'Toole. 1975. Acetylene reduction assay at ambient Po_2 of field and forest soils: laboratory and field core studies. In: W. D. P. Stewart (ed.), Nitrogen Fixation by Free-living Microorganisms, pp. 285–294. Cambridge University Press, London.

Lee, K. K., T. Castro, and T. Yoshida. 1977. Nitrogen fixation throughout growth and varietal differences in nitrogen fixation by the rhizosphere of rice planted in pots. Plant Soil 48:613–619.

Masterson, C. L., and P. M. Murphy. 1976. Application of the acetylene reduction technique to the study of nitrogen fixation by white clover in the field. In: P. S. Nutman (ed.), Symbiotic Nitrogen Fixation in Plants, pp. 299–316. Cambridge University Press, London.

Pate, J. S. 1976. Physiology of the reaction of nodulated legumes to environment. In: P. S. Nutman (ed.), Symbiotic Nitrogen Fixation in Plants, pp. 335–360. Cambridge University Press, London.

Sloger, C., D. Bezdicek, R. Milberg, and N. Boonkerd. 1975. Seasonal and diurnal variations in N_2 (C_2H_2)-fixing activity in field soybeans. In: W. D. P. Stewart (ed.), Nitrogen Fixation by Free-living Microorganisms, pp. 271–284. Cambridge University Press, London.

Sprent, J. I. 1976. Nitrogen fixation by legumes subjected to water and light stresses. In: P. S. Nutman (ed.), Symbiotic Nitrogen Fixation in Plants, pp. 405–420. Cambridge University Press, London.

Subrahmanyam, P., and C. S. Prabhakar. 1975. Mobilization and exudation of photo-assimilated ^{14}C by developing pods of groundnut (*Arachis hypogaea L.*). Plant Soil 43:687–690.

Wheeler, C. T. 1969. The diurnal fluctuation in nitrogen fixation in the nodules of *Alnus glutinosa* and *Myrica gale*. New Phytol. 68:675–682.

Nitrogen Fixation, Volume II
Edited by W. E. Newton and W. H. Orme-Johnson

Nitrogen Fixation Associated with Roots of Sorghum and Wheat

R. V. Klucas and W. Pedersen

Nitrogen fixation by free-living microorganisms in the root environment of plants is of interest because of its tremendous potential for contributing fixed nitrogen to the biosphere. Considerable research using nitrogen-fixing bacteria as soil inoculants showed crop yield increases in some cases, but nitrogen fixation did not appear to be the cause of the observed increases. Interest was renewed by the availability of a simple and sensitive assay for nitrogen fixation and by recent research, especially that of Döbereiner (1968), demonstrating that specific associations existed between *Azotobacter paspali* and *Paspalum notatum*. Subsequently, Döbereiner and Day (1974a) reported associations between a number of plant species and gram-negative nitrogen-fixing bacteria initially identified as *Spirillum lipoferum*. Recently, from biochemical and DNA homology studies, the generic name *Azospirillum* was proposed for this microorganism, with species of *lipoferum* and *brasilense* (Krieg, 1977). These associations were of interest because: 1) the microorganism was closely associated with roots with a possible internal location; 2) the efficiency (N_2 fixed per g of substrate consumed) of the free-living cultured *Spirillum* was excellent; and 3) the nitrogenase activity as measured by acetylene reduction in root segments was high (Döbereiner and Day, 1974a). There is no evidence for specificity between this microorganism and plants such as that observed between

This research was supported by contract AID/ta-C-1093 from the Agency for International Development and constitutes paper No. 5565, Journal Series, Nebraska Agricultural Experiment Station.

legumes and rhizobia but, since microbial differences are apparent (Krieg, 1977), bacterial strain specificity may exist.

The basic unsolved problems involved with associative symbiotic nitrogen fixation are: 1) the supply of an adequate energy source; 2) the maintenance of environmental conditions that are conducive for the process; and 3) the translocation of inorganic and organic compounds between plants and specific bacteria. The supply and efficient utilization of energy sources must be sufficient if associative nitrogen fixation is to contribute a significant amount of nitrogen to plants. Factors affecting plant root exudation, including plant species, plant age, temperature, light, plant nutrition, microorganisms, soil moisture, and root damage (Rovira, 1969), are generally the same factors that affect associative symbiotic nitrogen-fixing relationships (Döbereiner and Day, 1974b; Day et al., 1975; Balandreau et al., 1977). The quantities of organic compounds exuded from roots are low and generally represent less than 0.4% of the total carbon fixed during photosynthesis (Rovira, 1969), but the total loss of carbon from roots is much greater (Barber and Martin, 1976; Martin, 1977). For example, wheat plants exposed to $^{14}CO_2$ for periods from 3 to 8 weeks lost from roots 14% to 23% of the total ^{14}C recovered from plants or 29% to 44% of the ^{14}C translocated to roots. Martin (1977) proposed that a major loss of root carbon resulted from autolysis of the root cortex. The extent of root exudate and lysate utilization by nitrogen-fixing bacteria remains to be determined. The energy requirement for biological nitrogen fixation is substantial and, for legumes, 60 to 100 mg of N_2 are fixed per g of carbohydrate consumed (Hardy, 1977). For free-living nitrogen-fixing bacteria, the efficiencies generally are much lower and depend on the microorganism (aerobic, microaerophilic, facultative anaerobic, or anaerobic), nutritional status and environmental conditions. Values ranging from zero for non–nitrogen-fixing conditions, such as high pO_2, to 39 mg of N_2 fixed per gram of mannitol consumed at a pO_2 of 0.02 atm for a continuous culture of *Azotobacter chroococcum* (Dalton and Postgate, 1969) were reported. Efficiencies to 12.5 (Okon et al., 1977) and 66 (Döbereiner and Day, 1974a) mg of N_2 fixed per g of malate consumed were reported for *S. lipoferum* grown under microaerophilic conditions. The efficiency for a nitrogen-fixing mutant of *Klebsiella pneumoniae* under non-growth, ammonia-excreting, and optimal anaerobic conditions for cell maintenance and nitrogen fixation was about 20 mg of N_2 fixed per g of sugar consumed (Anderson and Shanmugam, 1977). With better utilization of the carbohydrate, such as with low pO_2 conditions (Hill, 1976), the efficiency may be twofold higher and higher still if hydrogen evolution by nitrogenase can be controlled. Additionally, the efficiency in terms of energy utilization is higher than indicated by carbohydrate consumption measurements because, in a root-soil ecosystem, products derived from fermentation are subsequently used by other microorganisms or plants.

Besides environmental factors affecting plants, microenvironmental conditions of the rhizosphere or roots, such as pO_2, moisture, and growth factors, are important for bacterial nitrogen fixation. Although no conclusive evidence exists concerning the location(s) of nitrogen-fixing bacteria in associative symbiotic systems, there are areas in the outer cortex of *Digitaria*, maize, and wheat that exhibit tetrazolium-reducing bacteria, and these bacteria appear to be *Spirillum* (Döbereiner and Day, 1974a; Döbereiner, 1977a). *Panicum virgatum*, a temperate grass, was reported to possess nitrogenase activity in the rhizosphere (Tjepkema, 1975), but the causative agent was not determined. An interesting symbiotic bacterial species, tentatively identified as *Erwinia herbicola*, was shown to associate with *Pan. virgatum* (Lewis and Crotty, 1977). Bacteria were not found inside the plant cells but rather on the surface of and buried in the outer walls of the more basal elongating epidermal cells. It was not reported whether this strain of *E. herbicola* possessed nitrogenase genes, but other *E. herbicola* were reported to fix nitrogen (Neilson and Sparell, 1976; Pedersen et al., 1978a).

Research with associative symbioses is complicated by uncertainties of assay procedures. Acetylene reduction to ethylene is a well established, convenient, and reliable indication of nitrogenase activity but it is not necessarily valid for quantitating N_2 fixed in situ. Generally, nitrogenase assays for associative symbiosis are performed on root segments, soil-root cores, or minimally disturbed plants. Considerable concern was expressed regarding the relevance of isolated root segment assays as described by Döbereiner and Day (1974a) to nitrogen fixation, primarily because of lengthy incubations required before nitrogenase activity was detected and lack of correlation in comparable less-damaging assays with maize and a sorghum (Tjepkema and Van Berkum, 1977) and several grasses (Eskew and Ting, 1977). However, Döbereiner (1977a) reported excellent correlation between the two assay procedures for wheat, *Digitaria*, and *Panicum*. Obviously, additional research is required to resolve this question. Nevertheless, the root segment procedure, which may be of limited use for quantitating N_2 fixed, is valuable because it not only provides evidence for the presence of nitrogen fixers in roots but also provides a means for obtaining these microorganisms. As with enriched soils experiments, root segment procedures demonstrate the potential for nitrogen fixation and not the natural status. Data derived from soil-root cores and intact plant assay procedures are probably more valid measurements of in situ nitrogenase activities because these procedures require minimum manipulations of the plants. However, these procedures are more cumbersome and time lags are generally observed before nitrogenase activity is observed.

Assessments of associative symbiotic relationships in which nitrogen fixation is correlated with positive plant responses are not particularly promising. Long term experiments by research groups at Rothamsted (Day et al., 1975) suggested that considerable amounts of nitrogen (34 kg of nitrogen per ha per annum) were gained by nonsymbiotic nitrogen fixation. Short term experiments with *Pas. notatum* by Döbereiner and Day (1974b) also supported the accumulation of nitrogen, but the results did not eliminate blue-green algae as the source. Recently, De-Polli et al. (1977) and Döbereiner (1977b), using $[^{15}N]N_2$, confirmed nitrogen fixation in intact soil-cores containing *Pas. notatum* or *Digitaria decumbens* by demonstrating ^{15}N incorporation into plants. The results were important not only because acetylene reduction assays for associative nitrogen fixation were verified but also because fixed nitrogen, albeit in meager amounts, was incorporated into roots, rhizomes, and upper portions of the plants. Field studies to test inoculation effects of *S. lipoferum* on *Pennisetum americanum* and *Panicum maximum* revealed increased dry matter with certain levels of nitrogen fertilizer if inoculated plots were compared to uninoculated ones, but no significant difference in forage nitrogen content (Smith et al., 1976). Barber et al. (1976) examined the effects of inoculation with *S. lipoferum* on sorghum and corn breeding lines in field and greenhouse experiments. Although root segments from corn inoculated with *S. lipoferum* reduced acetylene after 16 hr of incubation at low pO_2 yield and nitrogen content of inoculated plants were comparable to uninoculated plants. Other trials involving field- and Biotron-grown maize (Albrecht, Okon, and Burris, 1977; Okon, Albrecht, and Burris, 1977) also provided no evidence for beneficial effects by *Spirillum* inoculations. Sixteen growth conditions, involving four temperatures and four light intensities, did not show any significant benefit either in dry weight or total nitrogen with inoculations, and rates of ethylene production by replicate root samples were low and variable and did not correlate with measurements of total nitrogen.

We are involved in studying nitrogen fixation associated with winter wheat, sorghum (Pedersen et al., 1978a), and perennial grasses (Pedersen et al., 1978b) in Nebraska. Our approach is to survey multiple locations and plant genotypes for associated symbiotic nitrogen-fixing systems using acetylene reduction by root segments, roots, soil-root cores, or intact plants as an indicator. From materials exhibiting nitrogenase activity, nitrogen-fixing bacteria are isolated and identified. Selected plant genotypes and nitrogen-fixing bacterial isolates are used for field and growth chamber experiments. Sorghum, wheat, and perennial grasses were selected for the study because of differences in photosynthetic and cultural properties as well as the availability of diverse genetic material. Wheat samples from approximately 150 sites and 2400 samples of either root segments or soil-plant cores collected over a period of 2 years throughout Nebraska had very low or no nitrogenase activity. Values ranged from 0 to 3.1 nmol of C_2H_4

produced per hr per g dry weight of root segment or core. However, a commercial field of variety Scout 66 did exhibit some activity with root segments (290 nmol of C_2H_4 produced per hr per g dry wt of root segments). The activity in the roots appeared to be localized, with 85% of the activity being found in 2-cm segments below the crown. The roots were extremely fibrous and the root systems below the crown were unusually large in diameter. Sorghums (400 lines and crosses) were also surveyed for associative nitrogen fixation. Grain sorghums generally exhibited higher nitrogenase activity than forage sorghums or winter wheats, with acetylene reduction rates ranging from 0 to 1100 nmol of C_2H_4 per hr per core. The predominant nitrogen-fixing bacteria isolated from root segments of wheat or sorghum exhibiting the highest acetylene reduction activity were identified primarily as members of Enterobacteriaceae. Of the 58 identified isolates, 50 were *K. pneumoniae*, 6 were *Enterobacter cloacae* and 2 (both from wheat) were *E. herbicola*.

Research described in this paper concerns field experiments to test effects of inoculations with nitrogen-fixing bacteria on winter wheat (*Triticum aestivum* L. em Thell) and grain sorghum (*Sorghum bicolor* L. Moench). Nitrogen-fixing bacteria (*E. herbicola* or *K. pneunomiae*) were used as seed inoculants and were isolates from winter wheat roots that exhibited nitrogen fixation.

METHODS

Winter wheat inoculation studies were conducted at the High Plains Research Station, Sidney, Nebraska, and the Panhandle Research Station, Scottsbluff, Nebraska, using the winter wheat cultivar Scout 66. The field plots at Sidney were in a crop rotation of alternating winter wheat and fallow without irrigation during this study. The plots at Scottsbluff followed a crop of sugarbeets and were irrigated prior to planting to ensure adequate germination. Soil tests showed 8 and 12 μg of nitrogen per g of soil at Sidney and at Scottsbluff, respectively. Treatments included bacterial isolate W-8 (live and heat killed), Chappell wheat field soil, and western wheatgrass soil. Isolate W-8 was previously identified as *E. herbicola* and was obtained from root segments of Scout 66 at Chappell, Nebraska (Pedersen et al., 1978a). Seeds of Scout 66 were inoculated with isolate W-8 using the following procedure. The bacterial isolate was grown to a concentration of 10^8 cells per ml in nutrient broth (Difco) at 30°C under aerobic conditions. Seeds were treated with 10% Clorox for 3 min, followed by a 5-min rinse in distilled water, and were then placed in the bacterial suspension under a vacuum of 30 cm Hg for 3 min. The excess liquid was drained from seeds and they were allowed to air dry before planting. Approximately 10^4 cells per seed of isolate W-8 were recoverable after 1 week of storage at 20°C. The same inoculation procedure was used with a bacterial suspension of W-8 that had been autoclaved for 15 min at 121°C. Soil from a wheat field at Chappell in which Scout 66 had shown acetylene reduction activity in 1976 and soil from native wheatgrass pastures were applied to the field plots at a rate of 228 kg per ha. The soil was applied at the time of planting using a granular fertilizer applicator attached to the grain drill. The experiment utilized a randomized complete block design with four replications and plots of 2.1 × 10.6 m. The number of crown roots and tillers and amount of fresh weight of forage and roots per m were measured at monthly intervals from October through March. Acetylene reduction levels were determined five times from May through June using the washed root and root-soil core assays (Pedersen et al., 1978a). Grain yields were determined by harvesting a subplot of 2.1 × 7.6 m.

Two $3^2 \times 5$ factorial experiments were designed to test the effect of: 1) three levels of nitrogen (0, 57, and 114 kg per ha) or three levels of phosphorus (0, 46, and 91 kg per ha; 2) three lines of sorghum (CK60A, CK60B, and B517); and 3) five bacterial inoculants (W-8, W-8 [heat killed], W-10, W-10 [heat killed], and an uninoculated control) on acetylene reduction levels, forage yields, forage nitrogen levels, grain yields, and grain nitrogen levels. The selection of lines CK60A and B517 was based on acetylene reduction assays from 1976 (Pedersen et al., 1978a). Line CK60B was the same as line CK60A (male-sterile A cytoplasm) except for the presence of a normal "B" cytoplasm. The sorghum seeds were inoculated with isolates W-8 (*E. herbicola*) and W-10 (*K. pneumoniae*) as previously described for wheat. Isolate W-10 was also obtained from root segments of the wheat cultivar Scout 66 at Chappell, Nebraska, and was capable of reducing acetylene (80–100 nmol per hr per ml of sodium malate broth). The field plots had been cropped to soybeans for the past 4 years and had received no nitrogen fertilizer. Soil tests estimated 5 μg of nitrogen and 8 μg of phosphorus per g of soil at the beginning of the study. The plots were 1.5 × 6.1 m and replicated three times. Supplemental water was applied via furrow irrigation three times during the study.

Acetylene reduction levels were assayed four to six times from July through September. The assay procedures (Pedersen et al., 1978a) involved digging the plant from the field, removing any loose soil, cutting the stem just above the crown, and placing the excised root in a 1-quart wide-mouth glass jar. The jar was then sealed with a rubber serum stopper. Methane as an internal standard and acetylene (0.1 atm) were injected into jars and gas samples were assayed for ethylene production after 16 to 24 hr. Forage yields were determined by harvesting five randomly selected plants (excluding root systems) at the dough stage and determining the total dry weight. Grain yields involved the harvesting of 3.1-m rows of mature plants. Nitrogen in samples from forage and grain yields was determined by Kjeldahl analysis.

Twelve lines of sorghum were assayed for acetylene reduction. They were selected on the basis of their acetylene reduction activity in 1976 (Pedersen et al., 1978a). A randomized complete block design with six replications was used. No fertilizer or supplemental irrigation was added during the experiment.

RESULTS AND DISCUSSION

Winter wheat field trials were performed to test the effects of nitrogen-fixing bacteria native to the area as seed inoculants. *E. herbicola* was selected because it was a nitrogen-fixing isolate from wheat roots that exhibited nitrogenase activity previously (Pedersen et al., 1978a). Soil from the field in which these roots were obtained was used as an inoculant to supply unidentified microorganisms. Soil from western wheatgrass land was used also to supply unidentified nitrogen fixers because soils from areas planted in wheatgrass appeared to accumulate fixed nitrogen, suggesting possible nitrogen fixation. Very significant differences were observed between the two sites, Scottsbluff (SB) and Sidney (S), for: 1) crown roots (58–66 at S; 102–115 at SB); 2) tillers (65–78 at S; 111–130 at SB); 3) total forage (24–30 g fresh wt at S; 69–79 g fresh wt at SB); 4) total roots (5.2–6.3 g fresh wt at S; 12.6–13.9 g fresh wt at SB); and 5) grain yield (1568–1786 kg per ha at S: 3177–3429 kg at SB). All values except yields represent the ranges of averages from five samplings and are expressed in terms of per meter of row. From these parameters, it is evident that the growth characteristics of plants from the two sites were quite different. Significant differences were not observed in any of these parameters among the five treatments at either site. Rates of acetylene reduction obtained either with soil-root cores (0 to 2.6 nmol of C_2H_4 formed per hr per core) or with washed root segments (0 to 3.1 nmol per hr per g dry wt of root) taken on five different dates (May through June) and representing between 96 and 168 samples at each date were very low both in treated areas and between sites, with no significant differences among treated areas.

In the two $3^2 \times 5$ factorial experiments with sorghum, statistical analysis of the acetylene reduction assays did show significant differences in nitrogenase activity for all factors (fertilizers, lines, and inoculants) and for all interactions for both experiments, except the line factor in the phosphorus experiment (Table 1). Supplemental nitrogen gave a decrease in nitrogenase activity for all three lines (Table 2), which was consistent with its known effects on nitrogen-fixing bacteria (e.g., Döbereiner and Day, 1974a; Döbereiner, 1977b; Pedersen et al., 1978a). Increased phosphorus did increase nitrogenase activity when the fertilizer factor was analyzed independently ($F = 9.15^{**}$, Table 1). The mean levels for 0, 46, and 91 kg of phosphorus per ha of the combined three lines were 135.6, 133.7, and 189.2 nmol C_2H_4 produced per hr per root sample, respectively (Table 2),

Table 1. Summary of F values using a three-way analysis of variance for the five parameters measured on both nitrogen and phosphorus experiments

Experiment	Source of variation	Acetylene reduction	Forage yield	Forage nitrogen	Grain yield	Grain nitrogen
Nitrogen	Fertilizers[b]	73.96**[a]	1.72	23.40**	0.30	5.91**
	Lines[c]	51.57**	1.27	103.90	34.33**	91.65**
	Inoculants[d]	7.74**	0.66	0.28	1.37	0.58
	Fert. × lines	31.57**	0.18	0.40	0.67	1.11
	Fert. × inoc.	2.99*	2.03*	1.16	0.50	0.50
	Lines × inoc.	2.71*	0.91	1.25	0.77	2.56*
	Fert. × lines × inoc.	4.18**	1.04	1.09	0.55	0.96
	C.V.[e]	55.4%	12.4%	5.6%	18.0%	6.6%
Phosphorus	Fertilizers[f]	9.15**	1.35	3.45*	0.16	5.02**
	Lines	1.05	6.29**	80.37**	42.38**	89.12**
	Inoculants	2.52*	1.44	1.67	1.24	0.36
	Fert. × lines	4.52	0.56	0.29	0.10	0.17
	Fert. × inoc.	3.48*	0.83	0.92	0.79	1.69
	Lines × inoc.	4.50*	0.49	1.64	0.71	1.23
	Fert. × lines × inoc.	2.09*	0.79	1.17	0.34	0.67
	C.V.	68.9%	13.3%	6.2%	15.7%	6.1%

[a] Significant at the 95% level (*) and 99% level (**).

[b] Fertilizers = 0, 57, and 114 kg of nitrogen per ha.

[c] Lines = CK60A, CK60B, and B517.

[d] Inoculants = W-8, W-8 (heat killed), W-10, W-10 (heat killed), and an uninoculated control.

[e] C.V. = coefficient of variability.

[f] Fertilizers = 0, 46, and 91 kg of phosphorus per ha.

with the highest level being significantly different according to Duncan's new multiple range test ($p = 0.05$). The interaction between lines and inoculants was significant. However, only CK60B showed an increase in activity at the highest phosphorus level (Table 2), in contrast to the negative effect observed previously (Pedersen et al., 1978a) with CK60A, in which nitrogenase activity was lower at high phosphorus levels. The other interactions were also significant but no clear linear relationship was observed. The coefficients of variability for the acetylene reduction assays in both the nitrogen and the phosphorus experiments were very high (Table 1), making analysis very difficult.

No significant differences were obtained by the inoculant factor in the analysis of forage yield, forage nitrogen, grain yield, and grain nitrogen (Table 1). Both nitrogen and phosphorus did have an effect on forage nitrogen and grain nitrogen (Table 3). Unlike the decrease observed in nitrogenase activity, the highest levels of nitrogen and phosphorus did exhibit the highest levels of both forage nitrogen and grain nitrogen. The response was important because, although a relatively insensitive measure,

Table 2. Nitrogenase activity of root samples from three sorghum lines

Treatment		Sorghum lines[a]		
		6CK60A	CK60B	B517
Nitrogen:	0 kg/ha	482.0 A[b]	131.0 A	119.9 A
	57 kg/ha	153.7 B	124.7 A	75.9 B
	114 kg/ha	53.8 C	77.1 B	23.1 C
Phosphorus:	0 kg/ha	167.7 X	102.2 X	136.9 X Y
	46 kg/ha	212.8 X	103.1 X	85.2 X
	91 kg/ha	194.5 X	207.2 Y	166.1 Y

[a] Measured in nmol of C_2H_4 produced per hr per root sample. Values represent means from the fertilizers × lines interaction with four replications.

[b] Numbers in each vertical column that are followed by the same letter are not significantly different according to Duncan's new multiple range test (p = 0.05).

it shows that increases in forage and grain nitrogen would have been detected with inoculant treatments if significant amounts of nitrogen were fixed. Differences were observed between lines with CK60A having the highest levels of forage yield, forage nitrogen, grain yield, and grain nitrogen for both experiments. The interactions fertilizer × lines, fertilizer × inoculation, lines × inoculations, and fertilizer × lines × inoculations were not significant for either experiment with two exceptions. The fertilizer × inoculations interaction (nitrogen experiment) for forage yield did show an increased yield in response to the bacterial inoculants (Table 4). At the lowest level, 0 kg of nitrogen per ha, there was no difference between the live and heat-killed inoculants. However, combined inoculations gave sig-

Table 3. The effect on mean forage and grain nitrogen levels of three levels of nitrogen and phosphorus fertilizers

Fertilizer (kg/ha)		Nitrogen level (mg/g dry wt)	
		Forage	Grain
Nitrogen:	0	14.31 A[a]	19.50 A
	57	15.17 B	19.88 B
	114	15.48 C	20.44 C
Phosphorus:	0	14.47 A	19.21 A
	46	14.38 A	19.27 A
	91	14.85 B	19.93 B

[a] Numbers in each vertical column that are followed by the same letter are not significantly different according to Duncan's new multiple range test (p = 0.05).

nificantly higher yields than the uninoculated control. A possible explanation for this increase is that nutrients applied with the bacteria were utilized by the plant or by native microorganisms that stimulated growth, but this is not likely because only trace amounts of nutrients were added to seeds. When the level of nitrogen increased to 57 kg of nitrogen per ha, W-8 live had a significantly higher yield than W-8 heat killed, as did W-10 at 114 kg of nitrogen per ha. This apparent need for nitrogen by the microorganisms has been previously observed (Smith et al., 1976). The other exception, the line × inoculation interaction for grain nitrogen (nitrogen experiment), showed that the highest nitrogen levels for CK60A and B517 were with W-8 (live). However, CK60B had the highest nitrogen level with W-10 (heat killed). There were no significant differences between the rest of the inoculations or the uninoculated control.

Significant differences were observed among 12 lines of sorghum tested for acetylene reduction with roots (Table 5). These lines were selected because previous work (Pedersen et al., 1978) suggested high nitrogenase activities or because they represented near-isogenic B (normal) lines as compared to A (cytoplasmic male-sterile) lines. However, because multiple sites and different samples were collected, the two studies are not directly comparable. Lines B517, CK60A, and CK60B were studied both in the inoculation study and this line study and are comparable. The first two lines exhibit

Table 4. The difference in forage yield for live and heat-killed inoculants W-8 and W-10 at three nitrogen levels

Fertilizer level (kg per ha)	Inoculants	Forage yield difference (g dry wt)
0	W-8[a] vs. W-8 (h.k.)[b]	−24.7
	W-10[c] vs. W-10 (h.k.)	0.9
	All[d] vs. control[e]	48.8**[f]
57	W-8 vs. W-8 (h.k.)	44.4**
	W-10 vs. W-10 (h.k.)	−23.1
	All vs. control	15.1
114	W-8 vs. W-8 (h.k.)	−2.5
	W-10 vs. W-10 (h.k.)	51.2**
	All vs. control	−15.1

[a] W-8 is *Erwinia herbicola*.

[b] Heat-killed (h.k.) bacterial inoculant.

[c] W-10 is *Klebsiella pneumoniae*.

[d] All includes W-8 and W-10 (live and heat killed) inoculants.

[e] Control was not inoculated with live or heat-killed cells.

[f] Significant at the 99% level (**).

Table 5. Nitrogenase (C_2H_4) activity of roots from sorghum plots

Sorghum line	Nitrogenase activity[a] (nmol of C_2H_4 produced per hr per plant)
WCT 116	180 A[b]
Wheatland B	210 A
R-8192	280 A B
B517	290 A B
R8191	390 B
Redland A	410 B
KS56A	570 C
CK60A	620 C
Wheatland A	860 D
KS56B	970 D
R-8193	1430 E
CK60B	1440 E

[a] Mean activity from 4–6 replications.

[b] Numbers followed by the same letter are not significantly different according to Duncan's new multiple range test. ($p = 0.05$).

about the same nitrogenase activity in both studies, but CK60B activity was about tenfold greater in the varieties trials in which it exhibited the greatest activity. Comparisons of CK60A with CK60B, KS56A with KS56B, and Wheatland A with Wheatland B reveal that, of the first two A and B lines, the B lines exhibited significantly higher activity, but with Wheatland the opposite was observed. An interesting variable that should have been included is the comparison of multiple sources of a single line. Although inconsistencies are obvious if sorghum lines are compared from different growth sites or seasons, overall the data support the existence of genotypic differences among plants regarding associative symbiotic nitrogen fixation.

CONCLUSIONS

Undoubtedly, associative symbiotic relationships between plant roots and microorganisms do result in nitrogen fixation, but the efficacy of these associations still is questionable. Most of the difficulties discussed by Jensen (1965) still exist (Burris, 1977). Data in the literature suggest that plant genotypes, microorganisms, and environmental conditions are important and that the bacterial location may be in the roots or in the rhizosphere of the roots. For a number of reasons, including energetics, parasitism, and translocation, those associations that resemble true symbiotic systems, like rhizobia and legumes, probably will be the most effective, but this does not mean that useful levels of nitrogen cannot be derived from the "loose" associations. The dilemma concerning the supply and efficient use of energy sources continues to be a major problem. Nitrogen-fixing bacteria under nongrowth conditions appear to be most suitable for optimal associations, but such bacteria may not be competitive among soil microflora unless protected by the plant or microorganisms. Only additional research will reveal the importance and magnitude of associative symbiotic relationships.

ACKNOWLEDGMENTS

We thank E. Kerr and M. Crocker for their assistance.

REFERENCES

Albrecht, S. L., Y. Okon, and R. H. Burris, 1977. Effects of light and temperature on the association between *Zea mays* and *Spirillum lipoferum*. Plant Physiol. 60:528–531.

Anderson, K., and K. T. Shanmugam. 1977. Energetics of biological nitrogen fixation: Determination of the ratio of formation of H_2 to NH_4^+ catalyzed by nitrogenase of *Klebsiella pneumoniae in vivo*. J. Gen. Microbiol. 103:107–122.

Balandreau, J. P., C. R. Millier, P. Weinhard, P. Ducerf, and Y. R. Dommergues. 1977. A modelling approach of acetylene reducing activity of plant-rhizosphere diazotroph systems. In: W. E. Newton, J. R. Postgate, and C. Rodriguez-Barrueco (eds.), Recent Developments in Nitrogen Fixation, pp. 523–529. Academic Press, London.

Barber, D. A., and J. K. Martin. 1976. The release of organic substances by cereal roots into soil. New Phytol. 76:69–80.

Barber, L. E., J. D. Tjepkema, S. A. Russell, and H. J. Evans. 1976. Acetylene reduction (nitrogen fixation) associated with corn inoculated with *Spirillum*. Appl. Environ. Microbiol. 32:108–113.

Burris, R. H. 1977. A synthesis paper on nonleguminous N_2-fixing systems. In: W. E. Newton, J. R. Postgate, and C. Rodriguez-Barrueco (eds.), Recent Development in Nitrogen Fixation, pp. 487–511. Academic Press, London.

Dalton, H., and J. R. Postgate. 1969. Growth and physiology of *Azotobacter chroococcum* in continuous culture. J. Gen. Microbiol. 56:307–319.

Day, J. M., D. Harris, P. J. Dart, and P. Van Berkum. 1975. The Broadbalk experiment. An investigation of nitrogen gains from non-symbiotic nitrogen fixation. In: W. D. P. Stewart (ed.), Nitrogen Fixation by Free-Living Micro-organisms, pp. 71–84. Cambridge University Press, Cambridge.

De-Polli, H. E. Matsui, J. Döbereiner, and E. Salati. 1977. Confirmation of nitrogen fixation in two tropical grasses by $^{15}N_2$ incorporation. Soil Biol. Biochem. 9:119–123.

Döbereiner, J. 1968. Non-symbiotic nitrogen fixation in tropical soils. Pesqui. Agropecu Bras. 3:1–6.

Döbereiner, J. 1977a. N_2 fixation associated with non-leguminous plants. In: A. Hollaender (ed.), Genetic Engineering for Nitrogen Fixation. Basic Life Sciences, Vol. 9, pp. 451–461. Plenum Publishing Corp., New York.

Döbereiner, J. 1977b. Physiological aspects of N_2 fixation in grass-bacteria associations. In: W. Newton, J. R. Postgate, and C. Rodriguez-Barrueco (eds.), Recent Developments in Nitrogen Fixation, pp. 513–522. Academic Press, London.

Döbereiner, J., and J. M. Day. 1974a. Associative symbiosis in tropical grasses: Characterization of microorganisms and dinitrogen-fixing sites. In: W. E. Newton and C. J. Nyman (eds.), Proceedings of the 1st International Symposium on Nitrogen Fixation, pp. 518–538. Washington State University Press, Pullman.

Döbereiner, J., and J. M. Day. 1974b. Potential significance of nitrogen fixation in rhizosphere associations of tropical grasses. In: E. Bornemisza and A. Alvarado (eds.), Soil Management in Tropical America, pp. 197–210. North Carolina State University Press, Raleigh.

Eskew, D. L., and I. P. Ting. 1977. Comparison of intact plant and excised root assays for acetylene reduction in grass rhizospheres. Plant Sci. Lett. 8:327–331.

Hardy, R. W. F. 1977. Increasing crop productivity: Agronomic and economic considerations on the role of biological nitrogen fixation. In: A. Hollaender (ed.), Report on the Public Meeting on Genetic Engineering for Nitrogen Fixation, pp. 77–107, U.S. Government Printing Office, Washington, D.C.

Hill, S. 1976. Influence of atmospheric oxygen concentration on acetylene reduction and efficiency of nitrogen fixation in intact *Klebsiella pneumoniae*. J. Gen. Microbiol. 93:335–345.

Jensen, H. L. 1965. Nonsymbiotic nitrogen fixation. In: W. W. Bartholomew and F. E. Clark (eds.), Soil Nitrogen. Agronomy Monograph 10, pp. 436–480. American Society of Agronomy, Inc. Madison, Wisc.

Krieg, N. R. 1977. Taxonomic studies of *Spirillum lipoferum*. In: A. Hollaender (ed.), Genetic Engineering for Nitrogen Fixation. Basic Life Sciences, Vol. 9, pp. 463–472. Plenum Publishing Corp., New York.

Lewis, R. F., and W. J. Crotty. 1977. The primary root epidermis of *Panicum virgatum* L. II. Fine structural evidence suggestive of a plant-bacterium-virus symbiosis. Am. J. Bot. 64:190–198.

Martin, J. K. 1977. Factors influencing the loss of organic carbon from wheat roots. Soil Biol. Biochem. 9:1–7.

Neilson, A. H., and L. Sparell. 1976. Acetylene reduction (nitrogen fixation) by Enterobacteriaceae isolated from paper mill process waters. Appl. Environ. Microbiol. 32:197–205.

Okon, Y., S. L. Albrecht, and R. H. Burris. 1977. Methods for growing *Spirillum lipoferum* and for counting it in pure culture and in association with plants. Appl. Environ. Microbiol. 33:85–88.

Okon, Y., J. P. Houchins, S. L. Albrecht, and R. H. Burris. 1977. Growth of *Spirillum lipoferum* at constant partial pressures of oxygen, and the properties of its nitrogenase in cell-free extracts. J. Gen. Microbiol. 98:87–93.

Pedersen, W. L., K. Chakabarty, R. V. Klucas, and A. K. Vidaver. 1978a. Nitrogen fixation (acetylene reduction) associated with roots of winter wheat and sorghum in Nebraska. Appl. Environ. Microbiol. 35:129–135.

Pedersen, W. L., R. C. Shearman, E. J. Kinbacher, and R. V. Klucas. 1978b. Associative nitrogen fixation studies with Park and Nugget Kentucky bluegrass. 16th Nebraska Turfgrass Conference Proceeding, pp. 5–8.

Rovira, A. 1969. Plant root exudates. Bot. Rev. 35:35–57.

Smith, R. L., J. H. Bouton, S. C. Schank, K. H. Quesenberry, M. E. Tyler, J. R. Milam, M. H. Gaskins, and R. C. Littell. 1976. Nitrogen fixation in grasses inoculated with *Spirillum lipoferum*. Science 193:1003–1005.

Tjepkema, J. 1975. Nitrogenase activity in the rhizosphere of *Panicum virgatum*. Soil Biol. Biochem. 7:179–180.

Tjepkema, J., and P. Van Berkum. 1977. Acetylene reduction by soil cores of maize and sorghum in Brazil. Appl. Environ. Microbiol. 33:626–629.

Section III
Cyanobacteria and their Associations

Nitrogen Fixation, Volume II
Edited by W. E. Newton and W. H. Orme-Johnson

Heterocyst Differentiation and Nitrogen Fixation in Cyanobacteria (Blue-Green Algae)

R. Haselkorn, B. Mazur, J. Orr, D. Rice,
N. Wood, and R. Rippka

It is now evident that nitrogen fixation in cyanobacteria can occur in either of two modes. Under anaerobic conditions, some unicellular strains and many filamentous strains, both heterocystous and nonheterocystous, can reduce acetylene. It is possible, but not yet proven, that nitrogenase is located in all the cells of an induced anaerobic culture. Under aerobic conditions, a very few unicellular strains and nearly all heterocystous filamentous strains can reduce acetylene. In the latter case, nitrogenase is confined to the heterocysts. The study of heterocyst differentiation is therefore the study of the programmed conversion of a CO_2-fixing, O_2-evolving vegetative cell to an anaerobic factory for N_2 reduction. The differentiation is additionally remarkable because the heterocyst is connected by cytoplasmic bridges to neighboring undifferentiated vegetative cells. Consequently, we can expect that the development of transport systems regulating the flow of small molecules between heterocyst and vegetative cell will be an important feature of differentiation (Haselkorn, 1978).

COMPARISON OF HETEROCYSTS AND VEGETATIVE CELLS

The biochemical activities that distinguish heterocysts from vegetative cells, particularly those involved in photosynthesis, were recently reviewed by

---

This work was supported by research grant GM 21823, a postdoctoral traineeship to B. M. (GM 7190), and predoctoral traineeships to J. O. and D. R. (GM 780), all from the National Institute of General Medical Sciences.

Stewart, Rowell, and Apte (1977). Heterocysts lack photosystem II activity and have very little phycobiliprotein, which serves as the exclusive light-harvesting pigment for photosystem II. In addition, heterocysts contain only 10% of the bound Mn^{2+} of vegetative cells, presumably a manifestation of some alteration in a photosystem II component (Tel-Or and Stewart, 1977). All of the components of the photosynthetic electron transport chain are present in heterocysts (Tel-Or and Stewart, 1977). There are 50% fewer photosystem I reaction centers in heterocysts and each photosynthetic unit has only 35%–50% of the amount of antenna chlorophyll found in vegetative cells (Alberte and Tel-Or, 1977).

Carbon metabolism in heterocysts is strikingly different from that in vegetative cells. Heterocysts do not fix CO_2. They lack ribulose bisphosphate carboxylase activity and antigen (Winkenbach and Wolk, 1973; Codd and Stewart, 1977). On the other hand, they have very high levels of glucose-6-phosphate (G-6-P) dehydrogenase and 6-phosphogluconate dehydrogenase (Winkenbach and Wolk, 1973). The latter enzymes are thought to provide reduced pyridine nucleotides that somehow donate electrons to nitrogen reduction. Indeed, with newer preparations of isolated heterocysts, it is now possible to demonstrate a G-6-P–dependent reduction of acetylene (Peterson and Wolk, 1978).

The form in which carbon is transported from vegetative cells to heterocysts is unknown, although the question is being pursued actively. Glucose is not a likely candidate because heterocysts do not have a sufficiently active hexokinase to process it for generation of reductant. A pulse-labeling study using $^{14}CO_2$ suggested that maltose and UDP-sugars were the most abundant recipients of newly fixed carbon in heterocysts (Jüttner and Carr, 1976). However, these experiments have been repeated with conflicting results (see Wolk, this volume).

The path of electron transfer from pyridine nucleotide to nitrogenase is believed to involve a ferredoxin, but the evidence is quite indirect. Heterocysts of *Anabaena cylindrica* contain a ferredoxin-$NADP^+$ oxidoreductase that can catalyze a G-6-P–dependent reduction of cytochrome *c*, but the rates observed, thus far, are very low (Apte, Rowell, and Stewart, 1978). The newest magic bullet in the field is a herbicide called metronidazole, which appears to be a specific inhibitor of ferredoxin-mediated electron transfer. This compound inhibits acetylene reduction in *A. cylindrica* without affecting O_2 production, CO_2 fixation, respiration, or light- or O_2-dependent H_2 uptake. It is concluded that the electron transfer proteins associated with nitrogen fixation are distinct from the ferredoxin pool that serves photosystem I (Tetley and Bishop, 1978).

With respect to nitrogen metabolism, heterocysts from free-living cyanobacteria differ from vegetative cells in three important ways: they contain nitrogenase, they have a higher level of glutamine synthetase (GS), and they have a lower level of glutamine-oxoglutarate amidotransferase (GOGAT). The particular role of GS in the control of heterocyst differentiation and nitrogen fixation (*nif*) gene expression is considered in detail subsequently. Here, we need only mention briefly the lovely experiments of Wolk and collaborators that indicate that glutamine is the carrier of fixed nitrogen from the heterocyst to vegetative cells (Wolk et al., 1976; Thomas et al., 1977). $[^{13}N]N_2$ was administered for very brief periods either to intact filaments of *Anabaena* or to isolated heterocysts. In the former case, the flow of ^{13}N was observed to be from N_2 to NH_3, from NH_3 to the amide of glutamine, and then to the α-NH_2 of glutamate and finally to other amino acids. In isolated heterocysts, the ^{13}N only proceeds as far as glutamine, which is exported to the medium. These results are interpreted in terms of the regulated transport of both nitrogen and carbon in Figure 1 (Haselkorn, 1978). As mentioned previously, the identification of maltose as the major form of carbon transported into the heterocyst has not been confirmed.

Uptake hydrogenase, which can recycle the H_2 produced by nitrogenase and protect all components from O_2, was shown to be present in *Anabaena*, inhibited by CO, and repressed by NH_3 (Bothe et al., 1977). As expected, it is found only in heterocysts (Peterson and Wolk, 1978).

The heterocyst differs from the vegetative cell in another very obvious way. It is surrounded by a multilayered envelope exterior to the vegetative cell wall, which confers resistance to both lysozyme and mechanical stresses and limits diffusion of gases into the cell. The chemical description of the envelope layers, provided by Wolk's laboratory, has been reviewed (Wolk, 1975; Stanier and Cohen-Bazire, 1977; Haselkorn, 1978). From the point of view of understanding the control of differentiation, it would be useful to have assays for the enzymes required to make the unique glycolipids and oligosaccharides found in the envelope. These enzymes must be synthesized or activated during the first hours of heterocyst development, but they have not yet been identified.

HETEROCYSTS AND AEROBIC NITROGEN FIXATION

Although our subject is heterocysts, some experiments with *Plectonema boryanum*, a nonheterocystous cyanobacterium, are instructive because very similar results are obtained with nonheterocyst-forming mutants of *Anabaena*. These, like *Plectonema*, express nitrogenase activity only under anaerobic conditions (Rippka and Stanier, 1978) and offer an explanation for the necessity for heterocysts for aerobic nitrogen fixation. The experiments exploit a method described by Rippka and Waterbury (1977) for the efficient and rapid induction of nitrogenase in nonheterocystous cyanobacteria. The cells are first starved for nitrogen under argon/O_2/CO_2, during

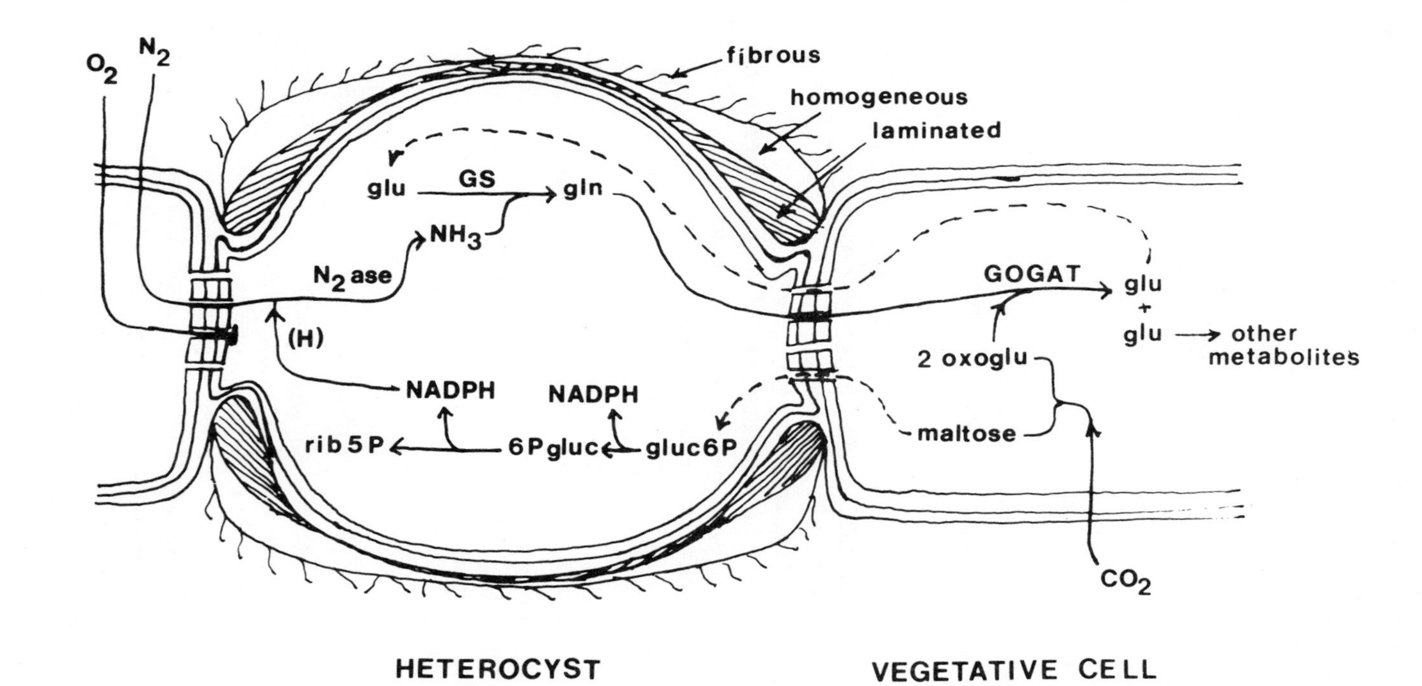

Figure 1. Model of the flow of carbon and nitrogen between heterocyst and vegetative cell in nitrogen-fixing cyanobacteria. Enzymes: N_2ase = nitrogenase, GS = glutamine synthetase, GOGAT = glutamine-oxoglutarate amidotransferase. Maltose should probably be replaced by sucrose; see Wolk, this volume. NADPH probably reduces a ferredoxin. An alternative pathway for ferredoxin reduction by pyruvate is not shown. The heterocyst envelope (fibrous, homogeneous, laminated) is believed to limit diffusion of N_2, CO_2, and O_2. Microplasmodesmata are shown connecting the two cells, but the polar bodies of the heterocyst are omitted. The heterocyst, lacking photosystem II, neither fixes CO_2 nor produces O_2. (Haselkorn, 1978).

which time they accumulate glycogen. The gas phase is then switched to argon/CO_2/C_2H_2 and DCMU is added to inhibit photosynthetic oxygen evolution. Nitrogenase synthesis is detectable within an hour. Very few proteins are synthesized under these conditions. Figure 2 shows a fluorogram of the proteins labeled with $^{35}SO_4^{2-}$ for various times after changing the gas phase. Figure 3 shows the pattern of stained protein bands after 6 hr of induction. Bands corresponding in molecular weight to nitrogenase components I and II are present only in induced, anaerobic cultures. If the anaerobic culture induced for 6 hr is then exposed to oxygen, the bands identified as nitrogenase disappear (Figure 4). The conclusion drawn from these results is that, without the oxygen-protective mechanism of the heterocysts, oxygen-inactivated *nif* gene products are cleared from the cells by protease action. This conclusion applies to at least two other proteins, of unknown function, in addition to components I and II.

PROTEASES ACTIVE DURING HETEROCYST DIFFERENTIATION

How many genes are differentially expressed in heterocysts? Assigning, for the sake of argument, half a dozen to the enzymes for envelope synthesis, an equal number for reductant, and 15 for the *nif* operons, glutamine synthetase, and uptake hydrogenase, there must be at least 25–30 genes turned on or turned up in the heterocyst. An unknown, but large, number must be turned off. To obtain a qualitative picture of these differences, we began to characterize the proteins of heterocysts and vegetative cells by polyacrylamide gel electrophoresis (Fleming and Haselkorn, 1973, 1974). Even with one-dimensional analyses, it was clear that the number of proteins differentially synthesized in heterocysts and vegetative cells was very large. Heterocysts synthesize many fewer proteins than vegetative cells. Moreover, many vegetative cell proteins are missing from heterocysts. This observation and the knowledge that, until nitrogen fixation begins, protein turnover must supply amino acids for heterocyst-specific protein synthesis, prompted us to look for specific proteases active during heterocyst differentiation.

The first such enzyme found in *Anabaena* 7120 is Ca^{2+}-dependent, soluble, and active against the synthetic collagen substrate Azocoll. It is not found in NH_4^+-grown cultures and its appearance under nitrogen-free conditions is blocked by chloramphenicol. In vitro, it appears to have the specificity required for an enzyme that is responsible for turnover of vegetative cell proteins in vivo. Figure 5 (Wood and Haselkorn, 1979) shows an electrophoretic analysis of: 1) vegetative cell proteins; 2) an extract of vegetative cell proteins digested with the Ca^{2+}-dependent enzyme in vitro; and 3) heterocyst proteins labeled with $^{35}SO_4^{2-}$ prior to the induction of differentiation, to show only those proteins of the vegetative cell that do *not* turn over during differentiation. The important result is that not all vegetative cell pro-

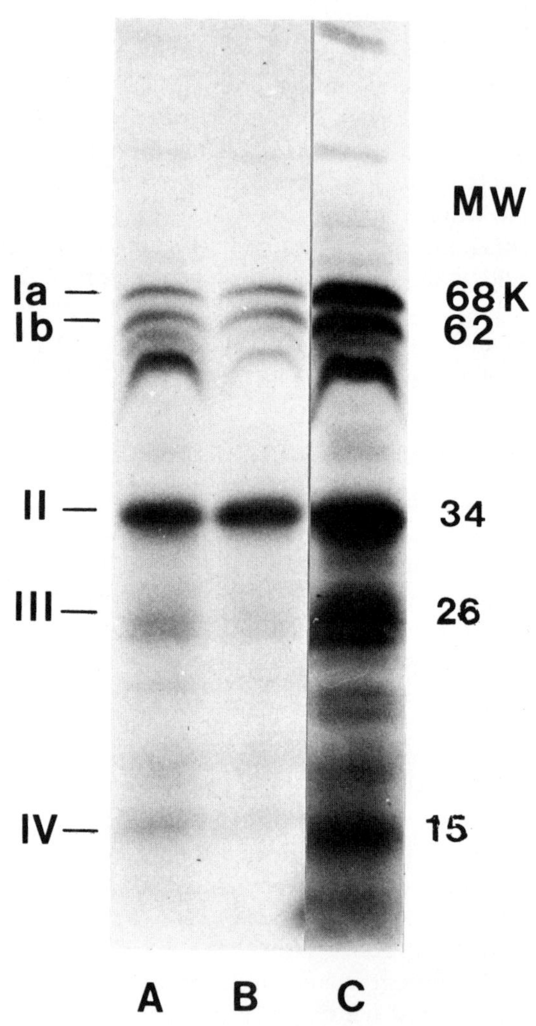

Figure 2. Proteins synthesized during anaerobic induction of nitrogenase in *Plectonema boryanum*, a nonheterocystous cyanobacterium. Cells were starved for nitrogen under Ar/O_2, as described by Rippka and Waterbury (1977), and then switched to an anaerobic gas phase. Nitrogenase activity is detected within 1 hr and is maximal within 6 hr. Cultures were labeled with $^{35}SO_4^{2-}$ for various times following the switch in the gas phase and the total cellular protein was subjected to SDS–acrylamide gel electrophoresis. The figure shows an autoradiograph of the dried gel containing samples labeled: (A) 0–2 hr, (B) 2–4 hr, and (C) 0–6 hr. The numbered bands, whose molecular weights based on coelectrophoresis with standards are also shown, are not labeled under aerobic conditions or in the presence of NH_4^+.

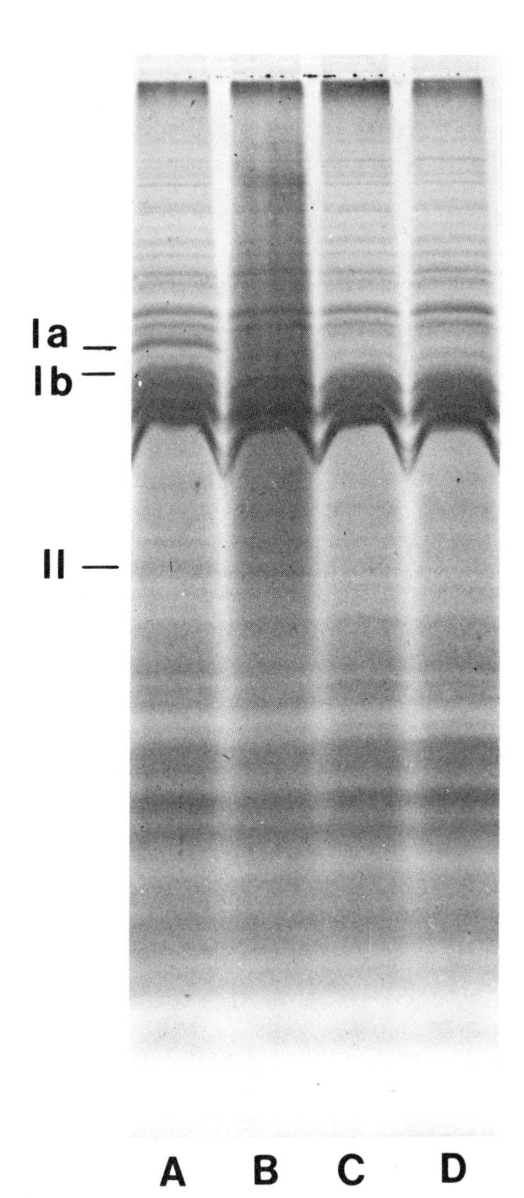

Figure 3. Synthesis of nitrogenase components in *Plectonema* requires anaerobic induction. Bands corresponding to nitrogenase components I and II in molecular weight can be seen in stained gels following SDS–acrylamide gel electrophoresis of whole cell protein. Gels are (A) extract from cells induced anaerobically for 6 hr; (B) the same but with 15 mM NH_4^+; (C) as in (A) but with 50 μg/ml chloramphenicol; (D) as in (A) but with 20% O_2 in the gas phase.

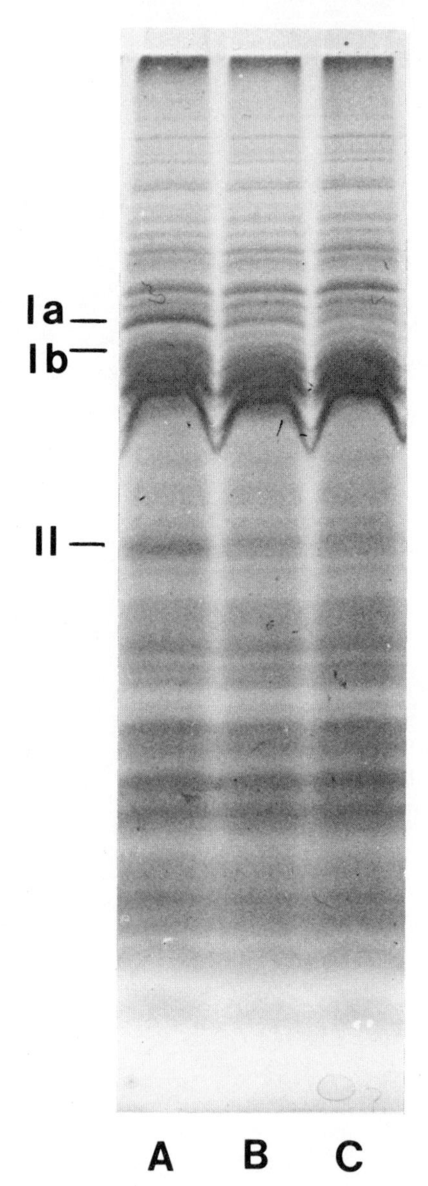

Figure 4. Disappearance of nitrogenase components from anaerobically induced *Plectonema* following admission of O_2. SDS–acrylamide gels, stained as in Figure 3, of total protein from cells (A) induced anaerobically for 6 hr; (B) same as (A) after 1 hr in 20% O_2; and (C) same as (A) after 2 hr in 20% O_2. Bands identified as nitrogenase components are specifically removed under aerobic conditions.

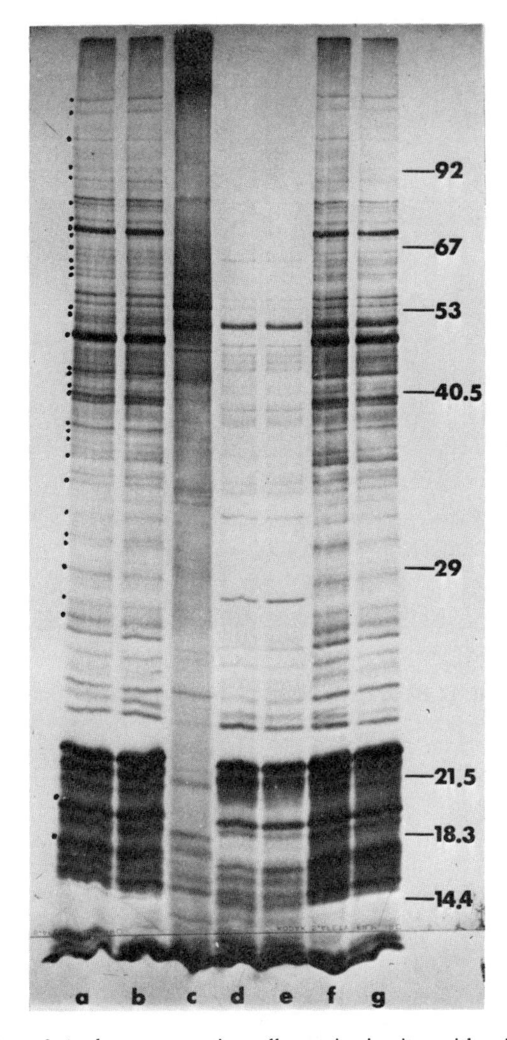

Figure 5. Digestion of *Anabaena* vegetative cell proteins in vitro with soluble protease from nitrogen-starved cells. Lysozyme extracts of nitrate-grown *Anabaena* 7120 labeled with $^{35}SO_4^{2-}$ were incubated with a portion of the soluble extract from vegetative cells starved for nitrate for 24 hr. Reactions were carried out in 0.1 M Tris (pH 7.2), 10 mM KCl, and 2 mM $CaCl_2$. Each reaction mixture contained protease activity capable of releasing 1.0 A^{520} units of Azocoll/hr (Wood and Haselkorn, 1979). A portion of each reaction mixture was analyzed by electrophoresis on a 30-cm polyacrylamide slab gel (25% acrylamide:0.083% bisacrylamide), impregnated with PPO, dried between two sheets of cellophane, and exposed to Kodak XR-5 film at −70°C. Each gel slot contained 28,000 cpm. (a) Substrate alone, no incubation; (b) substrate alone, 9-hr incubation; (c) proteins of heterocysts isolated after 22 hr of differentiation; (d) and (e) 4- and 9-hr incubation in the presence of enzyme and 2 mM Ca^{2+}; (f) and (g) 4- and 9-hr incubation in the presence of enzyme and 2 mM EDTA. Molecular weight markers are shown on the right. Dots on the left indicate polypeptides present in vegetative cell extracts, missing from mature heterocysts, and degraded in vitro.

teins are equally susceptible to the Ca^{2+}-dependent enzyme in vitro. Although more than 30 proteins are degraded, many of those found in heterocysts are not. The pattern of protein turnover in vivo is not mimicked perfectly in vitro, however. The very abundant phycobiliproteins with molecular weights of 15,000–20,000, which are absent from heterocysts, are largely untouched by the Ca^{2+}-dependent enzyme in vitro.

This result led to experimentation that identified a proteolytic activity in the membrane fraction of *Anabaena* specifically degrading phycobiliproteins in vitro (Wood and Haselkorn, 1977, 1979). The activity can be released from cell membranes with non-ionic detergents and is assayed by the release of acid-soluble radioactivity from purified ^{35}S-labeled phycobilisomes. The latter are particulate assemblages of phycobiliproteins that are found attached to the photosynthetic lamellae in vivo and purified intact by centrifugation. Unlike the Ca^{2+}-dependent activity, the phycobiliproteinase is present in cells grown on NH_4^+. However, the activity increases six- to tenfold following induction of heterocyst differentiation. We have used the disappearance of phycocyanin absorption in vivo and the increase in phycobiliproteinase activity in vitro as indicators of early events in heterocyst differentiation (Wood and Haselkorn, 1979).

CONTROL OF HETEROCYST SPACING PATTERN

Heterocysts develop at regularly spaced intervals along *Anabaena* filaments following transfer to nitrogen-free medium. Fogg (1949) originally proposed that NH_4^+ or a simple derivative of it inhibited heterocyst differentiation. Wolk (1967) subsequently suggested that mature heterocysts were the source of the inhibitor, which diffused through vegetative cells. For stability in the heterocyst spacing pattern, the inhibitor must be destroyed in vegetative cells and differentiating heterocysts must destroy, sequester, or export the inhibitor faster than they manufacture it. Establishment of the normal heterocyst spacing pattern does not require nitrogen fixation. It is observed: 1) before nitrogenase activity can be detected; 2) in an atmosphere of argon/CO_2, in which nitrogen fixation cannot occur; and 3) in some *nif*$^-$ mutants of *Anabaena*. The proposal that glutamine is the diffusible inhibitor is still consistent with these facts (Haselkorn, 1978). Protease activation, which is among the earliest responses to nitrogen starvation, should produce glutamine by proteolysis and a glutamine gradient could be responsible for the initial determination of heterocyst spacing.

We have attempted to test the proposal that the level of glutamine specifically, rather than the level of amino acids in general, is critical to the regulation of heterocyst differentiation. The proposition is a difficult one to establish unequivocally. To begin, we found it easier to refute the converse (that lowering the level of an amino acid other than glutamine provokes differentiation) by using methionine auxotrophs of *Anabaena* (Currier, Haury, and Wolk, 1977). With the mutant *met* 38 we asked whether starvation for methionine, in the presence of NH_4^+, would provoke differentiation and found that it does not. The data in Table 1 show that neither phycocyanin breakdown nor phycobiliproteinase activation occur in methionine-starved cultures as long as NH_4^+ is present. When NH_4^+ is withdrawn as well, differentiation, phycocyanin breakdown, and the rise in phycobiliproteinase occur. The important conclusion is that general signals of the sort provided by stringent strains of *Escherichia coli*, i.e., the manufacture of ppGpp by ribosomes when the level of aminoacyl tRNA falls, do not play a role in regulating heterocyst differentiation. Indeed, this conclusion was already foreshadowed by Adams et al. (1977), who were unable to find ppGpp and related compounds in *Anabaena* shifted to nitrogen-free medium.

The specific role of glutamine, however, has been indicated by a number of experiments involving the analog methionine sulfoximine (MSX). This compound is a very poor inhibitor of *Anabaena* glutamine synthetase in vitro, although, at 1 μM concentration, it irreversibly inhibits the cyanobacterial enzyme in vivo (Orr et al., 1978). When MSX is administered to *Anabaena*, heterocyst differentiation and nitrogenase activity are induced, even in the presence of NH_4^+ (Stewart and Rowell, 1975; Orr et al., 1978). We can then ask whether glutamine represses heterocyst differentiation in cultures containing MSX. The interpretation of this experiment is not straightforward because glutamine effectively prevents the transport of MSX into *Anabaena* (J. Orr, unpublished results). We therefore allowed 6 hr for MSX to enter the cells and inhibit glutamine synthetase before adding glutamine. The results shown in Table 2 indicate that glutamine, but not NH_4^+, reverses the induction of heterocyst differentiation by MSX. We conclude that glutamine, or a metabolite of glu-

Table 1. Phycocyanin degradation following starvation of *Anabaena variabilis* Met 38 for methionine and/or ammonia

		$-$Met, $+NH_3$		$+$Met, $-NH_3$		$-$Met, $-NH_3$	
Assay	0 hr	21 hr	45 hr	21 hr	45 hr	21 hr	45 hr
Phycocyanin[a]	100.0	100.0	90.0	73.0	—[c]	49.0	13.0
PBPase[b]	1.0	1.8	1.8	4.3	9.4	3.2	7.9

[a] Measured as absorbance at 620 nm. 100 = absorbance of 0.18.

[b] Phycobiliproteinase was assayed as follows: at the indicated times, cells from 10 ml of culture were collected by centrifugation and broken by sonication, and a membrane fraction was collected by centrifugation at 15,000 rpm for 30 min. The resulting pellet was suspended in 0.2 ml of 20 mM Tris (pH 7.8) containing 0.2% NP-40. A preparation of ^{35}S-labeled phycobilisomes containing 14,000 cpm was added and the mixture incubated for 10 hr at 37°C. An activity of 1.0 corresponds to the release of 23 cpm/hr in TCA-soluble form.

[c] This culture was differentiated and growing under N_2-fixing conditions by 24 hr.

tamine other than NH_4^+, is a component of the system that represses heterocyst differentiation.

GLUTAMINE SYNTHETASE AND NITROGEN FIXATION

Although these experiments indicate that glutamine is part of a negative regulatory system repressing differentiation, they do not bear on the possible role of glutamine synthetase (GS) itself as a positive regulatory element in *Anabaena*. *Anabaena* GS is a dodecameric double disc with structure similar to that of *E. coli* GS, although it is not subject to adenylylation control (Orr et al., 1978). The level of enzyme activity increases severalfold during heterocyst differentiation. The enzyme has been purified to homogeneity

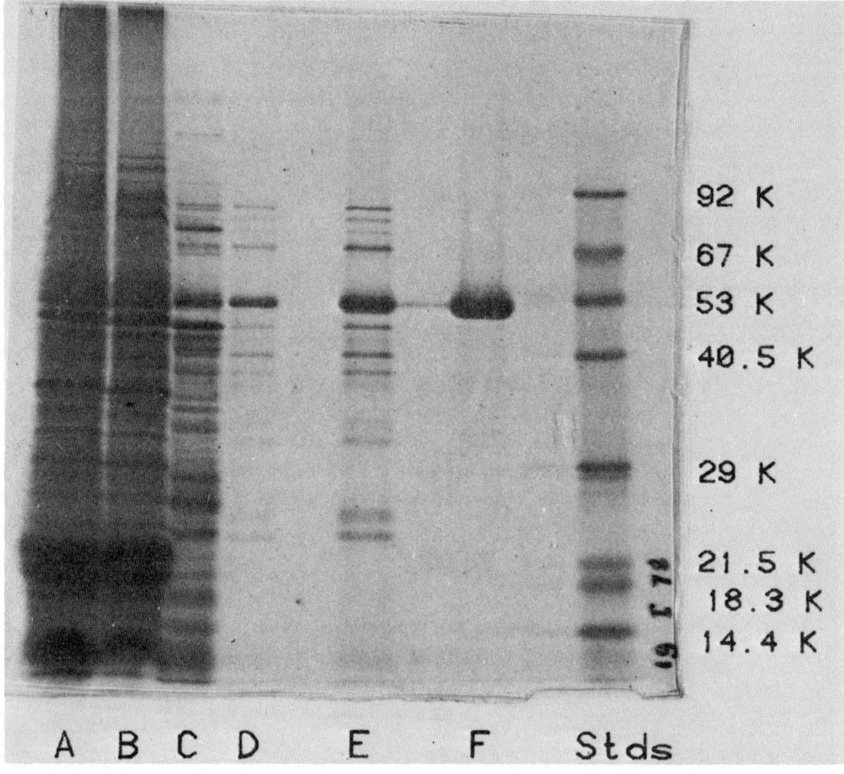

Figure 6. SDS–polyacrylamide gel showing purification of glutamine synthetase from *Anabaena* 7120. The slots contained: (A) crude extract, (B) streptomycin sulfate supernatant, (C) 31%–52% saturated ammonium sulfate precipitate, (D) agarose A 1.5-m gel column purification. Samples (A)–(D) were loaded at constant GS activity. (E) 3 × sample D. (F) blue Sepharose-4B column-purified material, overloaded. The standards slot contained protein markers with the molecular weights shown. The main gel was 15% acrylamide: 0.087% bisacrylamide, run at pH 8.8.

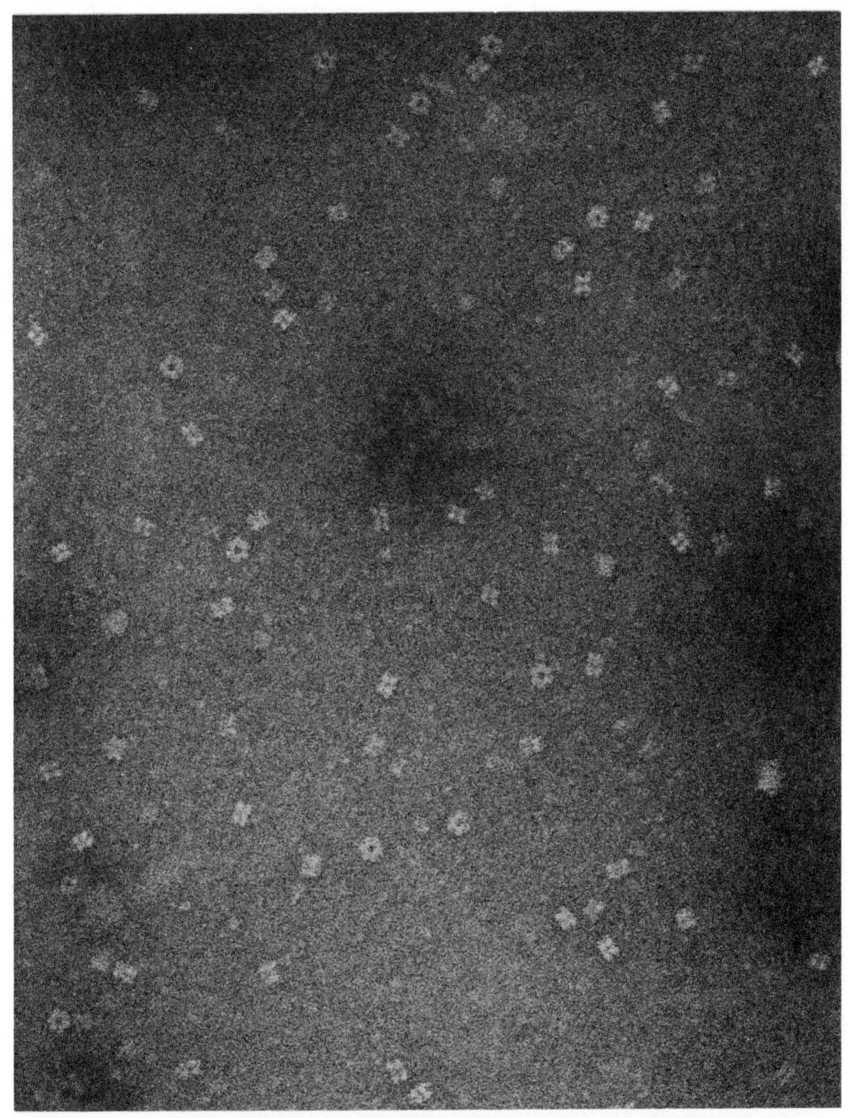

Figure 7. Glutamine synthetase from *Anabaena* 7120, purified as described in the legend to Figure 6, negatively stained with 2% PTA. The enzyme contains two hexameric discs with an edge-to-edge distance of ~150 Å. We are grateful to Thomas Wellems for the electron micrograph.

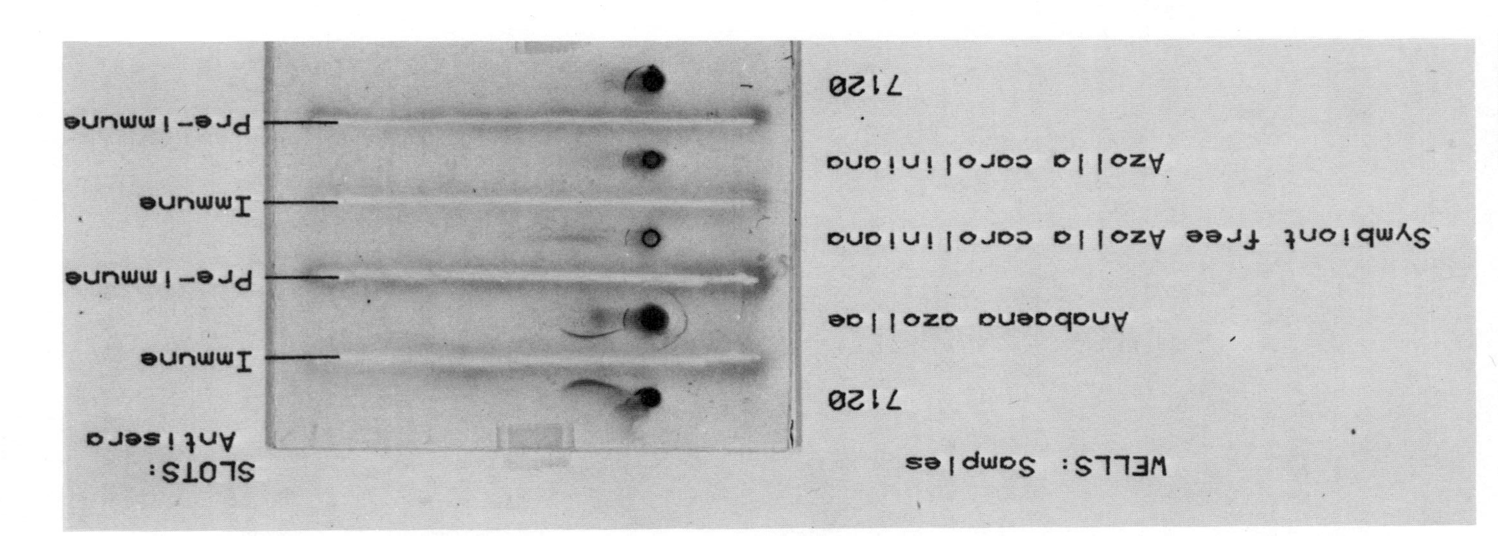

Figure 8. Immunoelectrophoresis in agarose gels. Crude extracts were placed in the wells, as labeled, and electrophoresed at pH 8.5 with the positive electrode on the right. Slots were then cut in the gel and antiserum to purified *Anabaena* 7120 GS was placed in the slots and allowed to diffuse. After 24 hr of development the gels were washed free of nonprecipitated proteins and stained with Coomassie Brilliant Blue. The reaction with 7120 extract shows the antibody to be monospecific. The reaction with *Anabaena azollae* extract shows cross-reaction with only one molecular species. Absence of a precipitin line in the reaction between GS antibody and *Azolla* extract shows that the latter contains less than 5% of the GS antigen present in free-living *Anabaena azollae*.

Table 2. Repression of heterocyst differentiation by glutamine[a]

Culture conditions	% Prohets	% Hets	Cells/ml $\times 10^{-6}$
Nitrogen-free	2.6	8.8	4.0
+2 mM NH_4Cl	0.9	3.6	5.6
+2 mM Gln	0.8	4.5	5.1
+5 μM MSX, +2 mM Gln	0.5	4.1	4.9
+5 μM MSX	3.4	12.3	3.2
+5 μM MSX, +2 mM NH_4Cl	3.9	11.8	3.7

[a] *Anabaena variabilis* 29413 was grown in nitrogen-free medium C (Kratz and Myers, 1955) to a density of around 2.5×10^6 cells/ml. The cells were collected by filtration, washed, suspended in the same medium, and distributed among flasks to which the indicated additions were made. In the fourth flask, glutamine was added 6 hr after the methionine sulfoximine. In all cases, cells were counted and the fraction of heterocysts and proheterocysts determined 48 hr later. We are grateful to Kent Fedde for these data.

(Figures 6 and 7) and used to prepare antibody in rabbits (Orr et al., 1978), which has permitted us to determine the level of glutamine synthetase antigen in several interesting situations. The first is the *Anabaena* cells treated with 1 μM MSX. Such cells actively synthesize nitrogenase, presumably by using glutamine released by proteolysis. However, the level of GS antigen remains constant 24 hr after the addition of MSX, which means that the inactivated enzyme is not itself degraded by protease and that glutamine starvation does not induce synthesis of new GS.

The second experiment made possible by the antiserum to GS is the measurement of antigen in the association of the water fern *Azolla caroliniana* with its symbiotic cyanobacterium *Anabaena azollae*. In this association, discussed in detail in this volume by Peters, the fern derives all the NH_3 it needs for growth from the symbiont. It does so by regulating the GS activity of the symbiont, causing the latter to excrete newly fixed nitrogen for the host's benefit. Stewart (1977) suggested that the fern inactivates the cyanobacterial GS with a compound acting like methionine sulfoximine, although regulation at the level of synthesis of GS was not excluded. Figure 8 shows that this association, although vigorously fixing nitrogen, contains no GS antigen detectable by immunoelectrophoresis. The test was made more quantitative by labeling pure GS from *Anabaena* 7120 with ^{125}I and using a radioimmunoassay based on binding of immune complexes to *Staphylococcus aureus* A. In such assays, the fern-symbiont association contained less than 5% and about 10%, in two experiments, of the GS antigen found in free-living N_2-fixing *A. azollae*, normalized to nitrogenase activity. In a single experiment, we found the GS antigen preferentially concentrated in the young leaves of the fern, which contain dividing, non-N_2-fixing vegetative cells of the symbiont (Hill, 1977). Thus, it is possible that, with continued dissection of the fern, we can determine whether GS antigen is required as a positive activator of *nif* gene expression in *Anabaena*.

CONTROL OF TRANSCRIPTION DURING HETEROCYST DIFFERENTIATION

The extensive work on *E. coli* indicates that these studies will require genetic analysis, including the isolation and characterization of mutants and the preparation of DNA probes to detect in vivo transcripts and to serve as templates for transcription in vitro. The work of Currier, Haury, and Wolk (1977) and of Wilcox, Mitchison, and Smith (1975) indicates that the

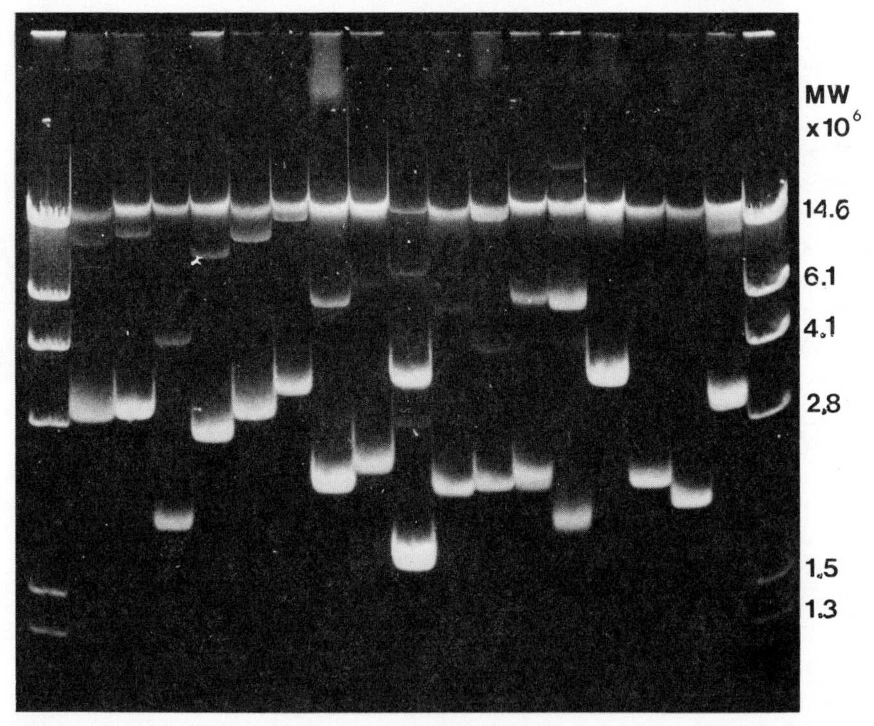

Figure 9. Agarose gel electrophoresis of crude plasmid DNA preparations from independent clones containing *Anabaena* 7120 DNA inserted into the plasmid pBR322. The two outside slots contain a set of molecular weight markers generated by *Hind*III restriction of bacteriophage λ DNA, with the molecular weights shown. The central slot contains the parent plasmid pBR322. Each of the other slots contains a recombinant DNA plasmid. The fastest moving (lowest) band in each slot corresponds to closed circular DNA; the fainter middle band is nicked circular plasmid DNA; the slowest band, migrating roughly with the λ DNA fragment of 14.6×10^6, is chromosomal DNA contaminating the crude plasmid DNA preparation. The plasmid in the fifth slot from the right end contains an insert of *Anabaena* DNA of about 3×10^6 in molecular weight. Other clones have been shown by electron microscopy to contain inserts up to 12×10^6 in molecular weight.

necessary developmental mutants can be isolated in *Anabaena*. Sorely missing is a system for complementation and for recombinational analysis. Although phages for *Anabaena* are known, none has been shown to transduce host markers. Transformation and conjugation are also unknown in *Anabaena*.

We are attempting two remedies for this situation. The first involves recombinant DNA, using the cloning systems developed for *E. coli*. We have prepared a collection of clones using as vector the small plasmid pBR322, which carries two antibiotic resistance markers. In one cloning method, pBR322 was cleaved with the restriction endonuclease *Pst*I, which cuts once in the gene for ampicillin resistance. This enzyme creates 3′—OH termini to which poly(dT) tails were added with terminal transferase after treatment with λ-exonuclease. *Anabaena* DNA was sheared to uniform size, briefly digested with λ-exonuclease, and poly(dA) tails were added with terminal transferase. The *Anabaena* DNA was then annealed with the vector and used to transform *E. coli*, selecting for resistance to tetracycline. Figure 9 shows an agarose gel in which crude plasmid DNA preparations from a number of independent clones were electrophoresed. Comparison with pBR322 by electrophoresis and by electron microscopy suggests that the recombinant plasmids contain inserts up to 18 Kb.

Plasmids containing DNA coding for proteins active in heterocyst differentiation have not yet been identified. To determine which recombinant plasmids contain *Anabaena nif* genes, two methods are being tried. One uses the plasmids containing *Klebsiella nif* genes prepared by Cannon, Reidel, and Ausubel (1977). These can be labeled with ^{32}P by nick translation and used in colony hybridization tests to locate colonies with the heterologous *Anabaena nif* genes. The second method is to prepare *nif* messenger RNA, either from *Anabaena* heterocysts or from anaerobically induced *Plectonema*, in which we expect *nif* mRNA to be extremely abundant. The RNA can be labeled *in vitro* and also used as a colony hybridization probe.

The alternative approach to cloning in *E. coli* is to develop a cloning system in *Anabaena* itself. The advantages of such a system are that it would permit complementation analysis and would avoid such problems as restriction and promoter recognition. We have found, as has Simon (1977), that *Anabaena* strains contain cryptic plasmids. There are many size classes, of which the smallest is around 1.5 μ (Figure 10). We are attempting to open this small plasmid with restriction endonucleases in such a way that wild-type *Anabaena* DNA can be inserted without damaging its capacity for replication. If the inserted fragment carries wild-type *Anabaena met* or *ura* genes, we may be able to select the recombinant plasmid by transformation of the Currier *met* or *ura* mutants. Only time will tell if this approach can succeed.

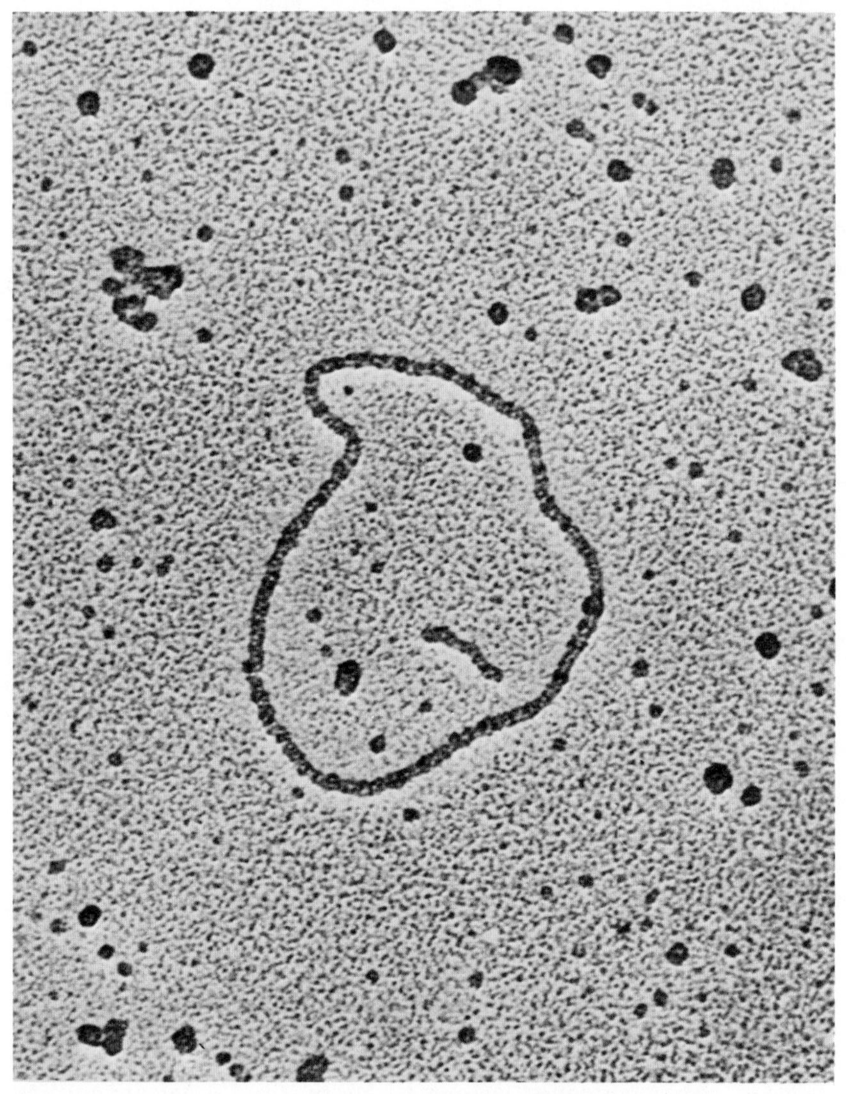

Figure 10. Small cryptic plasmid from *Anabaena* 7120. Closed circular DNA was prepared by the method of Sharp et al. (1972) and spread by the aqueous basic protein film technique described by them. Some of the molecules are nicked after collection from the CsCl-EtBR gradient and appear relaxed in the electron microscope. The molecule shown has a contour length of 1.5 μ. Others are found with lengths up to 18 μ.

REFERENCES

Adams, D. G., D. O. Phillips, J. M. Nichols, and N. G. Carr. 1977. The presence and absence of magic spot nucleotide modulation in cyanobacteria undergoing nutritional shiftdown. FEBS Lett. 81:48–52.

Alberte, R. S., and E. Tel-Or. 1977, Characteristics of the photosynthetic apparatus of heterocysts and vegetative cells of *Nostoc* and *Anabaena*. Plant Physiol. 59(Suppl.):129.

Apte, S. K., P. Rowell, and W. D. P. Stewart. 1978. Electron donation to ferredoxin in heterocysts of the nitrogen-fixing alga *Anabaena cylindrica*. Proc. R. Soc. Lond. [Biol.] 200:1–25.

Bothe, H., J. Tennigkeit, G. Eisbrenner, and M. G. Yates. 1977. The hydrogenase-nitrogenase relationship in the blue-green alga *Anabaena cylindrica*. Planta 133:237–242.

Cannon, F. C., G. E. Reidel, and F. M. Ausubel. 1977. Recombinant plasmid that carries part of the nitrogen fixation (*nif*) gene cluster of *Klebsiella pneumoniae*. Proc. Natl. Acad. Sci. USA 74:2963–2967.

Codd, G. A., and W. D. P. Stewart. 1977. Ribulose-1,5-bisphosphate carboxylase in heterocysts and vegetative cells of *Anabaena cylindrica*. FEMS Microbiol. Lett. 2:247–249.

Currier, T. C., J. F. Haury, and C. P. Wolk, 1977. Isolation and preliminary characterization of auxotrophs of a filamentous cyanobacterium. J. Bacteriol. 129:1556–1562.

Fleming, H., and R. Haselkorn. 1973. Differentiation in *Nostoc muscorum:* Nitrogenase is synthesized in heterocysts. Proc. Natl. Acad. Sci. USA 70:2727–2731.

Fleming, H., and R. Haselkorn. 1974. The program of protein synthesis during heterocyst differentiation in nitrogen-fixing blue-green algae. Cell 3:159–170.

Fogg, G. E. 1949. Growth and heterocyst production in *Anabaena cylindrica* Lemm II. In relation to carbon and nitrogen metabolism. Ann. Bot. 13:241–259.

Haselkorn, R. 1978. Heterocysts. Annu. Rev. Plant Physiol. 29:319–344.

Hill, D. J. 1977. The role of *Anabaena* in the *Azolla-Anabaena* symbiosis. New Phytol 78:611–616.

Jüttner, F., and N. G. Carr. 1976. The movement of organic molecules from vegetative cells into the heterocysts of *Anabaena cylindrica* In: G. A. Codd and. W. D. P. Stewart (eds.), Proceedings of Second International Symposium on Photosynthetic Procaryotes, pp. 121–123. University of Dundee, Scotland.

Kratz, W. A., and J. Myers. 1955. Nutrition and growth of several blue-green algae. Am. J. Bot. 42:282–287.

Orr, J., N. D. Toan, M. Ohtsuki, T. Wellems, and R. Haselkorn. 1978. Glutamine synthetase from the cyanobacterium *Anabaena* 7120. Fed. Proc. 37:1431.

Peterson, R. B., and C. P. Wolk, 1978. Localization of an uptake hydrogenase in *Anabaena*. Plant Physiol. 61:688–691.

Rippka, R., and R. Y. Stanier. 1978. The effects of anaerobiosis on nitrogenase synthesis and heterocyst development by nostocacean cyanobacteria. J. Gen. Microbiol. 105:83–94.

Rippka, R., and J. B. Waterbury. 1977. The synthesis of nitrogenase by non-heterocystous cyanobacteria. FEMS Microbiol. Lett. 2:83–86.

Sharp, P. A., M.-T. Hsu, E. Ohtsubo, and N. Davidson. 1972. Electron microscope heteroduplex studies of sequence relations among plasmids of *E. coli*. J. Mol. Biol. 71:471–497.

Simon, R. D. 1977. Determination of the range of extrachromosomal DNA found in species of filamentous cyanobacteria. Abstract of the Annual Meeting of the

American Society for Microbiology, p. 172. American Society for Microbiology, Washington, D.C.

Stanier, R. Y., and G. Cohen-Bazire. 1977. Phototrophic prokaryotes: The cyanobacteria. Annu. Rev. Microbiol. 31:225:274.

Stewart, W. D. P. 1977. A botanical ramble among the blue-green algae. Br. Phycol. J. 12:89–115.

Stewart, W. D. P., and P. Rowell. 1975. Effects of L-methionine-DL-sulfoximine on the assimilation of newly fixed NH_3, acetylene reduction and heterocyst production in *Anabaena cylindrica*. Biochem. Biophys. Res. Commun. 65:846–856.

Stewart, W. D. P., P. Rowell, and S. K. Apte. 1977. Cellular physiology and the ecology of N_2-fixing blue-green algae. In: W. Newton, J. R. Postgate, and C. Rodriguez-Barrueco (eds.), Recent Developments in Nitrogen Fixation, pp. 287–307. Academic Press, London.

Tel-Or, E., and W. D. P. Stewart. 1977. Photosynthetic components and activities of nitrogen-fixing isolated heterocysts of *Anabaena cylindrica*. Proc. R. Soc. Lond. [Biol.] 198:61–86.

Tetley, R. M., and N. I. Bishop. 1978. Differential inhibition of photosynthesis, nitrogenase activity and H_2 uptake by metronidazole in bluegreen algae. Plant Physiol. 61(suppl.):2.

Thomas, J., J. C. Meeks, C. P. Wolk, P. W. Shaffer, S. M. Austin, and W.-S. Chien. 1977. Formation of glutamine from [^{13}N]ammonia, [^{13}N]dinitrogen and [^{14}C]glutamate by heterocysts isolated from *Anabaena cylindrica*. J. Bacteriol. 129:1545–1555.

Wilcox, M., G. J. Mitchison, and R. J. Smith. 1975. Mutants of *Anabaena cylindrica* altered in heterocyst spacing. Arch. Mikrobiol. 130:219–223.

Winkenbach, F., and C. P. Wolk. 1973. Activities of enzymes of the oxidative and the reductive pentose phosphate pathways in heterocysts of a blue-green alga. Plant Physiol. 52:480–483.

Wolk, C. P. 1967. Physiological basis of the pattern of vegetative growth of a blue-green alga. Proc. Natl. Acad. Sci. USA 57:1246–1251.

Wolk, C. P. 1975. Differentiation and pattern formation in filamentous blue-green algae. In: P. Gerhardt, H. Sadoff, and R. Costilow (eds.), Spores VI, pp. 85–96. American Society for Microbiology, Washington, D.C.

Wolk, C. P., J. Thomas, P. W. Shaffer, S. M. Austin, and A. Galonsky. 1976. Pathway of nitrogen metabolism after fixation of ^{13}N-labeled N_2 by *Anabaena cylindrica*. J. Biol. Chem. 251:5027–5034.

Wood, N. B., and R. Haselkorn. 1977. Protein degradation during heterocyst differentiation in the blue-green alga *Anabaena* 7120. Fed. Proc. 36:886.

Wood, N. B., and R. Haselkorn. 1979. Proteinase activity during heterocyst differentiation in nitrogen-fixing cyanobacteria. In: G. N. Cohen and H. Holzer (eds.), Limited Proteolysis in Microorganisms, pp. 159–166. U.S. D.H.E.W. Publication No. (NIH) 79-1591.

Nitrogen Fixation, Volume II
Edited by W. E. Newton and W. H. Orme-Johnson

Heterocysts, ^{13}N, and N_2-fixing Plants

C. P. Wolk

Aerobic fixation of dinitrogen by many cyanobacteria is dependent on the differentiation of a percentage of their cells into heterocysts and on interactions between the heterocysts and the remaining vegetative cells. Nitrogen fixation by the heterocysts is dependent on products of photosynthesis that they receive from the vegetative cells; in return, they supply the vegetative cells with products of N_2 fixation.

We view work on these organisms as having potential bearing on certain practical problems, and have sought to identify the mechanisms and intercellular interactions by which nitrogenase is protected against inactivation by O_2 and supplied with reductant and ATP, and the ammonia produced is removed. Such analysis would provide a model of conditions required for N_2 fixation within the tissues of nonleguminous higher plants and might permit development of isolated heterocysts as self-maintaining photocatalysts for the large-scale production of ammonia, with carbohydrate (or H_2) as electron donor. We have also tried to develop the genetics of N_2-fixing cyanobacteria. Genetic techniques would facilitate analysis of the biochemical controls governing N_2 fixation and cellular differentiation in these organisms, *might* permit the construction of N_2-fixing endosymbioses between cyanobacterial cells and the cells of nonleguminous crop plants, and *could* perhaps permit the construction of cyanobacterial strains that could be used as self-maintaining, stable "photocatalysts" of bioproduction of NH_3 and H_2 with H_2O as the source of electrons.

This work was supported by the U.S. Department of Energy under contract EY-76-C-02-1338.

NITROGEN FIXATION IN HETEROCYSTS

The first, pivotal question to be asked is: Is nitrogen fixation restricted to heterocysts (Fay et al., 1968)? Many indirect and sometimes ingenious experiments have been interpreted as indicating that nitrogenase activity in aerobically grown filaments is restricted to heterocysts. One reason to remain skeptical was that relatively little of the in vivo nitrogenase activity

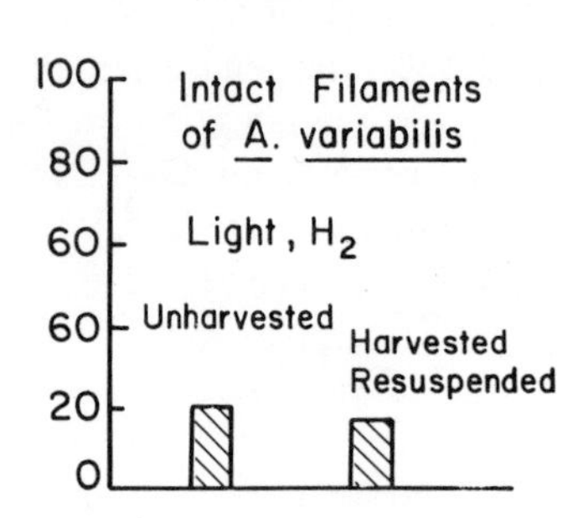

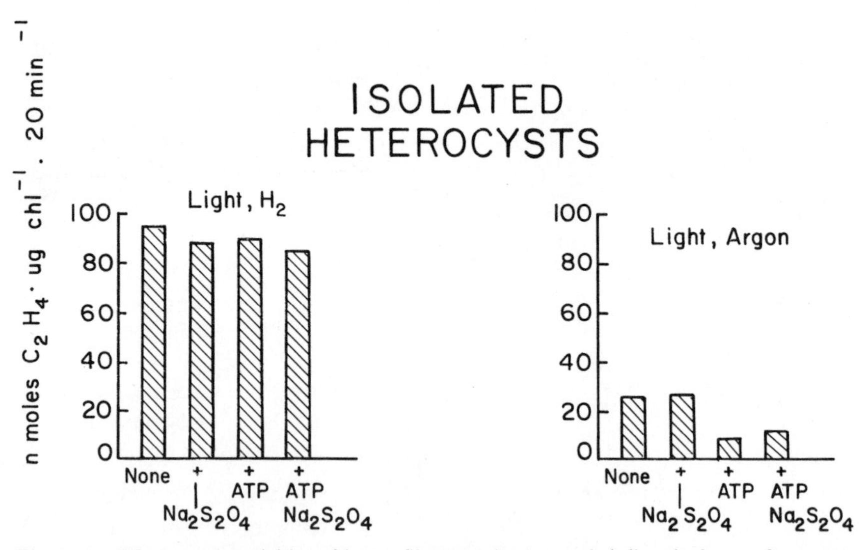

Figure 1. Nitrogenase activities of intact filaments (top) sampled directly from a fermentor, and following centrifugation under argon and resuspension; and of isolated heterocysts under H_2 (bottom left) and Ar (bottom right), in all cases in the presence of 7% C_2H_2. The activities shown, about 5 μmol mg of Chl^{-1} min^{-1}, are typical rather than maximal, are largely dependent on the presence of H_2, and are not stimulated by dithionite or ATP (and an ATP-generating system). In this experiment the chlorophyll concentration in the fermentor was 0.84 μg of Chl ml^{-1}, and the light intensity during assay was 5.1×10^5 ergs cm^{-2} s^{-1}; the isolated heterocysts accounted for 63.5% of the activities of the intact filaments. There was no activity in the dark under hydrogen + 7% C_2H_2, in the presence of $Na_2S_2O_4$, ATP, and an ATP-generating system.

was recoverable in isolated heterocysts. We have recently observed that when, after incubation of filaments of *Anabaena variabilis* under H_2 for 0.5 hr in 30 mM HEPES, 30 mM PIPES, 1 mM $MgCl_2$, and 10 mM EDTA (pH 7.2) containing 1 mg of lysozyme/ml, vegetative cells are disrupted by cavitation in a sonic cleaning bath, heterocysts isolated by differential centrifugation have the following properties. They reduce acetylene in the light under H_2 at rates of up to 6.5 μmol of $C_2H_2 \cdot$mg of Chl $a^{-1} \cdot \text{min}^{-1}$, and have thereby accounted for an average of 60% (with a standard deviation, for 13 experiments, of 12%) of the nitrogenase activity of the parent filaments. These high rates of reduction remain constant for 1.5 to 3 hr. In the dark, these isolated heterocysts express an O_2- and H_2-dependent nitrogenase activity, but have extremely low activity in the dark with ATP and dithionite (see Figure 1 and its legend; Peterson and Wolk, unpublished observations).

One may object, however, that, because the uptake hydrogenase activity in our cultures is restricted to the heterocysts (Peterson and Wolk, 1978), we may have favored the detection of nitrogenase localized in heterocysts. Dr. Richard Peterson, in my laboratory, has therefore performed the following experiments. He labeled filaments with ^{55}Fe and made extracts of filaments, of isolated heterocysts, and of lysate released from vegetative cells during isolation of heterocysts, all under anaerobic conditions. The extracts were subjected to gel electrophoresis under nondenaturing, anaerobic conditions, and the gels were sliced and the radioactivity in the slices was measured. Approximately 7 major bands of radioactivity were observed in the gels (Figure 2). Bands 1 and 4 were identified as the MoFe and Fe protein bands of nitrogenase, respectively, on the basis of the following evidence: 1) these bands, and the corresponding bands of protein, disappeared in ammonium-grown filaments; 2) only these two bands stained quickly with bathophenanthroline sulfonate, a reaction characteristic of nonheme iron proteins; 3) as visualized by counts of ^{55}Fe, by staining with bathophenanthroline sulfonate, or by staining with the protein stain Coumassie Brilliant Blue R-250, the former band was partially diminished in intensity and the latter disappeared completely, whereas other bands were not affected, when extracts were exposed to air for 45 min; 4) these bands, visualized in the same three ways, are highly prominent in our extracts of heterocysts, which are very active in reduction of acetylene); 5) ^{99}Mo labeled a protein that coelectrophoresed with the ^{55}Fe-containing protein of band 1; and 6) elution and SDS-gel electrophoresis of the proteins of bands 1 and 4 showed that those bands contain polypeptides of molecular weights approximately equal to the molecular weight of the subunits of the respective components of *Clostridium* nitrogenase. On a per-heterocyst basis, determined by quantification of ^{55}Fe, isolated heterocysts contained a mean

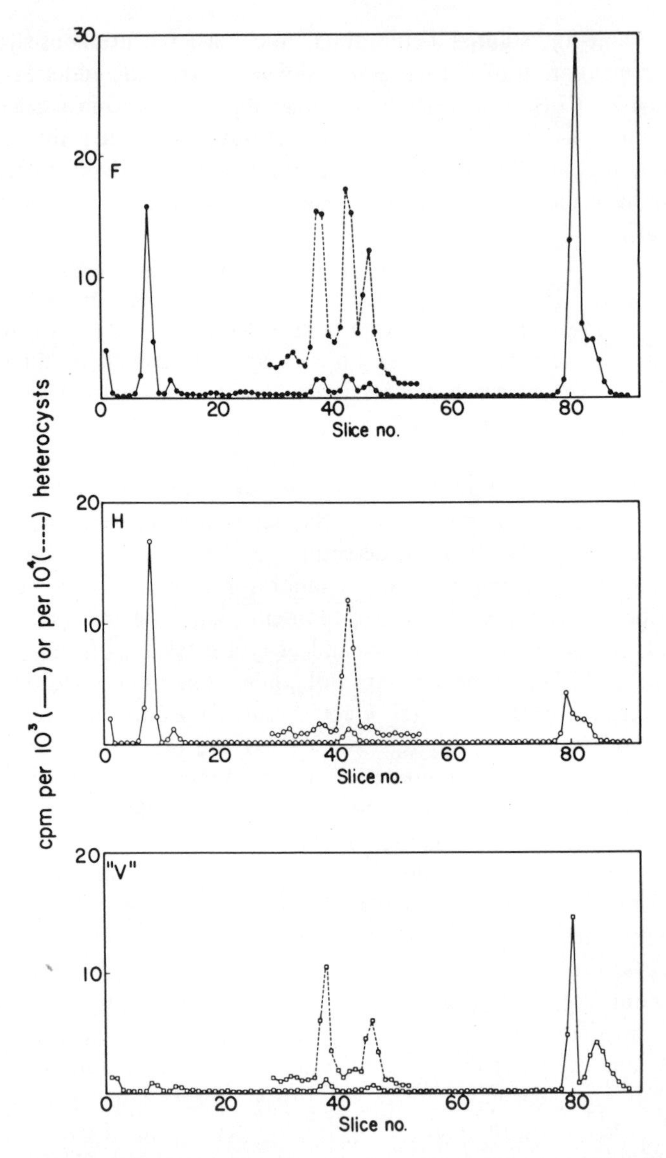

Figure 2. Nondenaturing, anaerobically run electrophoretograms of ^{55}Fe-labeled proteins from heterocysts (H) and residual fraction (V) derived from filaments (F) of *Anabaena variabilis*. Filaments were pretreated for 30 min with 30 mM HEPES, 30 mM PIPES, 1 mM $MgCl_2$, and 10 mM EDTA (pH 7.2) containing 1 mg of lysozyme/ml, and cavitated in a sonic bath to destroy vegetative cells; heterocysts were isolated by differential centrifugation. The centrifugal supernatant was used for (V), and the heterocysts and unincubated filaments, broken by probe sonication, for (H) and (F), respectively. Curves correspond to equal amounts of filaments at the start of the experiment. In this experiment heterocysts accounted for all of the MoFe protein and 81% of the Fe protein of the unincubated filaments.

of 91% of the MoFe protein and 70% of the Fe protein of intact filaments (88% and 80%, respectively, of the total recovered after lysozyme treatment), with the residuum of those proteins detected in the lysate produced during isolation of heterocysts. On the basis of these results, we accept for *A. variabilis* the conclusion proffered by others, that nitrogenase and nitrogenase activity of the aerobically grown filaments is completely, or nearly so, restricted to heterocysts.

What interactions take place between heterocysts and vegetative cells? Jüttner and Carr (1976) pretreated filaments of *Anabaena cylindrica* with lysozyme, pulse-labeled them with $H^{14}CO_3^-$ for 10 sec, and then isolated heterocysts in the cold. A disaccharide, tentatively identified as maltose, accounted for a majority of the radioactive constituents detected in the heterocyst extracts. Heterocysts of *A. cylindrica* have very high activities of enzymes of the oxidative pentose phosphate cycle, but very low activities of glyceraldehyde phosphate dehydrogenase as well as of ribulose diphosphate carboxylase (Winkenbach and Wolk, 1973; Lex and Carr, 1974; Rowell and Stewart, 1976; Codd and Stewart, 1977). With more porous, isolated heterocysts, glucose 6-phosphate can serve as electron donor both to oxygen (Peterson and Burris, 1976) and to substrates of nitrogenase (Lockau et al., 1978; Peterson and Burris, 1978). Thus, a disaccharide entering heterocysts from vegetative cells may be metabolized principally by the oxidative pentose phosphate cycle.

NITROGEN TRANSPORT TO VEGETATIVE CELLS

We now consider the identity of the substance or substances that move from heterocysts to vegetative cells. Using $[^{13}N]N_2$ and $^{13}NH_4^+$, we have been able to chart the initial metabolism of N_2 and NH_4^+ in *A. cylindrica* (Wolk et al., 1976; Meeks et al., 1977), and other unicellular and filamentous cyanobacteria, including organisms with and without heterocysts. In all except *Anacystis nidulans*, the glutamine synthetase–glutamate synthase pathway was the principal assimilatory pathway (in *Gloeocapsa*, the kinetics of assimilation was in accord with this pathway but methionine sulfoximine was without effect). Minor pathways were also determined. In nitrate-grown *Anacystis nidulans*, the formation of $[^{13}N]$glutamate from $^{13}NH_4^+$ was greatly increased by methionine sulfoximine, indicating that, in that organism, glutamine synthetase and glutamic acid dehydrogenase function coordinately in the assimilation of NH_4^+ (Meeks et al., 1978a). In addition, we have shown by use of $^{13}NH_4^+$ and $^{13}NO_3^-$ that cultured tobacco cells make use of the glutamine synthetase–glutamate synthase pathway, and also use glutamic acid dehydrogenase (Skokut et al., 1978).

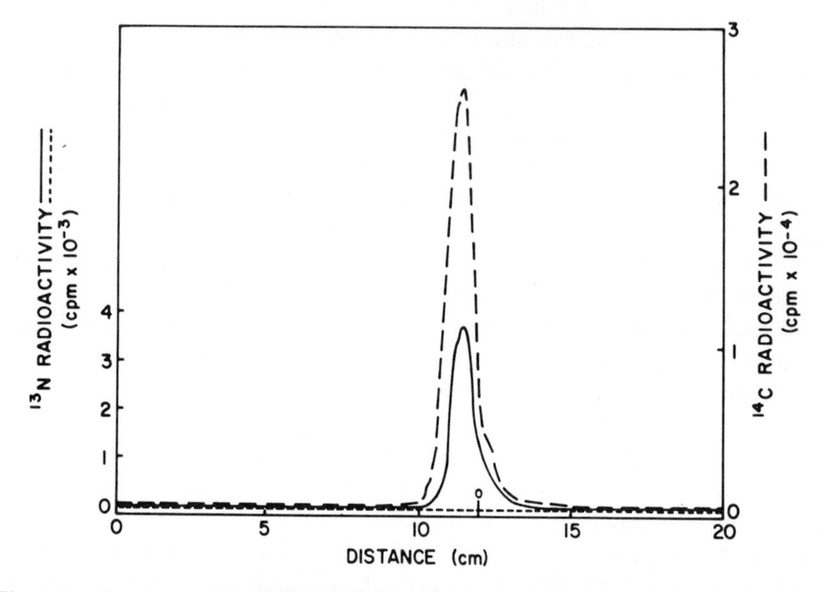

Figure 3. Distribution of radioactivity from ^{13}N in an electrophoretogram of compounds extracted from a suspension of heterocysts isolated from *Anabaena cylindrica* and allowed to fix $[^{13}N]N_2$ for 15 min in the light in the presence of 1 mM glutamate, 5 mM $MgCl_2$, 4 mM ATP, an ATP-generating system, 10 mM $Na_2S_2O_4$, and 24 mM TES buffer (pH 7.2). An 80% methanolic extract of the contents of the reaction vial, supplemented with unlabeled amino acids and $[^{14}C]$ glutamine as markers, was applied at position 0 of a cellulose thin layer plate, and was subjected to electrophoresis at 3000 V for 12 min in 70 mM sodium borate buffer (pH 9.2). The plate was scanned (——); rescanned after decay of the ^{13}N (----); and then, after removal of a thin layer of aluminum covering the detector, scanned for radioactivity from ^{14}C (— —). The plate was then dried and sprayed with a solution of ninhydrin to localize the stable marker amino acids. Glutamate was present at the 6.7-cm position after electrophoresis. From Thomas et al. (1977), with permission.

Using ^{13}N-labeled N_2, we demonstrated that the assimilation of N_2 by heterocysts isolated from *A. cylindrica* is coupled, via formation of NH_4^+, to formation of glutamine by glutamine synthetase (Figure 3). Using $^{13}NH_4^+$, we showed that this reaction is greatly stimulated by supplementary glutamate (Thomas et al., 1977). Of the enzymes catalyzing the two initial reactions for metabolism of N_2-derived NH_4^+, namely, glutamine synthetase and glutamate synthase, the latter is present only in vegetative cells, so that glutamate formation can take place only in vegetative cells (Thomas et al., 1977). We therefore conclude that the principal product of fixation of N_2 that moves from heterocysts to vegetative cells is *glutamine*, and that its formation is largely dependent on the movement of glutamate from vegetative cells to heterocysts. Differences in the relative kinetics of formation of glutamine and glutamate from $[^{13}N]N_2$ and $^{13}NH_4^+$ by a variety of heterocyst-forming cyanobacteria are also consistent with the idea that it is glutamine rather than NH_4^+ that moves from heterocysts to vegetative cells (Meeks et al., 1977; Meeks, et al., 1978a). Figure 4 summarizes the results of studies on intercellular interactions in species of *Anabaena*.

The finding that heterocysts produce glutamine bears on the investigation of the nature of the mechanism controlling the formation of heterocysts. Both mature and immature heterocysts inhibit the differentiation of nearby vegetative cells into heterocysts (Wolk, 1967; Wolk and Quine, 1975). The inhibition is mediated by a diffusible inhibitor or inhibitors produced by heterocysts (Wolk and Quine, 1975). Exogenously supplied ammonium also inhibits heterocyst formation (Fogg, 1949), and methionine sulfoximine, which inhibits glutamine synthetase and thereby inhibits the formation of glutamine from ammonium, prevents the inhibition by ammonium (Stewart and Rowell, 1975; Ownby, 1977). Heterocyst formation can also be inhibited by the addition of exogenous glutamine (Wolk, 1979a). We have therefore suggested that the substance that mediates the intercellular inhibition of heterocyst formation may be glutamine or a derivative of it (Thomas et al., 1977).

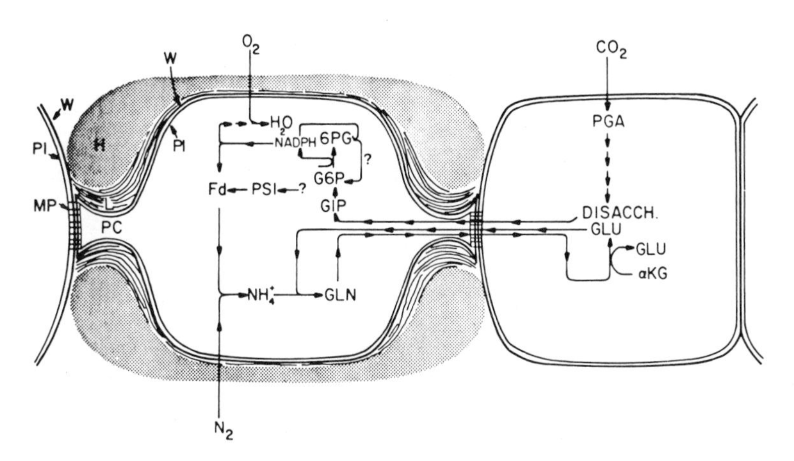

Figure 4. Diagram showing the principal known interactions between a heterocyst (at left) and a vegetative cell (at right). Outside of the wall (W) of the heterocyst is an envelope consisting principally of a laminated, glycolipid layer (L) and a homogeneous, polysaccharide layer (H). Microplasmodesmata (MP) join the plasma membranes (Pl) of the heterocyst and vegetative cell at the end of the pore channel (PC) of the heterocyst. A disaccharide formed by photosynthesis in the vegetative cells moves into heterocysts, and is thought then to be metabolized to glucose 6-phosphate, and oxidized by the oxidative pentose phosphate pathway. Pyridine nucleotide (NADPH) reduced by this pathway can react with O_2 to maintain reducing conditions within the heterocysts, and can reduce ferredoxin (Fd). Ferredoxin can also be reduced by photosystem I. Reduced ferredoxin can donate electrons to nitrogenase, which reduces N_2 to NH_4^+ with concomitant production of hydrogen. Glutamate produced by vegetative cells reacts in heterocysts with NH_4^+ to form glutamine. The glutamine moves into the vegetative cells, where it reacts with α-ketoglutarate (αKG) to form 2 molecules of glutamate. From Wolk (1979b), with permission.

N_2-FIXING PLANTS

Soybeans

In our studies directed at large N_2-fixing plants, we investigated the initial products of assimilation of $[^{13}N]N_2$ by detached soybean nodules and by intact soybean plants. Assimilation by detached nodules appears, from the kinetics, to involve the glutamine synthetase–glutamate synthase pathway, followed by transamination to form alanine (Figure 5). In attached nodules, however, an additional, unidentified compound becomes radioactive at early times (Figure 6). Because asparagine is very rich in the bleeding sap of nodules, it was assumed to be the principal form in which nitrogen is transported from the nodules. However, little or no ^{13}N was detected in asparagine or beyond the nodules themselves after 15 min of labeling, even though $[^{14}C]$asparagine was not degraded by our experimental procedures (Meeks et al., 1978b). To obtain greater transport out from nodules, it may be necessary to provide a source of water to their exterior.

Endosymbiont: Genetic Studies

We attempted to construct N_2-fixing endosymbioses between cyanobacteria and the cells of a dicot. In such an endosymbiosis, the cyanobacteria would

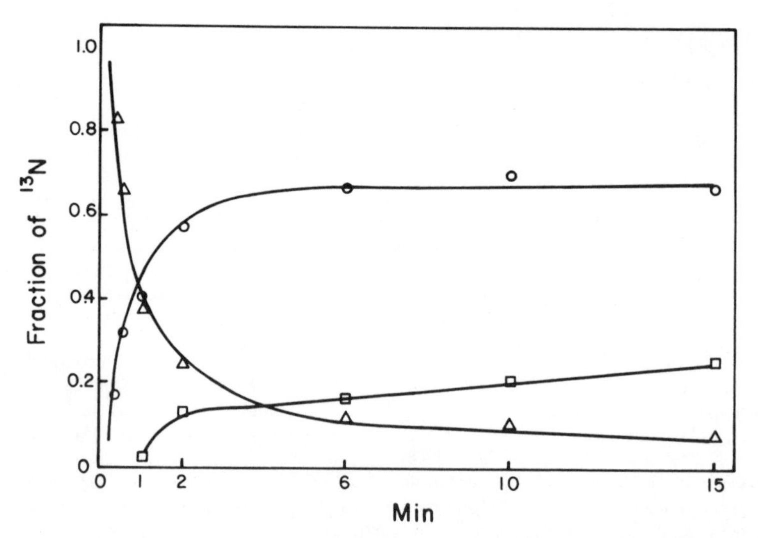

Figure 5. Distribution of ^{13}N in organic products extracted with 80% methanol after fixation of $[^{13}N]N_2$ for 0.33, 0.5, 1, 2, 6, 10, and 15 min by detached soybean nodules. The radioactivity in the constituents of extracts subjected to electrophoresis at pH 9.2 was quantified by integration of peaks in radioscans, with corrections applied for decay. Values presented are means, from two or three experiments, of the fraction of organic ^{13}N migrating with stable glutamine (△), glutamate (○), and alanine (□). From Meeks et al. (1978b), with permission.

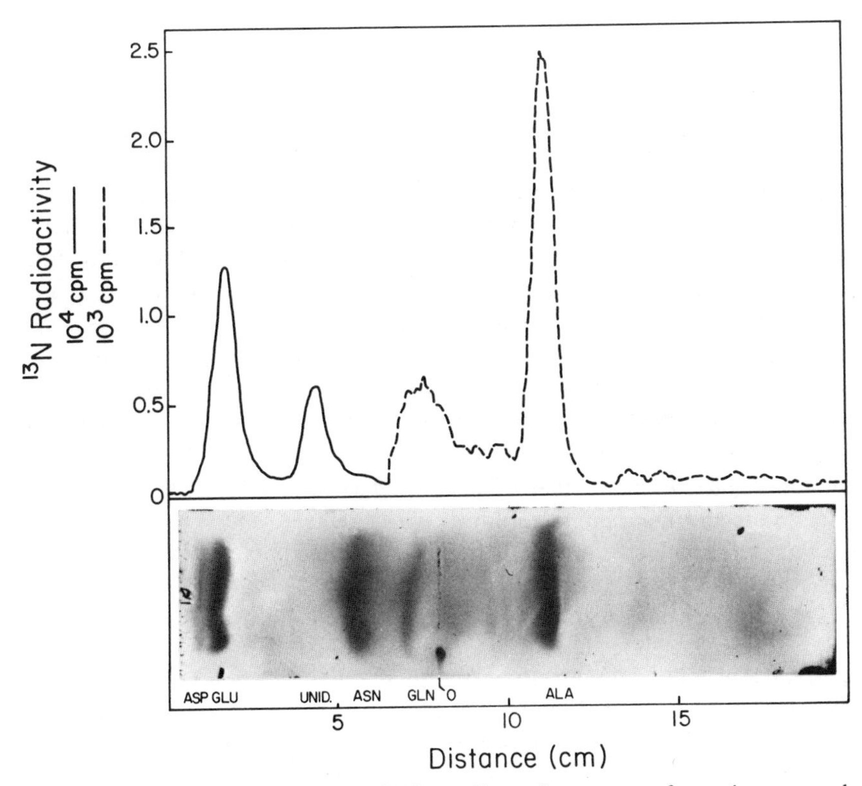

Figure 6. Scan of radioactivity from ^{13}N in an electrophoretogram of organic compounds extracted from a soybean nodule with 1.5 ml of 80% methanol after 6 min of fixation of ^{13}N-labeled N_2 by a seedling. The extract, supplemented with stable aspartate, glutamate, asparagine, glutamine, and alanine as markers, was applied to a cellulose thin layer plate in a thin strip (origin = 0). After displacement of lipids from the origin by chromatography, the extract was subjected to electrophoresis at 3 kV for 11 min in 70 mM sodium borate buffer (pH 9.2). The plate was scanned at 2 cm/min (τ = 3 sec) from + (at left) to −, and was then dried and sprayed with a solution of ninhydrin to localize the marker amino acids. The ^{13}N-containing substances detected that migrated with glutamate, an unidentified compound (unid.), glutamine, and alanine during electrophoresis were present in the ratio (time-corrected) 1.00:0.44:0.06:0.19. Less than 3% of the radioactivity was attributable to asparagine. From Meeks et al. (1978b), with permission.

photoreduce N_2 and, unlike *Azotobacter*, would not have to respire large amounts of host-derived photosynthate in order to protect nitrogenase from O_2. It seemed likely that the cyanobacterial cells would have to be "tailored" to such a function, i.e., might have to be dependent on some host-produced metabolite in order to grow in synchrony with the host.

The fact that the cells of filamentous cyanobacteria are joined together complicates attempts to obtain recessive mutants, such as auxotrophs. Three approaches have been taken to overcome this difficulty. Historically, the first approach was taken by R. N. Singh (1972), who introduced the use of cyanobacteria, e.g., *Anabaena doliolum* and *Nostoc linckia*, in which almost all of the cells form spores and, thus, separate and can be separately cloned. A second approach, taken by Wilcox, Mitchison, and Smith (1975), was to clone individual filaments selected on the basis of microscopic appearance. Our technique involves cavitation of vegetative filaments to rupture cells randomly and thereby fragment the filaments; by killing a fraction f of the vegetative cells, the average number of cells per fragment is reduced to approximately $1/f$. For *A. variabilis* (Tischer's strain A-37, isolated from a sewage oxidation pond, and later called *Anabaena flos-aquae* IUCC 1444), the resulting fragments, even when close to one cell in average length, can be plated quantitatively; the resulting clones can be replica-plated (Wolk and Wojciuch, 1973). By separating mutant and wild-type cells by cavitation and by enriching with penicillin, auxotrophic mutants and a large number of strains defective in aerobic N_2 fixation were isolated from this cyanobacterium (Currier, Haury, and Wolk, 1977). Some of the strains defective in N_2 fixation formed morphologically aberrant heterocysts. With two mutant strains, transfer to a solid medium lacking combined nitrogen led to differentiation of almost all of the vegetative cells into heterocysts. Because single cells of this organism can be plated heterotrophically, mutants defective in photoautotrophy could be isolated (Shaffer, Lockau, and Wolk, 1978). Recent isolates of potential relevance to studies of N_2 fixation include spontaneous mutants of *A. variabilis* in which the formation of heterocysts is not repressed by 2 mM NH_4^+, although nitrogenase activity is repressed under these conditions, and mutants that in the presence of 5 mM glutamate are resistant to 1 mM methionine sulfoximine.

The most thoroughly characterized of our mutants are two uracil auxotrophs, both of which lack only the first enzyme in the uridylic acid pathway, aspartic acid transcarbamylase (Currier and Wolk, unpublished data). The mutants impaired in photoautotrophy are of several types. One type, deficient in biliproteins and having low photosystem II activity at low light intensities, grows at normal rates only at high light intensities. A second mutant has a decreased growth rate and plating efficiency relative to the wild-type organism, especially at high light intensities (Shaffer, Lockau, and Wolk, 1978). Among the mutants defective in N_2 fixation, one has normal nitrogenase activity when cultured and assayed microaerobically (i.e., photosynthesizing under an O_2-free atmosphere), but very low nitrogenase activity when cultured aerobically and assayed aerobically; it is apparently defective in an O_2-protecting mechanism. Several of the mutants exhibit relatively normal nitrogenase activity if cultured microaerobically and if assayed anaerobically in the presence of dithionite, but not if assayed microaerobically; such mutants are apparently impaired in electron dona-

Table 1. Extracellular hydrolases of *Anabaena variabilis*[a]

Enzymatic activity	Method	Substrate hydrolyzed	Agar	Solution
Alkaline phosphatase	Healey and Hendzel (1975)	*p*-nitrophenyl phosphate	+	+
Protease	Smith, Gordon, and Clark (1946)	gelatin	+	
DNAse		DNA–methyl green	+	
DNAse	Currier (1976)	^{3}H-DNA		+

[a] Hydrolytic activity was demonstrated by streaking cyanobacteria onto substrate-containing, agarized growth medium and incubating in the light, or by incubation of solutions of substrate with culture filtrates. In the latter case, controls consisted of filtrates of uninoculated medium.

tion to nitrogenase (Haury and Wolk, unpublished data). We are currently trying to develop techniques for gene transfer in *A. variabilis*.

Cells of auxotrophic and wild-type strains of *A. variabilis* were introduced into tobacco protoplasts by means of polyethylene glycol, and the fate of the associations determined. Although about 25% of plant cells without internal cyanobacteria re-formed walls and grew into clones, no cell with an intracellular cyanobacterium divided. After several days, the host cell in most instances burst and died (Meeks, Malmberg, and Wolk, 1978). We have in the interim confirmed an earlier report that *A. variabilis* secretes phosphatase (Healey and Hendzel, 1975) and have, in addition, shown that it releases extracellular protease and deoxyribonuclease (Ladyman and Wolk, unpublished data; see Table 1). If derivative strains that do not secrete hydrolytic enzymes could be isolated and introduced into tobacco cells, they might not interfere with the growth of the host cells. The cyanobacterial cells that had been introduced into tobacco protoplasts did not undergo division, either. Therefore, either a more suitable host or derivative, resistant cyanobacterial strains may have to be used.

Isolated Heterocysts

The final type of large N_2-fixing "plant" toward which we have directed our efforts is an N_2-fixing *factory*. Now that we have identified certain of the principal interactions between heterocysts and vegetative cells, we are attempting to use this information to obtain prolonged nitrogenase activity of isolated heterocysts. If so, then there would be the possibility of deploying isolated heterocysts as industrial-scale catalysts of nitrogen fixation. Their virtues would be: 1) that as cells, they could potentially maintain their enzymatic activity; 2) as *nondividing* cells, they would not have the problems inherent in utilization of exponentially growing material; and 3) they can use sunlight as a source of energy. These experiments have just been initiated.

REFERENCES

Codd, G. A., and W. D. P. Stewart, 1977. Ribulose-1,5-diphosphate carboxylase in heterocysts and vegetative cells of *Anabaena cylindrica*. FEMS Microbiol. Lett. 2:247–249.

Currier, T. C. 1976. Isolation and characteristics of plastids determining pathogenicity in *Agrobacterium*. Ph.D. thesis, University of Washington.

Currier, T. C., J. F. Haury, and C. P. Wolk. 1977. Isolation and preliminary characterization of auxotrophs of a filamentous cyanobacterium. J. Bacteriol. 129:1556–1562.

Fay, P., W. D. P. Stewart, A. E. Walsby, and G. E. Fogg, 1968. Is the heterocyst the site of nitrogen fixation in blue-green algae? Nature 220:810–812.

Fogg, G. E. 1949. Growth and heterocyst production in *Anabaena cylindrica* Lemm. II. In relation to carbon and nitrogen metabolism. Ann. Bot. n.s. 13:241–259.

Healey, F. P., and L. L. Hendzel. 1975. Effect of phosphorus deficiency on two algae growing in chemostats. J. Phycol. 11:303–309.

Jüttner, F., and N. G. Carr. 1976. The movement of organic molecules from vegetative cells into the heterocysts of *Anabaena cylindrica*. In: G. A. Codd and W. D. P. Stewart (eds.), Proceedings of the 2nd International Symposium on Photosynthetic Prokaryotes, pp. 121–123. Federation of European Microbiological Societies, Dundee.

Lex, M., and N. G. Carr. 1974. The metabolism of glucose by heterocysts and vegetative cells of *Anabaena cylindrica*. Arch. Microbiol. 101:161–167.

Lockau, W., R. B. Peterson, C. P. Wolk, and R. H. Burris, 1978. Modes of reduction of nitrogenase in heterocysts isolated from *Anabaena* species. Biochim. Biophys. Acta 502:298–308.

Meeks, J. C., C. P. Wolk, J. Thomas, W. Lockau, P. W. Shaffer, S. M. Austin, W.-S. Chien, and A. Galonsky, 1977. The pathways of assimilation of $^{13}NH_4^+$ by the cyanobacterium, *Anabaena cylindrica*. J. Biol. Chem. 252:7894–7900.

Meeks, J. C., R. L. Malmberg, and C. P. Wolk. 1978. Uptake of auxotrophic cells of a heterocyst-forming cyanobacterium by tobacco protoplasts, and the fate of their associations. Planta 139:55–60.

Meeks, J. C., C. P. Wolk, W. Lockau, N. Schilling, P. W. Shaffer, and W.-S. Chien, 1978a. Pathways of assimilation of $[^{13}N]N_2$ and $^{13}NH_4^+$ by cyanobacteria with and without heterocysts. J. Bacteriol. 134:125–130.

Meeks, J. C., C. P. Wolk, N. Schilling, P. W. Shaffer, Y. Avissar, and W.-S. Chien, 1978b. The initial organic products of fixation of $[^{13}N]N_2$ by root nodules of soybean (*Glycine max*). Plant. Physiol. 61:980–983.

Ownby, J. D. 1977. Effects of amino acids on methionine-sulfoximine–induced heterocyst formation in *Anabaena*. Planta 136:277–279.

Peterson, R. B., and R. H. Burris, 1976. Properties of heterocysts isolated with colloidal silica. Arch. Microbiol. 108:35–40.

Peterson, R. B., and R. H. Burris. 1978. H_2 metabolism in isolated heterocysts of *Anabaena* 7120. Arch. Microbiol. 116:125–132.

Peterson, R. B., and C. P. Wolk. 1978. Localization of an uptake hydrogenase in *Anabaena*. Plant Physiol. 61:688–691.

Rowell, P., and W. D. P. Stewart. 1976. Alanine dehydrogenase of the N_2-fixing blue-green alga, *Anabaena cylindrica*. Arch. Microbiol. 107:115–124.

Shaffer, P. W., W. Lockau, and C. P. Wolk. 1978. Isolation of mutants of the cyanobacterium. *Anabaena variabilis*, impaired in photoautotrophy. Arch. Microbiol. 117:215–219.

Singh, R. N. 1972. Physiology and biochemistry of nitrogen fixation by blue-green algae (final technical report), 1967–1972. Banaras Hindu University, Varanasi, India.

Skokut, T. A., C. P. Wolk. J. Thomas, J. C. Meeks, P. W. Shaffer, and W.-S. Chien. 1978. The initial organic products of assimilation of $[^{13}N]$ammonium and $[^{13}N]$nitrate by tobacco cells cultured on different sources of nitrogen. Plant Physiol. 62:299–304.

Smith. N. R., R. E. Gordon, and F. E. Clark. 1946. Aerobic mesophilic sporeforming bacteria, p. 33. Misc. Publ. 559. U.S. Dept. of Agriculture, Washington, D.C.

Stewart, W. D. P., and P. Rowell. 1975. Effect of L-methionine-DL-sulphoximine on the assimilation of newly fixed NH_3, acetylene reduction and heterocyst production in *Anabaena cylindrica*. Biochem. Biophys. Res. Commun. 65:846–856.

Thomas, J., J. C. Meeks, C. P. Wolk, P. W. Shaffer, S. M. Austin, and W.-S. Chien. 1977. Formation of glutamine from $[^{13}N]$ammonia, $[^{13}N]$dinitrogen, and

[^{14}C]glutamate by heterocysts isolated from *Anabaena cylindrica*. J. Bacteriol. 129:1545–1555.

Wilcox, M., G. J. Mitchison, and R. J. Smith. 1975. Mutants of *Anabaena cylindrica* altered in heterocyst spacing. Arch. Microbiol. 103:219–223.

Winkenbach, F., and C. P. Wolk. 1973. Activities of enzymes of the oxidative and the reductive pentose phosphate pathways in heterocysts of a blue-green alga. Plant Physiol. 52:480–483.

Wolk, C. P. 1967. Physiological basis of the pattern of vegetative growth of a blue-green alga. Proc. Natl. Acad. Sci. USA 57:1246–1251.

Wolk, C. P. 1979a. Intercellular interactions and pattern formation in filamentous cyanobacteria. 37th Symposium of the Society for Developmental Biology, pp. 247–266.

Wolk, C. P. 1979b. Cyanobacteria (blue-green algae). In: P. K. Stumpf and E. E. Conn (eds.), The Biochemistry of Plants. A Comprehensive Treatise, Vol. 1. Academic Press, Inc., N.Y. In press.

Wolk, C. P., and M. P. Quine, 1975. Formation of one-dimensional patterns by stochastic processes and by filamentous blue-green algae. Dev. Biol. 46:370–382.

Wolk, C. P., J. Thomas, P. W. Shaffer, S. M. Austin, and A. Galonsky. 1976. Pathway of nitrogen metabolism after fixation of ^{13}N-labeled nitrogen gas by the cyanobacterium, *Anabaena cylindrica*. J. Biol. Chem. 251:5027–5034.

Wolk, C. P., and E. Wojciuch. 1973. Simple methods for plating single vegetative cells of, and for replica-plating, filamentous blue-green algae. Arch. Mikrobiol. 91:91–95.

Nitrogen Fixation, Volume II
Edited by W. E. Newton and W. H. Orme-Johnson

Azolla-Anabaena Association: Morphological and Physiological Studies

G. A. Peters, T. B. Ray, B. C. Mayne, and R. E. Toia, Jr.

The N_2-fixing symbiotic association between species of the eukaryotic water fern *Azolla* and its prokaryotic blue-green algal symbiont, an *Anabaena*, is found in temperate and tropical aquatic ecosystems throughout the world (Moore, 1969) and has recently undergone a transition from a botanical curiosity to a nitrogen source with ecological and agronomic significance. The potential of these associations as a nitrogen source in agriculture, especially in conjunction with rice culture, was indicated by Moore (1969) and has been substantiated by others (Talley, Talley, and Rains, 1977; Watanabe et al., 1977). We have initiated a program of studies on the *Azolla caroliniana–Anabaena azollae* association (Peters and Mayne, 1974a, 1974b; Peters, 1975, 1976, 1977; Peters, Evans, and Toia, 1976; Peters, Toia, and Lough, 1977; Peters et al., 1978; Ray et al., 1978) and the following is a synopsis of these studies.

MORPHOLOGY

The *Azolla* sporophyte consists of a branched floating stem bearing alternately arranged, deeply bilobed leaves and true roots. The ventral leaf lobes are thin, have little chlorophyll, and may be partially submerged. The dorsal lobes are aerial, chlorophyllous, and contain an internal cavity occupied by the symbiont (Figure 1).

The symbiont is associated with the dorsal leaf lobes throughout their development (Konar and Kapoor, 1972; Hill, 1975; Peters et al., 1978).

Relatively undifferentiated, actively dividing filaments are associated with the shoot apex (Figure 2). As development proceeds, the symbiont colonizes cavities formed in the ventral surface of the dorsal leaf lobes. Multicellular hairs, some of which exhibit transfer cell ultrastructure (Duckett, Toth, and Soni, 1975; Peters et al., 1978), are associated with all stages of development. Hill (1977) reported that, during leaf maturation in *Azolla filiculoides*, cell division of the symbiont is diminished and there is a concomitant cell enlargement and increased differentiation into heterocysts. Our observations indicate a comparable sequence in *Az. caroliniana* except for an increased differentiation into akinetes as well as heterocysts. In *Az. filiculoides*, the heterocyst frequency increases from near zero in the apex to above 30% in mature leaves (Hill, 1975). In *Az. caroliniana*, the frequency of cell types in the symbiont from all stages of leaf development is 62% vegetative cells, 21% heterocysts, and 17% akinetes (Peters, 1975). Although there appears to be a synchrony in the development of the host and symbiont, the factor(s) regulating the sequence are not known. However, the development of the leaf cavities and associated hair cells occurs in the absence of the symbiont (Peters, 1976).

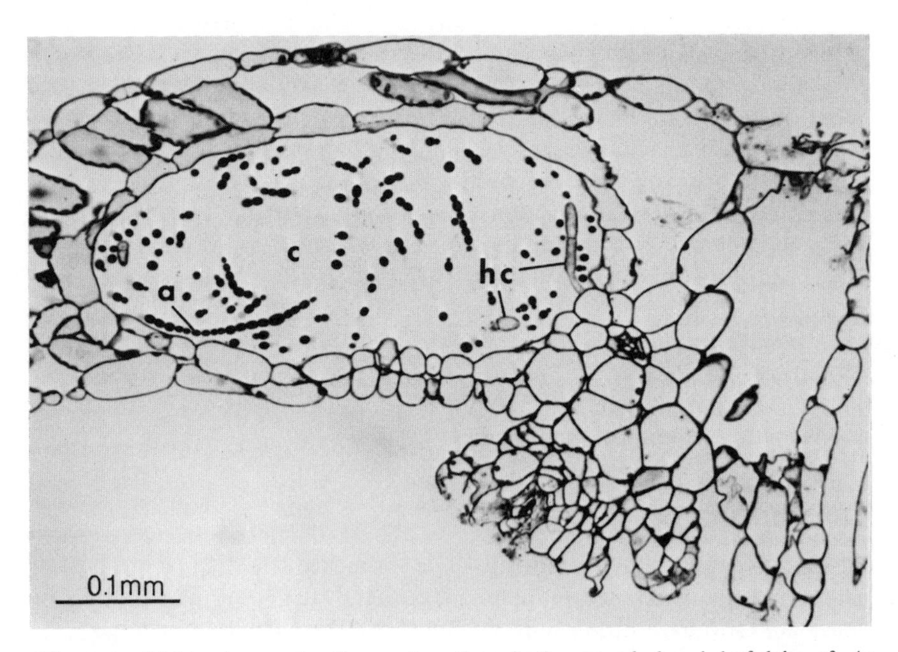

Figure 1. Light micrograph of a section through the second dorsal leaf lobe of *Az. caroliniana*. Filaments of the algal symbiont (a) and hair cells (hc) are shown within the leaf chamber (c).

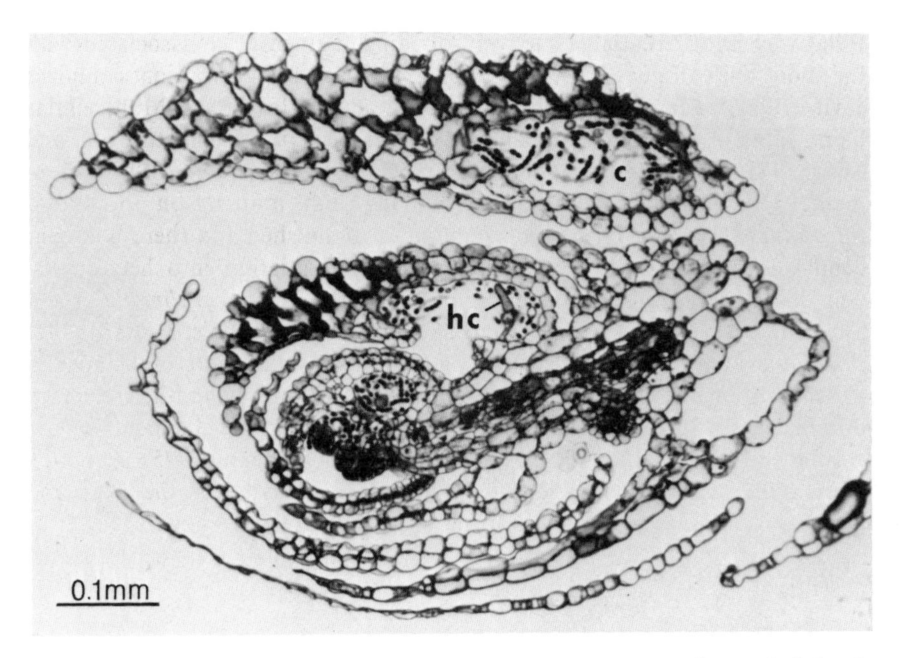

Figure 2. Light micrograph of a section through the stem tip of *Az. caroliniana* depicting the sequential development of leaf chambers in the dorsal lobes. Abbreviations as for Figure 1.

The leaf chambers in mature dorsal lobes can be isolated by digestion of the leaf tissue with cellulytic enzymes and have been termed algal packets or clumps (Peters, 1976; Peters et al., 1978). They exhibit the size and contours of the original leaf chambers and contain the symbiont and hair cells within a thin limiting envelope. A layer adjacent to the host plant cell wall that may correspond to this envelope can be observed in sections through intact leaves (Figure 3). Since there appears to be no direct contact between the vascular system and the leaf chamber (Duckett et al., 1975; Peters et al., 1978), in mature leaves the exchange of metabolites between the partners is presumed to occur through this limiting envelope and/or the hair cells. The physiological properties of the limiting envelope and hair cells have not been established.

PHYSIOLOGY

Nitrogen Fixation

In *An. filiculoides*, light-dependent C_2H_2 reduction by successive leaves from the apex paralleled the developmental pattern exhibited by the

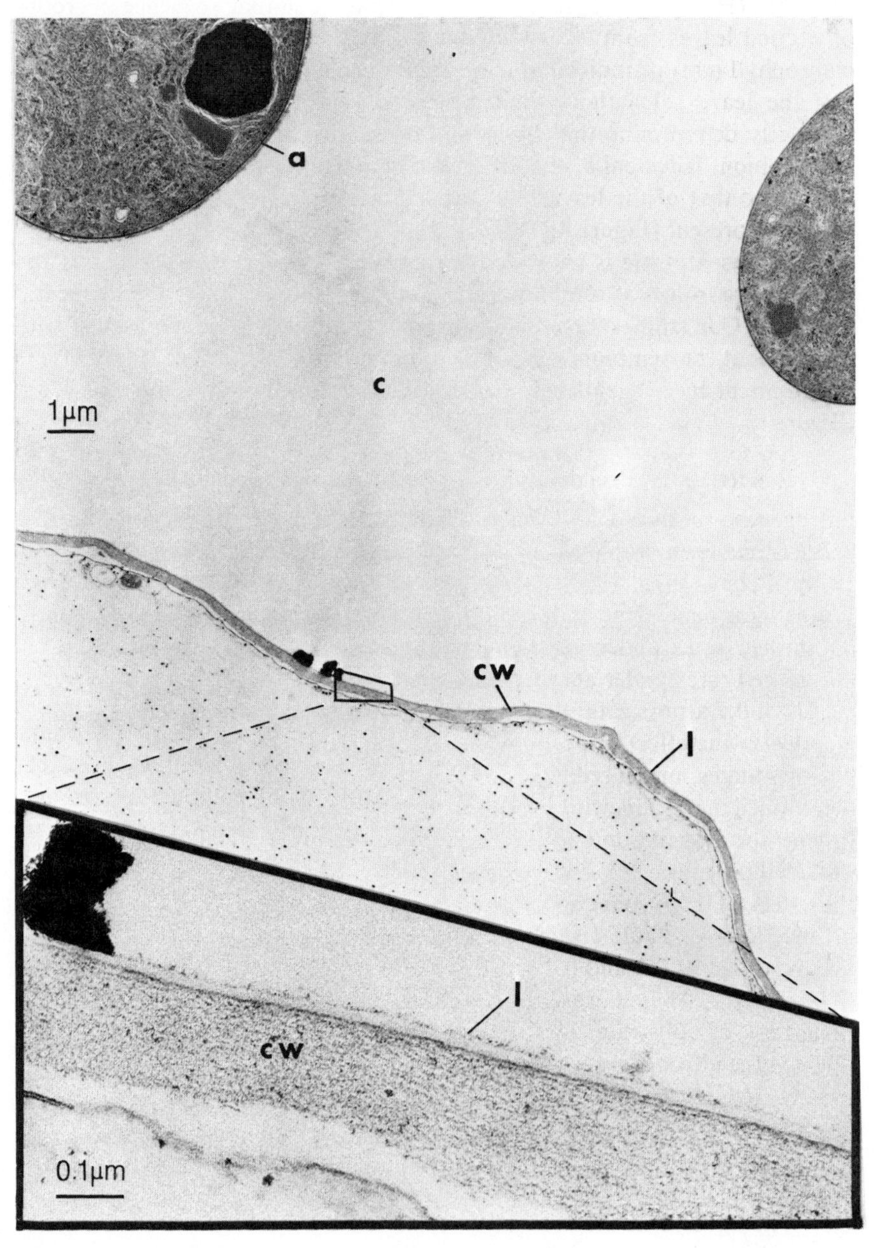

Figure 3. An electron micrograph of a thin section through the fourth dorsal lobe of *Az. caroliniana* illustrating the layer (l) adjacent to the plant cell wall (cw). This layer (inset) may correspond to the envelope surrounding algal packets obtained after enzymatic digestion of the host leaf tissue (Peters, 1976; Peters et al., 1978).

symbiont (Hill, 1977). Figure 4A demonstrates a similar sequence in groups of excised leaves from *Az. caroliniana*. The symbiont's contribution to the chlorophyll (chl) of individual leaves and the chl *a*:chl *b* ratios of the leaves and the leaves plus the symbiont were obtained by dissection and independently determining the chl content of the leaves and symbiont. Because the symbiont lacks chl *b*, the difference between the chl *a*:chl *b* ratio of the leaves and that of the leaves plus the symbiont is an index of the amount of symbiont present (Figure 4B).

At present, little is known concerning other physiological processes and the relative roles of the host and symbiont within this developmental sequence. Our studies have been concerned with intact *Azolla* plants, with and without the symbiont, and the symbiont isolated from all stages of leaf development by a "gentle" isolation procedure (Peters and Mayne, 1974b) (Figure 5).

C_2H_2 Reduction, H_2 Production, and $[^{15}N]N_2$ Fixation

Nitrogenase-catalyzed, ATP-dependent reduction of C_2H_2, protons, and $[^{15}N]N_2$ has been employed to characterize N_2 fixation (Peters and Mayne, 1974b; Peters, 1975, 1976, 1977; Peters et al., 1976, 1977). Substrate reduction is light dependent under microaerophilic conditions, occurs at slightly diminished rates under aerobic conditions in the light, and at appreciably diminished rates under aerobic dark conditions. In C_2H_2 reduction assays, a pC_2H_2 of 0.1 atm is saturating and inhibits H_2 production by 95% (Peters et al., 1977). The effect of pN_2 on H_2 production is shown in Figure 6. In companion studies, measurements of C_2H_4 production at saturating pC_2H_2 were determined in parallel with $[^{15}N]N_2$ fixed and H_2 evolved as a function of pN_2 in both the association and the symbiont (Peters et al., 1977). These studies provided ratios of C_2H_2 reduced:N_2 fixed as a function of pN_2, which reflects the effect of H_2 production on these ratios and an expression of electron balance in vivo. Table 1 shows the relationships obtained for the association and symbiont at 0.3 and 0.6 atm N_2. For the expression of electron balance, $C_2H_4/(3N_2 + H_2)$, it was assumed that total electron flow was constant, regardless of substrate, and that at a saturating pC_2H_2 all electron flow through the nitrogenase was used in its reduction. Hence, total electrons used in C_2H_2 reduction (a $2e^-$ process) should equal the sum of the electrons used in the reduction of N_2 (a $6e^-$ process) and protons (a $2e^-$ process) at any pN_2; i.e., moles C_2H_4 produced/(3 $\times$ moles N_2 fixed + moles of H_2) produced should equal unity.

When grown in the absence of combined nitrogen, rates of H_2 production under argon are variable in both the association and the symbiont but consistently less than rates of C_2H_2 reduction. The highest rates of H_2 production occur under 0.1 atm C_2H_2, 0.02 atm CO (Peters et al., 1976; 1977). These observations indicate unidirectional hydrogenase activity in the

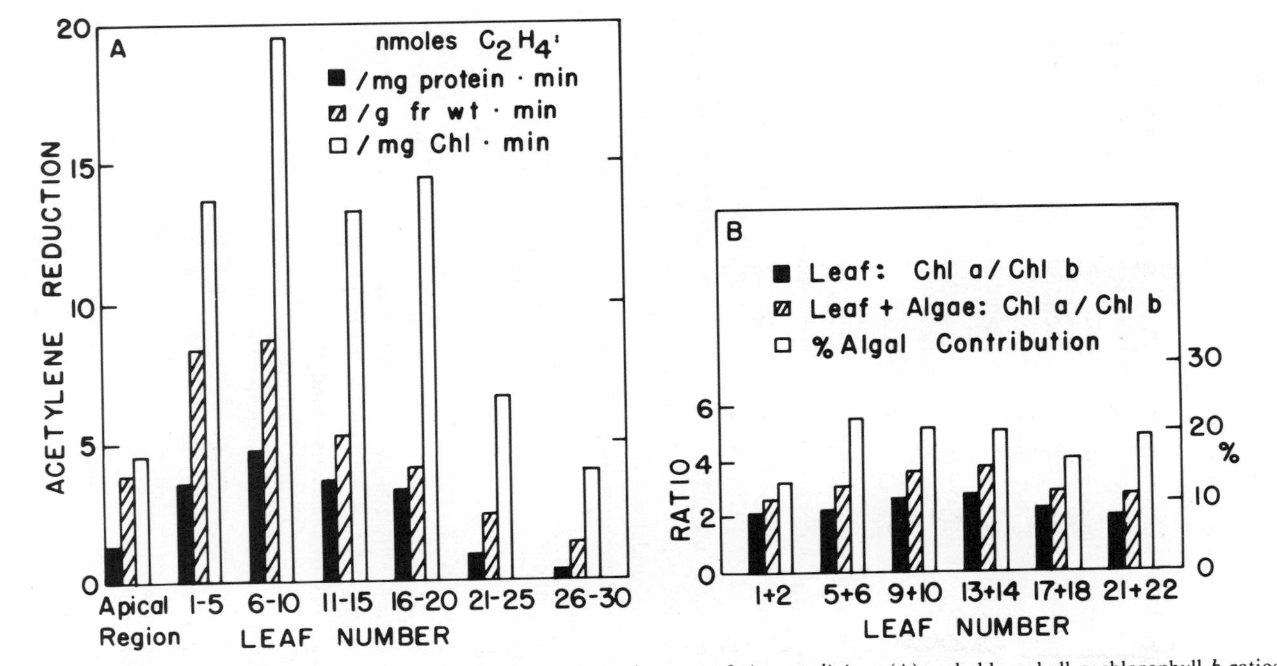

Figure 4. Light-dependent C_2H_2 reduction in successive leaves from the apex of *Az. caroliniana* (A) and chlorophyll *a*:chlorophyll *b* ratios of successive leaves from the apex with and without the symbiont as well as the percent of the total chlorophyll attributable to the symbiont (B).

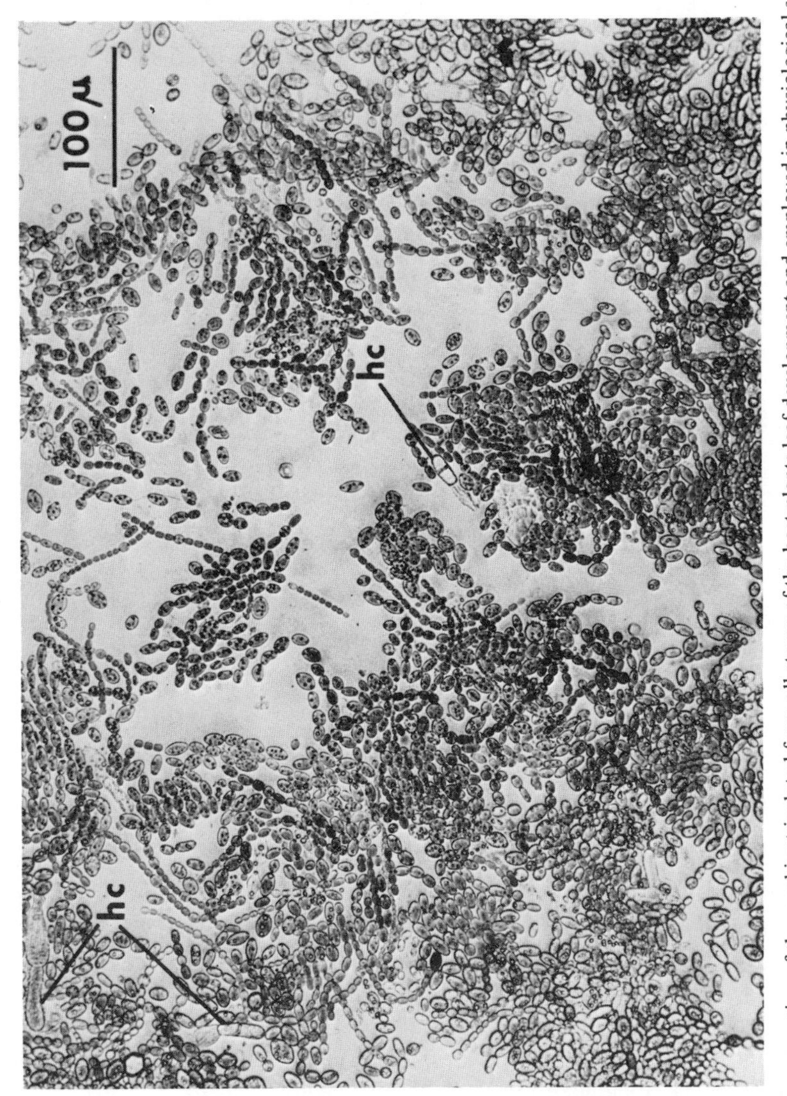

Figure 5. A preparation of the symbiont isolated from all stages of the host plants leaf development and employed in physiological studies. Some of the hair cells (hc) lining the leaf cavity are present in such preparations.

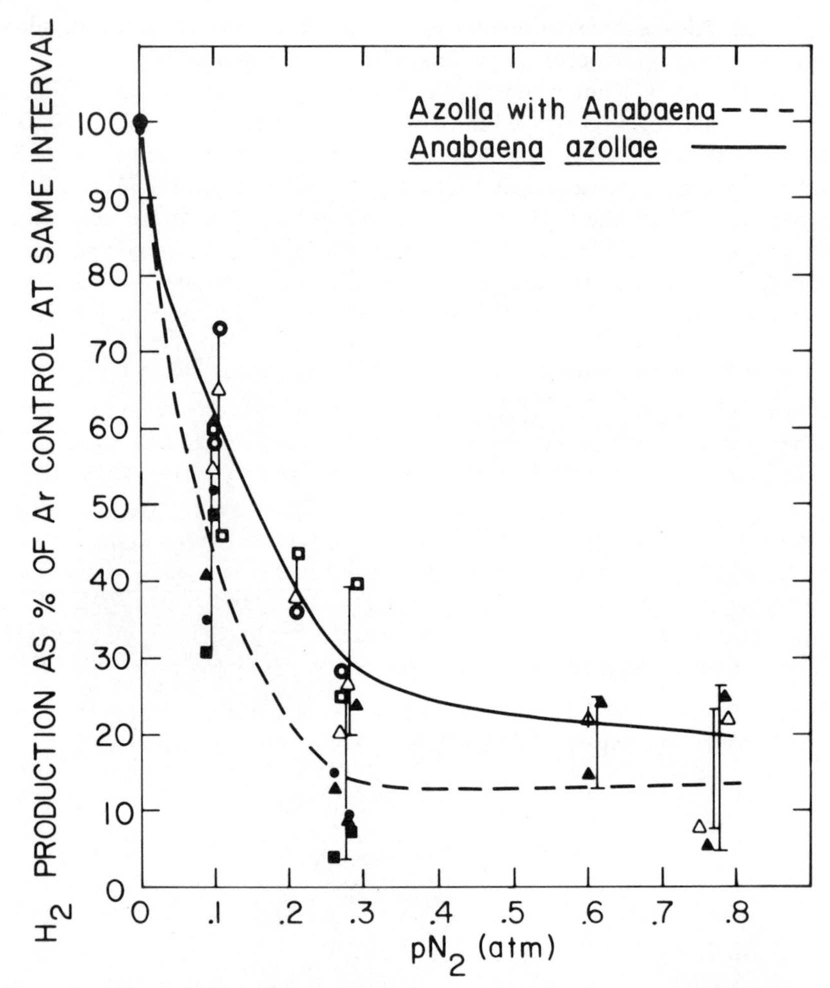

Figure 6. Effect of pN_2 on nitrogenase-catalyzed H_2 production in the *Azolla-Anabaena* association (---) and the isolated symbiont (——). Average rates of duplicate or triplicate samples at 3 hr (○/●), 5 or 6 hr (△/▲), and 9 or 10 hr (□/■) are expressed as a percentage of the rate of H_2 production under argon ($pN_2 < 0.005$ atm) controls. Marker lines indicate range at comparable partial pressures. Solid symbols represent the association, open symbols the symbiont. (From Peters et al., 1977).

symbiont (Smith, Hill, and Yates, 1976; Peters et al., 1977) with the measurable H_2 being diminished by its oxidation and a recycling of electrons into the system. Since this process cannot be 100% efficient, the expression of electron balance in vivo could presumably exceed a value of unity and still express a true balance. However, C_2H_4:N_2 values of less than 3.0 and expressions of electron balance of less than 1.0 are not readily explained. Although experimental error cannot be totally excluded, additional processes may occur in vivo that alter the stoichiometry.

In the study from which Table 1 was taken (Peters et al., 1978), a $C_2H_4:N_2$ value of 1.65 was attained for the association at 0.8 atm N_2, 0.2 atm O_2, 0.03 atm CO_2. Using this same gas phase, incubation periods of 30 and 60 min have subsequently yielded $C_2H_4:N_2$ values of 2.77 and 3.38, respectively. Thus, the $C_2H_4:N_2$ values obtained for *Azolla* are well within the range of values reported for other N_2-fixing organisms and agree remarkably well with the values of 1.6, 2.4, and 3.4 obtained by Watanabe et al. (1977) using C_2H_2 reduction and nitrogen increases per g fresh weight.

Effect of Combined Nitrogen

Appreciable nitrogenase activity is retained by the symbiont when the association is grown on nitrate or urea (Walmsley, Breen, and Kyle, 1973; Peters and Mayne, 1974b; Hill, 1975; Newton, 1976; Peters et al., 1976) and growth rates are similar whether the nitrogen source is N_2, combined nitrogen, or both (Peters et al., 1976). In contrast to the association grown on N_2, ATP-dependent H_2 production under argon can equal the rate of C_2H_2 reduction when the association is grown on nitrate (Newton, 1976; Peters et al., 1976), and there is no increase in H_2 production under C_2H_2-CO relative to that under argon (Peters et al., 1976; Peters, 1977).

Table 1. A comparison of C_2H_2 reduction, H_2 production under argon and 0.29 and 0.60 atm N_2, the C_2H_2:N_2 ratios and electron balance, i.e., $C_2H_4/(3N_2 + H_2)$, for the association and isolated symbiont[a]

Product (or ratio)	Association (nmol/mg of Chl·min)	Isolated symbiont (nmol/mg of Chl·min)
C_2H_4 (light)	13.96	92.97
C_2H_4 (dark)	0.06	0.35
H_2($pN_2 < 0.005$ atm)	8.58	89.97
H_2($pN_2 = 0.29$ atm)	2.02	35.90
H_2($pN_2 = 0.60$ atm)	1.31	20.21
N_2 fixed ($pN_2 = 0.29$ atm)	4.35	21.30
N_2 fixed ($pN_2 = 0.60$ atm)	6.91	30.76
$C_2H_2:N_2$ ($pN_2 = 0.29$ atm)	3.21	4.36
$C_2H_2:N_2$ ($pN_2 = 0.60$ atm)	2.02	3.02
$C_2H_4/(3N_2 + H_2)$ ($pN_2 = 0.29$ atm)	0.93	0.93
$C_2H_4/(3N_2 + H_2)$ ($pN_2 = 0.60$ atm)	0.63	0.83

From Peters, Toia, and Lough (1977).

[a] The incubation period was 5 hr at 500 ft-c, 28°C. The atom percent of excess ^{15}N in the alga was 0.413 at 0.29 atm N_2 and 0.386 at 0.60 atm N_2; for the association it was 0.219 at 0.29 atm N_2 and 0.258 at 0.6 atm N_2, based on an average of four to six determinations with and without carrier nitrogen.

Moreover, whereas atmospheric N_2 strongly inhibits H_2 production by N_2-grown plants, in nitrate-grown plants rates of H_2 production under air may be 40% of the rate of C_2H_2 reduction or H_2 production under argon (Newton, 1976; Peters et al., 1976). Although there is often an appreciable increase in H_2 production under C_2H_2-CO relative to that under argon in N_2-grown fronds, the rate of H_2 production is still significantly less than the rate of C_2H_2 reduction. These findings led to the suggestion that there may be two types of unidirectional hydrogenase in this association, one of which is inhibited by C_2H_2 and both of which are controlled, or inactivated, by a product of nitrate or nitrite metabolism (Peters, 1977). $[^{15}N]N_2$ fixation and studies of electron balance have not been conducted on the nitrate-grown association. However, since the symbiont retains nitrogenase activity and the normal, high heterocyst frequency (Hill, 1975) when the association is grown on nitrate, the increase in net H_2 production under these growth conditions may reflect a control mechanism.

Transfer of Fixed Nitrogen between Symbiont and Host

$[^{15}N]N_2$ studies have shown that, when removed from the leaf cavities, the symbiont releases up to 60% of the N_2 it fixes, predominantly as ammonia, with only small amounts of organic nitrogen (Table 2; Peters, 1977). Additional studies have compared ^{14}C labeling in the supernatants of the symbiont incubated with $^{14}CO_2$ under argon and N_2 atmospheres. These have shown NH_3 release under N_2 but not argon, and no significant differences in the ^{14}C content of the supernatants, both containing less than 5% of the incorporated ^{14}C. These data further demonstrate that the symbiont releases few or no amino acids. In related studies, the association was incubated under $[^{15}N]N_2$ for 3 hr followed by either immediate fractionation or fractionation after an additional incubation under room air for 1 and 2 hr. The distribution of the fixed ^{15}N among ammonia, ethanol-soluble, and ethanol-insoluble fractions after the three treatments has clearly demonstrated a sequential transfer of the fixed nitrogen into ethanol-soluble (amino acids) and then ethanol-insoluble (protein) fractions and indicated a relatively small free ammonia fraction accounting for about 6% of the total fixed ^{15}N. Although the label in the ammonia pool did not drop as quickly as expected during the chase periods with air, there was a direct correlation between the decrease in the ethanol-soluble and the increase in the ethanol-insoluble fractions. Newton and Cavins (1976) found that ammonia could account for up to half of the total pool nitrogen (based on amino acid and ammonia analysis data) when this N_2-grown association was extracted with boiling water. They obtained about 550 nmol of microdiffused NH_3 per g fresh wt. Using boiling 80% ethanol extraction followed by microdiffusion of ammonia and Kjeldahl digestion of both the remaining soluble and precipitated plant material, we obtain values

Table 2. Distribution of ^{15}N-labeled products during N_2 fixation by the isolated symbiont[a]

Sample	Atom % of excess ^{15}N	nmol of N_2 fixed	Distribution	Rate (nmol of N_2/mg of Chl·min)
Complete	0.345	119.52	—	10.23
Fractions				
(a) extracellular NH_3	18.756	37.11	34.72%	—
(b) intracellular NH_3	9.535	8.49	7.94%	—
(c) extracellular organic nitrogen	2.667	7.08	6.62%	—
(d) intracellular organic nitrogen	0.139	54.89	50.72%	—
Total fractions		106.89		10.39

[a] Experimental procedures essentially as described in Peters (1977), except for being based on a greater number of determinations.

between 650 and 1150 nmol of NH_3 per g fresh wt of tissue, which accounts for less than 1% of the total nitrogen and 5%–12% of the soluble nitrogen. (It should be noted that our soluble nitrogen fractions contain other compounds in addition to amino acids, whereas Newton and Cavins [1976] value of 50% is based solely on amino acid nitrogen.)

Thus, although it is not yet possible to state that fraction of the total fixed N_2 that is assimilated by the individual partners in the association, it is highly probable that ammonia released by the symbiont is the major nitrogen source for the host plant when grown in the absence of combined nitrogen.

Both the host and the symbiont have the capacity to metabolize ammonia to amino acids, as is apparent from the activities of the ammonia-assimilating enzymes. Glutamine synthetase (GS), glutamate synthase (GOGAT), and glutamate dehydrogenase (GDH) were measured in the association, the symbiont, and the algal-free plants (Table 3). The distribution of these enzymes between the host and symbiont was determined (Table 4) with respect to phycocyanin content. The symbiont was found to account for 16% of the association's protein and chlorophyll. Using these values, C_2H_2 reduction rates of the association were estimated on an algal chlorophyll or protein basis for comparison with the isolated symbiont (Table 5). The difference in rates of C_2H_2 reduction may reflect some filament breakage or the special microenvironment of the leaf chamber.

As noted earlier, changes in the symbiont's morphology and C_2H_2 reduction capacity as a function of leaf development in the host suggest the possibility of changes in other physiological and biochemical processes of the individual partners within this developmental sequence. The values shown in Tables 3 and 4 merely reflect an average of all stages of development, whereas GS and GDH activities may be associated with specific developmental stages. The difference in the affinities of these two enzymes for ammonia and their partitioning between host and symbiont as a function

Table 3. Specific activities of ammonia-assimilating enzymes in extracts of the *Azolla-Anabaena* association, the symbiotic endophyte, and algal-free *Azolla*

	Glutamine synthetase[a]			
Enzyme	Coupled biosynthetic activity	Transferase activity	GDH[a]	GOGAT[a]
Association	26	417	46	51
Symbiotic *An. azollae*	11	154	45	33
Algal-free *Azolla*	6	197	41	—

From Ray et al. (1978).

[a] nmole/min · mg of protein.

Table 4. Specific activities, total activities, and percent of GS (biosynthetic) and GDH in the *Azolla-Anabaena* association, the *Azolla* host, and the *Anabaena* symbiont (specific activities are averages of three determinations ± the standard deviation)

	Association	*Azolla* (host)	*Anabaena* (symbiont)
GS Activity			
Specific[a]	15 ± 5	16 ± 6	11 ± 4
total[b]	188 ± 35	168 ± 31	20 ± 13
% of total	100	89	11
GDH Activity			
Specific[a]	34 ± 15	24 ± 6	40 ± 10
total[b]	177 ± 16	219 ± 20	58 ± 22
% of total	100	79	21

From Ray et al. (1978).

[a] nmol/mg of protein · min.

[b] nmol/g fresh wt of association · min.

of leaf age could conceivably provide a regulatory mechanism for cross-feeding in this association.

Photosynthesis

The *Azolla-Anabaena* association and its components utilize C_3-type carbon fixation as demonstrated by the distribution of ^{14}C label shortly after adding $^{14}CO_2$ to photosynthesizing *Azolla-Anabaena*, algal-free *Azolla*, and the isolated *An. azollae*. The *Azolla-Anabaena* complex and the algal-free *Azolla* show the usual characteristics of photorespiration, i.e., a high CO_2 compensation point that is dependent on pO_2, and the inhibition of photosynthesis by oxygen. The isolated blue-green algal symbiont, however, does not show the usual characteristics of photorespiration, although it is also

Table 5. Acetylene reduction in the *Azolla-Anabaena* association, the isolated algae, and in the association when based on algal chlorophyll content

	nmol of C_2H_4/mg of chl · min		
Experiment	Association	Isolated algae	Association based on algal chlorophyll
1	11.9	79.6	131
2	19.9	87.7	199
3	21.6	109.0	137
Average	17.8 ± 5.2	92.1 ± 15.1	155 ± 37

From Ray et al. (1978).

the C_3 type, in agreement with Bidwell (1977) and Lloyd, Canvin, and Culver (1977).

Interactions: Photosynthesis and N_2 Fixation; Host-Symbiont Carbon Metabolism

As in free-living, N_2-fixing blue-green algae (Stewart, Rowell, and Apte, 1977), photosynthesis is the ultimate source of electrons and a direct source of ATP for nitrogenase activity in this association. Although dark, aerobic N_2 fixation occurs, it is dependent on endogenous reserves of photosynthate, as demonstrated by varying the length of light or dark aerobic incubation prior to assaying. Moreover, the diminished rates of aerobic dark C_2H_2 reduction as compared to those under aerobic light conditions imply that dark aerobic nitrogenase activity is ATP limited.

Although photosystem II activity is required to provide photosynthate for reducing power, noncyclic photophosphorylation is not a principal source of ATP for nitrogenase activity. Action spectra for O_2 evolution by the symbiont alone indicate that the phycobilins are the principal light-harvesting pigments for O_2 evolution (Ray, unpublished data). This process is inhibited by DCMU (an inhibitor of CO_2 fixation and noncyclic photophosphorylation) but there is no comparable direct effect on C_2H_2 reduction (Peters, 1975). This observation indicates that reductant provided by prior photosynthesis and ATP from cyclic photophosphorylation are the driving forces of nitrogenase activity. Moreover, simultaneous measurements of CO_2 fixation and C_2H_2 reduction *on the same sample* confirm the role of photosynthesis in providing ATP and reductant to nitrogenase in the association. Figure 7 (A and B) compares C_2H_2 reduction and CO_2 fixation after addition of DCMU to the association that had been previously maintained for 24 hr under aerobic-light conditions (Figure 7A) and 16.5 hr under aerobic-dark conditions (Figure 7B). The inhibition of CO_2 fixation by DCMU is the same regardless of the pretreatment. In the dark pretreatment, an inhibition of C_2H_2 reduction nearly parallels the inhibition of CO_2 fixation. The lessened inhibition of C_2H_2 reduction by DCMU in the plants receiving light pretreatment suggests that adequate reserves of photosynthate for reductant are available to maintain nitrogenase activity. The effect of DCMU on C_2H_2 reduction is thus an indirect one. These endogenous supplies of reductant are diminished by aerobic-dark pretreatment, resulting in a direct dependence of C_2H_2 reduction on CO_2 fixation. A rapid decline in C_2H_2 reduction when the light-pretreated association is placed in darkness (not shown) further demonstrates that cyclic photophosphorylation, and not oxidative phosphorylation, supplies the majority of the ATP for nitrogenase activity.

Indirect evidence indicates a possible interaction of fern-algal carbon metabolism, which includes the diminished rates of C_2H_2 reduction by the

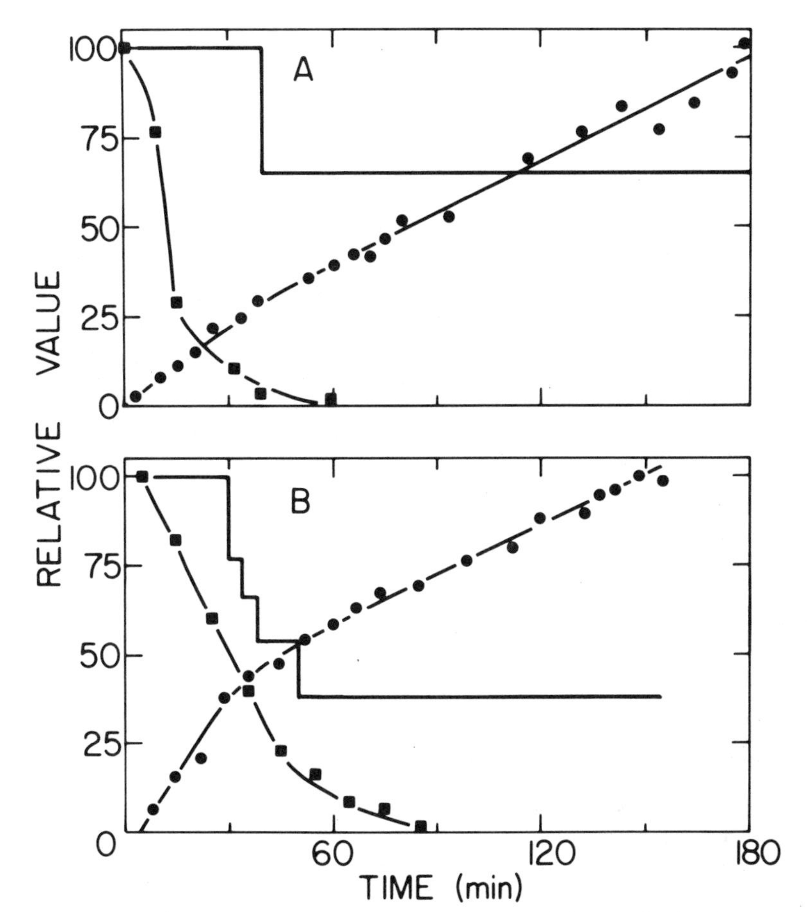

Figure 7. Simultaneous determinations of CO_2 fixation and C_2H_2 reduction in the same sample after addition of DCMU (10 μM) to the *Azolla-Anabaena* association, which had been maintained for either 24 hr under aerobic light conditions (A) or 16.5 hr aerobic dark conditions (B) prior to the addition of DCMU (t_0). The absolute values corresponding to 100% are: (A) 1.04 μmol of CO_2 fixed/mg of chl/min (■); 4.87 μmol of C_2H_4/mg of chl/min (●); and 35.7 nmol of C_2H_4/mg of chl/min (——); (B) 0.84 μmol of CO_2/mg of chl/min (■); 1.37 μmol of C_2H_4/mg of chl/min (●); and 16.70 nmol of C_2H_4/mg of chl/min (——).

algal symbiont after its removal from the host plant as compared to estimates of its C_2H_2 reduction in the leaf cavities (Ray et al., 1978) and the differences in the time required for DCMU inhibition of C_2H_2 reduction by the association and isolated algal symbiont (Peters, 1976). Furthermore, since heterocysts lack photosystem II capability (Stewart et al., 1977) and the heterocyst frequency of the symbiont can reach 30% in mature leaf cavities, it might be questioned whether or not the symbiont by itself has sufficient CO_2 fixation capability to generate the amount of reducing power required to maintain high levels of nitrogenase at these heterocyst frequencies. In free-living blue-green algae, the heterocyst frequencies seldom exceed 10% and are generally less. Thus, although there are at least 10 vegetative cells per heterocyst in free-living species, maximum C_2H_2 reduction is indicated to occur in leaf cavities where the vegetative cell: heterocyst ratio is only 2:3 (Hill, 1977; also see Peters, 1975). It is important to note that, where heterocyst frequencies comparable to those present in this association occur, namely, in the symbionts of liverworts (Rodgers and Stewart, 1977) and *Gunnera* (Silvester, 1975), the symbionts receive fixed carbon from the host plant. In contrast to the algal symbiont in *Azolla*, these algal symbionts exhibit no photosystem II activity and have little or no detectable phycobilins. Additional studies are required to determine the extent, if any, of the interaction of carbon metabolism by this association, in which both partners are photosynthetically competent.

ACKNOWLEDGMENTS

The authors wish to thank Dr. H. Calvert for the micrographs. This paper is contribution No. 624 from the Charles F. Kettering Research Laboratory.

REFERENCES

Bidwell, R. G. S. 1977. Photosynthesis and light and dark respiration in freshwater algae. Can. J. Bot. 55:809–818.

Duckett, J. G., R. Toth, and S. C. Soni, 1975. An ultrastructural study of the *Azolla, Anabaena azollae* relationship. New Phytol. 75:111–118.

Hill, D. J. 1975. The pattern of development of *Anabaena* in the *Azolla-Anabaena* symbiosis. Planta 122:178–184.

Hill, D. J. 1977. The role of *Anabaena* in the *Azolla-Anabaena* symbiosis. New Phytol. 78:611–616.

Konar, R. N., and R. K. Kapoor. 1972. Anatomical studies on *Azolla pinnata.* Phytomorphology 22:211–223.

Lloyd, N. D. H., D. T. Canvin, and D. A. Culver. 1977. Photosynthesis and photorespiration in algae. Plant Physiol. 59:936–940.

Moore, A. W. 1969. *Azolla:* Biology and agronomic significance. Bot. Rev. 35:17–34.

Newton, J. W. 1976. Photoproduction of molecular hydrogen by a plant-algal symbiotic system. Science 169:559–560.

Newton, J. W., and J. F. Cavins. 1976. Altered nitrogenous pools induced by the *Azolla–Anabaena Azolla* symbiosis. Plant Physiol. 58:798–799.

Peters, G. A. 1975. The *Azolla–Anabaena azollae* relationship. III. Studies on metabolic capabilities and further characterization of the symbiont. Arch. Microbiol. 103:113–122.

Peters, G. A. 1976. Studies on the *Azolla–Anabaena azollae* symbiosis. In: W. E. Newton and C. J. Nyman (eds.), Proceedings of the 1st International Symposium on Nitrogen Fixation, Vol. 2, pp. 592–610. Washington State University Press, Pullman.

Peters, G. A. 1977. The *Azolla–Anabaena azollae* symbiosis. In: A. Hollaender (ed.), Genetic Engineering for Nitrogen Fixation, pp. 231–258. Plenum Publishing Corp., New York.

Peters, G. A., and B. C. Mayne. 1974a. The *Azolla–Anabaena azollae* relationship. I. Initial characterization of the association. Plant Physiol. 53:813–819.

Peters, G. A., and B. C. Mayne. 1974b. The *Azolla–Anabaena azollae* relationship. II. Localization of nitrogenase activity as assayed by acetylene reduction. Plant Physiol. 53:820–824.

Peters, G. A., W. R. Evans, and R. E. Toia, Jr. 1976. *Azolla–Anabaena azollae* relationship. IV. Photosynthetically driven, nitrogenase-catalyzed H_2 production. Plant Physiol. 58:119–126.

Peters, G. A., R. E. Toia, Jr., and S. M. Lough. 1977. The *Azolla–Anabaena azollae* relationship. V. $^{15}N_2$ fixation, acetylene reduction and H_2 production. Plant Physiol. 59:1021–1025.

Peters, G. A., R. E. Toia, Jr., N. J. Levine, and D. Raveed. 1978. The *Azolla–Anabaena azollae* relationship. VI. Morphological aspects of the association. New Phytol. 80:583–593.

Ray, T. B., G. A. Peters, R. E. Toia, Jr., and B. C. Mayne. 1978. The *Azolla–Anabaena azollae* relationship. VII. Distribution of ammonia assimilating enzymes, protein and chlorophyll between host and symbiont. Plant Physiol. 62:463–467.

Rodgers, G. A., and W. D. P. Stewart. 1977. The cyanophyte-hepatic symbiosis. I. Morphology and physiology. New Phytol. 78:441–458.

Silvester, W. B. 1975. Endophyte adaptation in *Gunnera-Nostoc* symbiosis. In: P. S. Nutman (ed.), Symbiotic Nitrogen Fixation in Plants, pp. 521–538. Cambridge University Press, New York.

Smith, L. A., S. Hill, and M. G. Yates. 1976. Inhibition by acetylene of conventional hydrogenase in nitrogen-fixing bacteria. Nature 262:209–210.

Stewart, W. D. P., P. Rowell, and S. K. Apte. 1977. Cellular physiology and the ecology of N_2-fixing blue-green algae. In: W. E. Newton, J. R. Postgate, and C. Rodriguez-Barrueco (eds.), Recent Developments in Nitrogen Fixation, pp. 287–307. Academic Press, Inc., New York.

Talley, S. N., B. J. Talley, and D. W. Rains. 1977. Nitrogen fixation by *Azolla* in rice fields. In: A. Hollaender (ed.), Genetic Engineering for Nitrogen Fixation, pp. 259–281. Plenum Publishing Corp., New York.

Walmsley, R. D., C. M. Breen, and E. Kyle, 1973. Aspects of the fern-algal relationship in *Azolla filiculoides*. Limnol. Soc. S. Afr. 20:13–21.

Watanabe, I., C. R. Espinas, N. S. Berja, and B. V. Alimagno. 1977. Utilization of the *Azolla-Anabaena* complex as a nitrogen fertilizer for rice. IRRI Research Paper Series 11:1–15.

Nitrogen Fixation, Volume II
Edited by W. E. Newton and W. H. Orme-Johnson
Copyright 1980 University Park Press Baltimore

Azolla as a Nitrogen Source for Temperate Rice

S. N. Talley and D. W. Rains

Azolla is a genus of fast-growing floating aquatic ferns widely distributed throughout tropic and temperate fresh waters. It is one of only a few vascular plants now known to exist in intimate association with a blue-green alga capable of reducing molecular nitrogen (Peters, 1976). This attribute allows *Azolla* to grow without being limited by the availability of reduced or oxidized nitrogen ions in its immediate environment. For these reasons, *Azolla* is an attractive candidate for photosynthetic production of fertilizer nitrogen.

The tropical *Azolla pinnata* has been domesticated for several centuries in Vietnam, where it is now grown as a fertilizer crop on over 400,000 hectares of fallow rice fields, providing sufficient nitrogen to produce 5 metric tons/ha of paddy yield (Tran and Dao, 1973). Twenty years after its introduction into mainland China, *A. pinnata* is cultivated as the principal nitrogen source for approximately 1,300,000 hectares of subtropical and temperate rice. Yields are reported to be higher than those in Vietnam (FAO, 1977). Tropical *Azolla* culture has also been incorporated into rice culture in parts of Thailand (Moore, 1969) and Indonesia (Saubert, 1949) and is currently being investigated in India (Singh, 1977), the Republic of China (Lin, 1976), and the Philippines (Watanabe et al., 1977).

Dependence of commercial ammonia production on natural gas and the uncertain future cost and availability of this commodity have led to our

---

Research supported in part by grant NSF-RANN AER 77-07301 to D. W. Rains and in part by the Department of Agronomy and Range Science and the Plant Growth Laboratory of the University of California, Davis. Any opinions, findings, and conclusions or recommendations expressed in this publication are those of the authors and do not necessarily reflect the views of the National Science Foundation.

interest in *Azolla* cultivation in temperate regions. During the past year, our studies have pursued two main objectives: to define the potential of *Azolla* as a nitrogen source for temperate rice culture; and to define conditions necessary to optimize growth and nitrogen fixation by *Azolla*.

GROWTH AND NITROGEN FIXATION BY *AZOLLA* IN RICE FIELDS

Two species of *Azolla* are indigenous to the California rice growing region, *A. filiculoides* and *A. mexicana*. *A. filiculoides* is frost tolerant and could be cultivated as a fallow season (November through April) green manure crop that would be incorporated into the soil during spring field preparation. During summer, both *A. filiculoides* and *A. mexicana* could be grown as a companion crop with rice (Talley, Talley, and Rains, 1977).

Fallow season studies have been conducted during the springs of 1977 and 1978. Fifty grams fresh wt/m^2 of *A. filiculoides* (representing 1.5 kg of nitrogen/ha) were inoculated onto flooded rice paddies. This inoculum covered less than 10% of the water's surface, but within 15 days had attained 100% cover and increased its nitrogen content nearly ninefold (Figure 1). Nitrogen production by *Azolla* over the next 10 days averaged 2.5 kg/ha/day. This increase was paralleled by a significant increase in nitrogenase activity as measured by the acetylene reduction technique. Thirty-five days after inoculation onto the paddies, the *Azolla* cover was mature and possessed a biomass of 1700 kg dry wt/ha, which contained 52 kg/ha of nitrogen (Figure 1). Subsequent experiments revealed that mature *A. filiculoides* had lost the capacity for further growth and that there had been a precipitous decline in acetylene reduction activity. Maturation was followed by a gradual, almost linear decline in nitrogen content of the *Azolla* biomass between 35 and 70 days from the experiment's start.

Strong winds are common throughout the rice fallow season in California. In this experiment, *A. filiculoides* was repeatedly observed to compact along the boundaries of the 12-ft by 20-ft (22.4 m^2) study plots prior to formation of 100% cover, but was redistributed over the plot area during calm periods between storms. We believe this would not have occurred if the study areas had been several hectares in size. Studies conducted between March 24 and May 8, 1978, revealed that fallow season weeds and/or burned rice stubble from the previous season provided an effective wind barrier, suggesting that large areas could be planted with *A. filiculoides* with little, if any, advance field preparation. At maturity, the *A. filiculoides* biomass from this second fallow season experiment was 2276 kg dry wt/ha and contained 93 kg/ha of nitrogen. This *Azolla* cover has now been dried, incorporated into the soil, and is being used to grow rice. In a subsequent experiment, yield effects of 40 kg of nitrogen/ha of *A. filiculoides* incorporated into paddy soil during spring field preparation were the same,

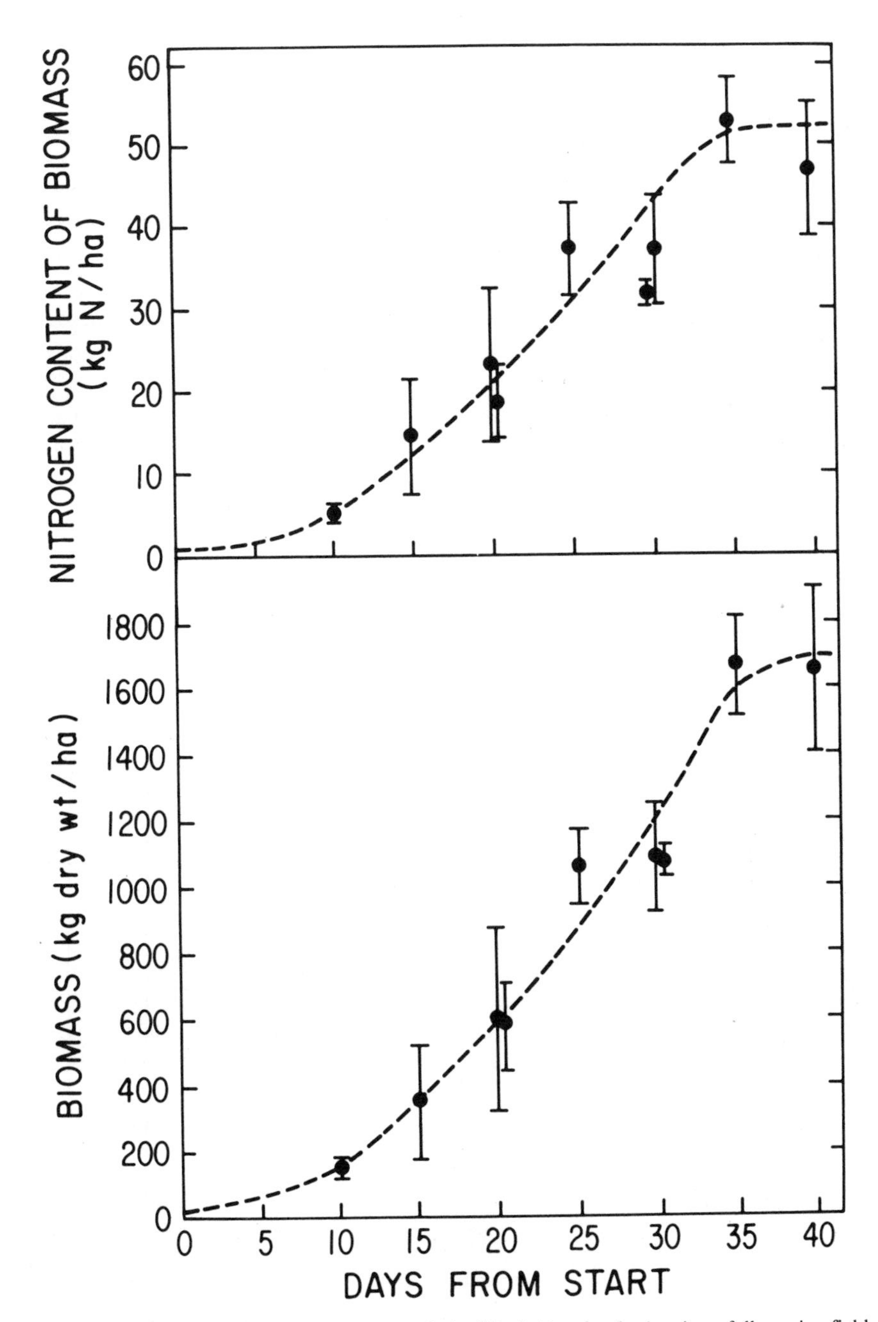

Figure 1. Biomass and nitrogen content of *A. filiculoides* developing in a fallow rice field (spring, 1977).

kilogram per kilogram, as ammonium sulfate (Figure 2), which suggests that *Azolla* is readily decomposed to ammonia, which is as available as the inorganic ammonia fertilizer.

Nitrogen contents of rice receiving dual crops of *A. filiculoides* and *A. mexicana* were 10 and 35 kg/ha, respectively, greater than unfertilized controls (Figure 2). This result represents an amount of nitrogen equivalent to 30% of the peak *A. filiculoides* biomass and 80% of the *A. mexicana* biomass (grown as two crops) associated with dual culture plots. Given the relative inefficiency of ammoniacal nitrogen transfer from water to rice (Mikkelsen and Finfrock, 1957), it is unlikely that the entire nitrogen increase resulting from *A. mexicana* dual culture could be derived from the fern cover.

Field and laboratory data suggest that ammoniacal nitrogen is released directly into the water from the developing *A. mexicana* cover. Although ammonium content of paddy water subject to dual culture is low (50 to 250

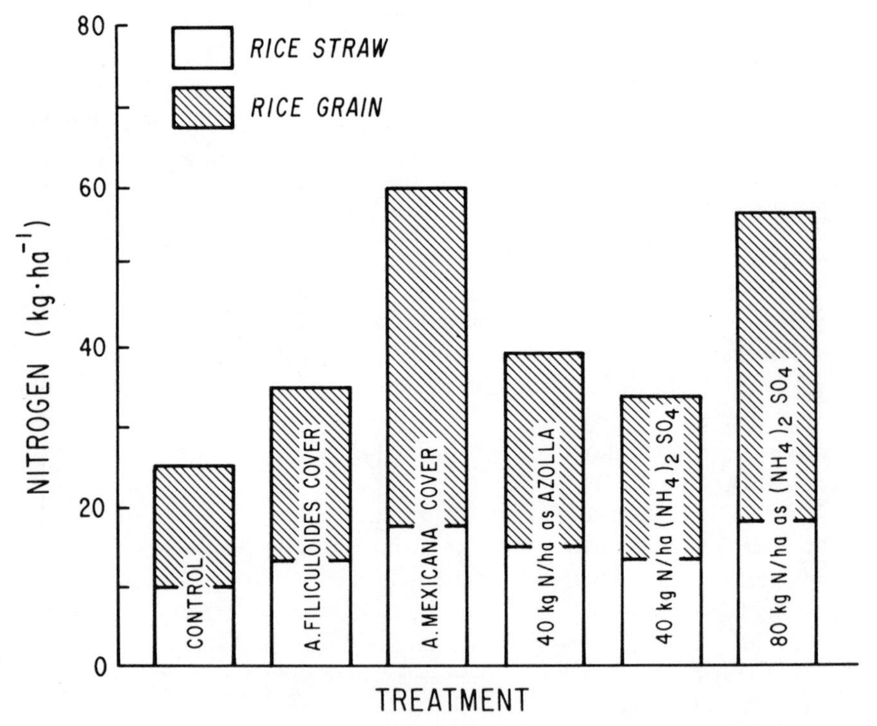

Figure 2. Nitrogen yield of rice straw and rice grain for unfertilized control, *A. filiculoides* cover (dual culture with rice), *A. mexicana* cover, *A. filiculoides* green manure, and ammonium sulfate incorporated into soil. Maximum nitrogen content of the *A. filiculoides* cover was 33 kg of nitrogen/ha. Cover for *A. mexicana* was 44 kg of nitrogen/ha as two successive crops (summer, 1977).

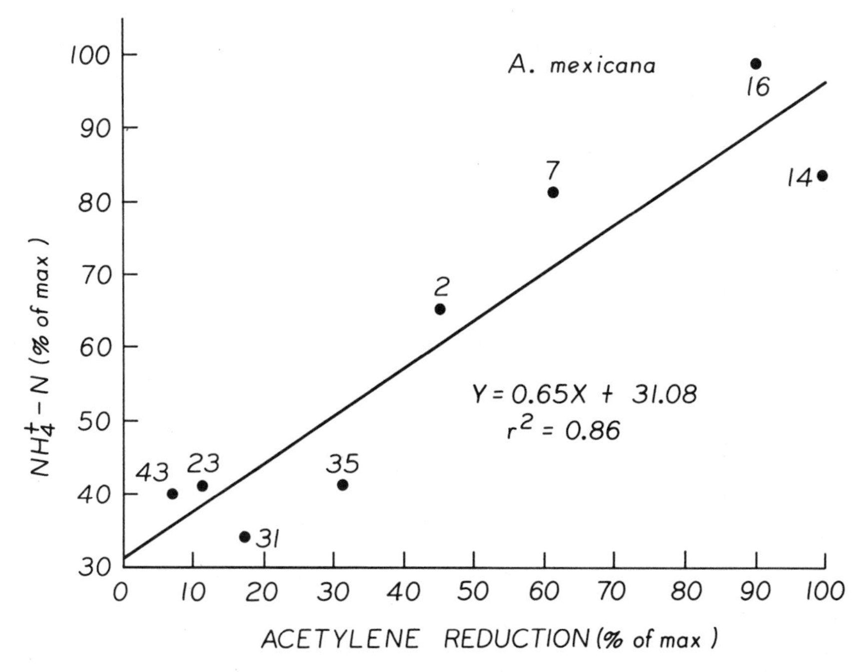

Figure 3. The relationship between total diurnal acetylene reduction activity of *A. mexicana* per unit area of paddy surface and average concentration of ammoniacal nitrogen in paddy water for the same periods. Numbers beside points indicate the days elapsed since *Azolla* was inoculated onto the paddy (summer, 1977).

μg of NH_4^+-N/liter), average concentration of ammoniacal nitrogen in these waters and total acetylene reduction activity per unit area of paddy surface exhibit strong positive correlation if these two parameters are summed over diurnal periods (Figure 3). Ammoniacal nitrogen in paddies with mature (31- or 35-day-old) or senescing (43-day-old) *A. mexicana* is very low (less than 50 μg of NH_4^+-N/liter).

Laboratory cultures of *A. mexicana* in closed, aerated flasks filled with sterile nitrogen-free nutrient media produce significant quantities of ammoniacal nitrogen (Figure 4). Ammonium release appears to be depressed at low light (130 μEinsteins$\cdot m^{-2}\cdot sec^{-1}$), but ranged between 2 and 3 mg NH_4^+-N/liter within 3 weeks for the high light (240 μEinsteins$\cdot m^{-2}\cdot sec^{-1}$) treatment (Figure 4). Results similar to the high light treatment were also obtained at light values of 330 and 540 μEinsteins$\cdot m^{-2}\cdot sec^{-1}$. Rates of increase for ammoniacal nitrogen were similar to exponential growth rates of the ferns and the period with the highest acetylene reduction rates coincides with the sharpest increases in ammoniacal nitrogen. Similar experiments conducted with *A. filiculoides* revealed only traces of ammoniacal nitrogen.

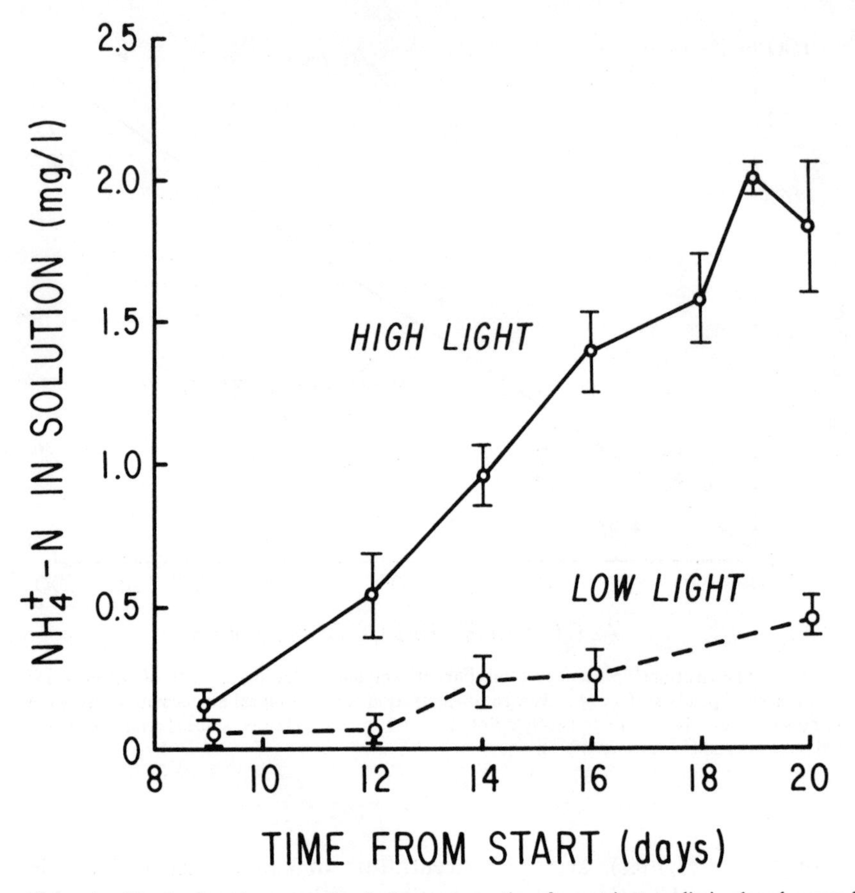

Figure 4. The increase in ammoniacal nitrogen over time for nutrient media in closed-aerated flasks into which *A. mexicana* was inoculated. The low light treatment was maintained at 130 μEinsteins$\cdot m^{-2} \cdot sec^{-1}$ (16-hr day). High light cultures were given 230 μEinsteins$\cdot m^{-2} \cdot sec^{-1}$.

Amino acid composition of the culture medium for the high light treatment in Figure 4 was 0.34 $\pm$ 0.03 mg of nitrogen/liter after 3 weeks. Amino acids present were glycine, leucine, and isoleucine. No protein was present. Detritus (determined by filtration) was 7.6 and 4.4 mg for high and low light cultures, respectively, accounting for 5.6 and 4.1% of the total *Azolla* dry weight. A total of 7.90 mg of nitrogen was contained in the flask for the high light experiment illustrated in Figure 4; 6.40 mg were in the *A. mexicana* biomass and the remaining 1.50 mg were in the culture medium, of which 1.10 mg existed as ammoniacal nitrogen and 0.21 mg was in amino acids. Assuming 2.5% nitrogen on a dry weight basis for dead or decomposing *Azolla*, another 0.19 mg, or 2% of the total, was in the detritus.

ENVIRONMENTAL CONSTRAINTS ON GROWTH AND NITROGEN FIXATION BY *AZOLLA*

Precise knowledge of the environmental tolerances of *Azolla* is central to its successful cultivation in rice paddies and will be critical to future efforts to quantitate the costs versus returns of solar ammonia production by *Azolla*. Temperature is the primary limiting factor for both species of temperate *Azolla* (Rains and Talley, 1978). *A. filiculoides* is frost tolerant but growth and nitrogen fixation are nearly zero at 10°C, increasing exponentially up to 25°C, at which temperature the biomass doubles every 60 hr and plants are 4% nitrogen on a dry weight basis. Above 30°C growth rate, and especially nitrogen fixation, decline rapidly. Prolonged exposure to 40°C days is lethal to *A. filiculoides*. However, *A. filiculoides* grown at 15°, 20°, and 30°C exhibits maximum acetylene reduction rates when exposed to temperatures of 30°, 35°, and 40°C, respectively, provided that this exposure is only for a few hours, as would occur during a normal diurnal cycle. Optimal growth and nitrogen fixation of *A. filiculoides* occur at light intensities above 800 μEinsteins $\cdot m^{-2} \cdot sec^{-1}$, with no depression at full sun (2000 μEinsteins-$\cdot m^{-2} \cdot sec^{-1}$). Temperature and light studies of *A. mexicana* have just commenced. As anticipated from field studies, this species has a very high temperature tolerance. At 30°–40°C and 12-hr, day-night thermoperiod and photoperiod, biomass of *A. mexicana* doubles every 48 hr.

Waters of the Sacramento Valley, which are used to supply rice fields, are usually deficient in phosphorus and iron with respect to *Azolla* growth (Rains and Talley, 1978). Preliminary studies on nitrogen-free nutrient media reveal a critical concentration of 20 μg of iron/liter as Fe-EDTA and 0.8 mg of phosphorus/liter for both *A. filiculoides* and *A. mexicana*. In the field, iron deficiency can be eliminated by addition of 50 g of iron/ha as Fe-EDTA every 10 days. Relief of phosphorus deficiency requires greater amounts of fertilizer during the fallow season than for companion cropping in summer, probably because of the colder temperatures prevailing at this time and longer maturation period of the *A. filiculoides* cover. Successful dual culture of *Azolla* with rice has been obtained by application of 30 kg of phosphorus/ha (as triple superphosphate) one day after inoculation or by 15 kg phosphorus/ha applied as a split treatment 1, 10, and 20 days after inoculation. Fallow season culture of *A. filiculoides* may require up to 60 kg of phosphorus/ha as one application or 30 kg of phosphorus/ha in a split treatment (Rains and Talley, 1978).

DISCUSSION

Data from this communication and studies of tropical *Azolla*-rice cultures (Singh, 1977) indicate a kilogram-per-kilogram equivalence of nitrogen as

Azolla or as commercial fertilizer (ammonium sulfate) when 50 kg of nitrogen/ha or less is incorporated into soil. Lower than anticipated rice yields, obtained from approximately 250 kg of nitrogen/ha as *A. pinnata* green manure in the Philippines (Watanabe et al., 1977), may have resulted because the *Azolla* cover was incorporated into the soil 40 days after seedling rice was transplanted. Current green manure studies are assessing the yield effects on rice when 93 kg of nitrogen/ha resulting from a single *A. filiculoides* crop is incorporated into soil. Given a favorable outcome for these studies, our next objective is to cultivate *A. filiculoides* as a green manure crop over a significant area of fallow rice paddy between mid-February and mid-April. The principal aim of this study is to quantitate both the fertilizer and energy inputs necessary to produce *Azolla* nitrogen on a significant scale in temperate (California) rice fields.

When a correction is made for the natural productivity of rice soils (yields of unfertilized controls), dual culture of approximately 40 kg of nitrogen/ha as *A. filiculoides* results in an increase in rice nitrogen content that is 70% of the response obtained when an equal amount of nitrogen is incorporated into the soil. Response of rice yield to a similar dual culture experiment in India, which utilized *A. pinnata*, was 64% (Singh, 1977). Singh and others (Moore, 1969) have concluded that the *Azolla* cover must die before its nitrogen becomes available to the developing rice crop. Laboratory studies by Saubert (1949) and Peters (1977) on *A. pinnata* and *A. caroliniana*, respectively, and our studies of *A. filiculoides* reveal little or no nitrogen in the culture media, further suggesting death and decomposition as the mode of nitrogen transfer.

The population of *A. mexicana* used in our dual culture studies appears to be unique among *Azolla* species thus far studied in that 10%–30% of the total nitrogen fixed is lost to the culture medium as ammoniacal nitrogen. The very high rice yields obtained from the dual culture experiments strongly suggest that ammoniacal nitrogen lost to the paddy water by the growing fern cover is available to the rice crop. This suggestion appears particularly reasonable because rice roots can remove ammoniacal nitrogen from solutions containing as little as 35 μg of NH_4^+-N/liter (Fried et al., 1965).

We believe leakage of ammoniacal nitrogen into the aquatic environment by the *A. mexicana–Anabaena azollae* pair is triggered by environmental conditions that result in a modest depression of carbon, but not nitrogen, fixation. Field experiments suggest that the quality and/or quantity of light prevailing under a canopy of rice (and in our growth chambers) might lead to ammonium loss. *Azolla mexicana* attains optimal biomass and nitrogen yield under conditions of high light (50% sun or more). Biomass approaches 1100 kg dry wt/ha and nitrogen content is about 38 kg/ha. The slightly reduced light under a canopy of rice growing without fertilizer results in an *A. mexicana* biomass of only 600 kg dry wt/ha, but nitrogen content is 32 kg/ha. Light levels under rice that has received considerable fertilizer (80 kg of nitrogen/ha) are very low and both carbon and nitrogen content of the *Azolla* cover is depressed (320 kg dry wt/ha and 16 kg of nitrogen/ha, respectively). The finding that equal intensities of green or white light were more effective than either red or blue light in supplying energy for nitrogen fixation by *Anabaena azollae* isolated from *A. mexicana* (Holst, 1977) may be critically related to this hypothesis.

Current dual culture research is aimed at evaluating the hypothesis that light quality and/or quantity is responsible for ammonium loss for growing *A. mexicana*. We are also attempting to define environmental conditions that will induce significant ammonium leakage in other species of *Azolla*.

REFERENCES

Food and Agriculture Organization of the United Nations. 1977. China: Recycling of organic wastes in agriculture. FAO Soils Bull. 40:29–38.

Fried, M., F. Zsoldos, P. B. Vose, and I. L. Shatokhin. 1965. Characterizing the NO_3 and NH_4 uptake process of rice roots by use of ^{15}N labeled NH_4NO_3. Physiol. Plant. 18:313–320.

Holst, R. W. 1977. Studies of the growth and nitrogen metabolism of the *Azolla mexicana–Anabaena azollae* symbiosis. Ph.D. dissertation, Southern Illinois University at Carbondale.

Lin, Pin-tung. 1976. Progress in the trial of an alternative to chemical fertilizer and its extension work. Sino-America Technical Cooperation Assn. 21(4):1–3.

Mikkelsen, D. S., and D. C. Finfrock. 1957. Availability of ammoniacal nitrogen to lowland rice as influenced by fertilizer placement. Agron. J. 49:298–300.

Moore, A. W. 1969. *Azolla:* Biology and agronomic significance. Bot. Rev. 35:17–34.

Peters, G. A. 1976. Studies of the *Azolla-Anabaena* symbiosis. In: W. E. Newton and C. J. Nyman (eds.), Proceedings of the 1st International Symposium on Nitrogen Fixation, Vol. 2, pp. 592–610. Washington State University Press, Pullman.

Peters, G. A. 1977. The *Azolla–Anabaena azollae* symbiosis. In: A. Hollaender (ed.), Genetic Engineering in Nitrogen Fixation, pp. 231–258. Plenum Publishing Corp., New York.

Rains, D. W., and S. N. Talley. 1978. The use of *Azolla* as a source of nitrogen for temperate zone rice culture. Proc. of INPUTS Project, 2nd Rev. Meeting, East-West Center, pp. 167–173.

Saubert, G. G. P. 1949. Provisional communication on the fixation of elementary nitrogen by a floating fern. Ann. Roy. Bot. Gard. Buitenzorg 51:177–197.

Singh, P. K. 1977. Multiplication and utilization of fern "*Azolla*" containing nitrogen-fixing algal symbiont as green manure in rice cultivation. Il Riso 36:125–137.

Talley, S. N., B. J. Talley, and D. W. Rains. 1977. Nitrogen fixation by *Azolla* in rice fields. In: A. Hollaender (ed.), Genetic Engineering in Nitrogen Fixation, pp. 259–281. Plenum Publishing Corp., New York.

Tran Quang Thuyet and Dao The Tuan. 1973. *Azolla:* A green manure. Agric. Prob. Vol. 4, Vietnamese Studies #38:119–127.
Watanabe, I., C. R. Espinas, N. S. Berja, and B. V. Alimagno. 1977. Utilization of the *Azolla-Anabaena* complex as a nitrogen fertilizer for rice. IRRI Research Paper Series #11.

Index

Actinomycete from *C. peregrina*
  cross-inoculability of, 223-224
  host specificity of, 223-224
  infection process of, 218, 221-223
  isolation and culturing of, 220-221
  nature of isolate, 217-218
  nodule formation by, 218
  sporangia formation by, 218, 220
  vesicle formation by, 219, 200
Actinomycetes, 135, 217-225
Aerobacter aerogenes, 35
Agrobacterium spp., 15, 140, 142, 143, 153, 157, 160, 161
Albizzia lophantha, 21
Alfalfa, 21, 75, 105, 108, 109, 165, 169, 176, 191, 199
Alnus crispa, 223
Alnus glutinosa, 22, 223, 224
Alnus rubra, 72, 223
Alnus rugosa, 220, 223
Aminolaevulinate dehydratase, in heme synthesis, 5
Aminolaevulinate synthetase, in heme synthesis, 5
Ammonia assimilation
  asparagine synthesis from, 39-46
  asparagine synthetase in, 19
  effect of glutamate synthase on, 9-15, 19-24, 304-305
    specific activity of, 35
  effect of glutamine synthetase on, 9-14, 19-24, 304-306
    specific activity of, 35, 304-305
  energy supply for, 46
  glutamate dehydrogenase in, 35, 304-305
    specific activity of, 35, 304-305
  glutamate metabolism in, 39-46
  regulation of, 9-14
  routes for, 33-34
  schemes for, in nodules, 20, 34, 35
Ammonia release
  by *Azolla,* 314-316, 318-319
  by rhizobia, 6-7
Anabaena 7120, 263-269, 275
Anabaena cylindrica, 260, 283-284
Anabaena doliolum, 288
Anabaena flosaquae, 288
Anabaena variabilis, 281, 283, 288, 290
Anacystis nidulans, 283
Arachis, *see* Peanut
Asparagine biosynthesis
  energy supply for, 46
  in legume nodules, scheme for, 44-46
Asparagine synthetase
  in nodules, 19, 39
  properties of, 43
Aspartate aminotransferase, properties of, 42
Associative symbiosis
  with sorghum roots, 249-253
    effect of fertilizer, 249-252
    effect of inoculant, 249-252
    effect of lines, 249, 252-253
  with wheat roots, 249
Azolla
  amino acid release by, 316

Azolla—continued
ammonia release by, 314–316, 318–319
biomass in rice fields, 312
comparison with $(NH_4)_2SO_4$ on rice, 314, 317–318
environmental constraints on, 317
as green manure, 311–319
iron requirement for, 317
nitrogen content of, in rice paddies, 312–314
phosphorus requirement for, 317
see also Azolla caroliniana/Anabaena azollae symbiosis
Azolla caroliniana/Anabaena azollae symbiosis
acetylene reduction by, 297–302
combined nitrogen effect on, 302
electron balance studies of, 297–302
fixed carbon transfer in, 306–308
fixed nitrogen transfer in, 303–304
glutamate dehydrogenase content of, 304–305
glutamate synthase content of, 304–305
glutamine synthetase content of, 272–274, 304–306
heterocyst frequencies in, 294
hydrogen evolution by, 297–302
matabolite exchange in, 297
morphology of, 293–297
nitrogen fixation in, 297–302
photosynthesis by, 306–308
photosynthesis and nitrogen fixation interaction in, 306–308
see also Azolla
Azolla filiculoides, 294, 297, 312–318
Azolla mexicana, 312–319
Azolla pinnata, 311, 318
Azospirillum brasilense, 181, 243
Azospirillum lipoferum, 243–246
Azotobacter chroococcum, 80, 244
Azotobacter paspali, 243
Azotobacter vinelandii, 8, 13, 14, 70, 173–175, 181

Bacillus amyloliquefaciens, 4
Bean, 21, 199
Blue-green algae, *see* Cyanobacteria; Heterocysts

Canavalia ensiformis, 167
Casuarina spp., 53, 219, 223
Ceanothus americanus, 220, 223
Ceanothus integerrimus, 224
Cell wall composition, *see* Rhizobium strain identification
Clover, 21, 74–76, 88, 97, 104, 108–112, 122, 135, 165, 168–171, 176–179, 182, 191, 199
Competition among rhizobia
environmental effects on, 136
host genotype effect on, 135
between introduced and indigenous populations, 132–133
among introduced populations, 133–134
other microorganisms' effect on, 135–136
Comptonia peregrina, 217–223
Cowpea, 18, 191, 206
Cyanobacteria
carbon transport in, 283
genetic studies of, 286–290
heterocyst differentiation in, 259–276, 285
nitrogen fixation in, 280–283
nitrogen transport in, 283–285
symbiosis involving higher plants, 286–290
see also Heterocysts
Cytochrome P-450, 59–60

Dactylis spp., 231, 236, 238
Digitaria spp., 245, 246

Elaeagnus angustifolia, 223
Elaeagnus umbellata, 221, 223
Enterobacter cloacae, 247
Erwinia herbicola, 245–249, 252
Escherichia coli, 11, 114, 115, 118, 269, 274, 275

Factors limiting nitrogenase activity, *see* Nitrogenase activity limitations
Free-living rhizobia
ammonia assimilation by, 36–37

control of nitrogenase synthesis and activity in, 6–8
excretion by, 36–37

Gloeocapsa spp., 283
Glutamate dehydrogenase, 19, 24, 35, 37, 44, 304–305
Glutamate synthase, 21, 22, 24, 35, 37, 284, 304–305
Glutamine synthetase, 7–14, 19–24, 35, 37, 44, 272–274, 283–285, 304–306
Glycine Max, see Soybean
Growth substances in nodule development, 95–97

H_2 evolution and uptake
by *Azolla,* 297, 302
by bacteroids, 16–19, 73–77
benefits of, 80–83
effect of environment on, 75–76
effect of plant host on, 75
efficiency of, 16–17, 71–73, 16–19, 69–76, 79–84
energy cost of, 18–19, 70–71
by free-living rhizobia, 77–79
by legume symbiosis, 71–72
by nodules, 16–19
reasons for, 16, 70
Heterocysts
aerobic nitrogen fixation in, 261–263, 280–283
communication between vegetative cells and, 283–285
comparison with vegetative cells, 259–261
differentiation of, 263–276
control of spacing pattern during, 268–280, 285
effect of glutamine on, 268–270
proteases active during, 263–268
transcriptional control during, 274–275
frequency in *Azolla/Anabaena* symbiosis, 294
hydrogenase location in, 261
as independent nitrogen-fixing catalysts, 290
nitrogen transport from, 283–285
nutrient flow into and out of, 262, 283
Hydrogenase, 16–19, 69–85, 297–302
see also H_2 evolution and uptake; Heterocysts

Infection
of cortical cells, 89–95
of root hairs, 88–90

Kalanchöe daigremomtiana, 157
Klebsiella pneumoniae, 118, 244, 247, 248, 252

Leghemoglobin
components of, 4
differences in, 4
effect of nitrate on, 5
effect of bacteroid ATP formation on, 57–59
effect on N_2 fixation of loss of, 5
function of, 63–64
heme biosynthesis for, 5
iron chromophore associated with, 61–63
location in nodules of, 3–4, 63–64
occurrence in non-legume symbioses of, 53–54
O_2 transfer to bacteroids by, 55–60
O_2 transport by, 54–55
Leghemoglobin-associated iron chromophore, 61–63
Legume nodules
ammonia assimilation in, 19–24, 33–49
cytochrome P-450 in, 60
export product of, 19–39
factors limiting nitrogen fixation in, 49
H_2 evolution and uptake by, 16–19
induction of ammonia-assimilating enzymes in, 37–44
nitrogen flow rates through, 46–49
nucleic acids in, 14–16
oxidases in, 55–61
physiology of, 3–24
Legume-rhizobium symbiosis
infection, 105–107

Legume-rhizobium symbiosis—*continued*
influence of host plant on, 109–113
influence of rhizobial genome on, 113–120
nature of, 103–105
nodule formation, 105–107
nodule function, 107–108
physiological and environmental effects on, 120–122
preinfection factors, 105
role of plasmids in, 119
sequence of events for, 104
symbiont specificity of, 108–109
Lemna spp., 22
Leucaena spp., 108
Lipopolysaccharide composition of rhizobial cell walls, *see Rhizobium* strain identification
Lotononis bainesii, 108
Lotus spp., 14
Lupin, 14, 15, 18, 19, 21, 37, 44, 191

Macroptilium atropurpurenum, 108
Maize, 231, 235–239, 245, 246
Medicago sativa, see Alfalfa
Myrica cerifera, 223
Myrica gale, 219, 220, 223

Nitrogenase
ammonia effect on, 5–8
control of activity of, 6–8
control of production of, 6–8
glutamine synthetase effect on, 7–8
leghemoglobin effect on, 5, 55
nitrite effect on, 5
O_2 effect on, 6–8
Nitrogenase activity limitations
field factors, 229–240
air temperature, 238
fruiting, competition by, 237–238
leaf dry weight, 235–236
light, 238
photosynthesis apparatus size, 235–236
soil temperature, 239–248
water content of soil, 238
water deficit of air, 239
in leguminous associations, 49

Nitrogen flow
in *Azolla,* 297–308
in cyanobacteria, 283–285
with heterocysts, 262, 283–285
through nodule, 46–49
transport compound from soybeans in, 286
Nodule formation and development
effect of growth substances during, 95–97
genetic control of, 98
infection and initial events during, 88–97
polyploid cell involvement in, 92–95
relationship of transfer cells to, 97–98
Nostoc linckia, 288
Nucleic acids in bacteroids, 14–16

Panicum maximum, 246
Panicum virgatum, 245
Parasponia parviflora miq., 53, 108
Paspalum notatum, 243, 246
Pea, 21, 75, 76, 79, 87–100, 109, 122, 157, 177, 180, 191, 199
Peanut, 231, 235–238
Pennisetum americanum, 246
Pennisetum purpureum, 238
Phaseolus spp., *see* Bean
Pisum spp., *see* Pea
Plasmids in rhizobia
comparative studies of, 153
"curing" attempts for, 153–157
detection and isolation of, 140–144
molecular weight determinations of, 142–153
Ti plasmid transfer, 157–160
Plectonema spp., 261, 275
Pseudomonas aeruginosa, 114

Rice, 231, 236, 238, 239
Rhizobial cell wall composition, *see Rhizobium* strain identification
Rhizobial competition, *see* Competition among rhizobia
Rhizobium cowpea, 77, 167, 206
Rhizobium japonicum, 5, 7, 9–15, 17, 21, 36, 40, 60, 72–75, 77,

79, 106, 110, 114, 116, 118, 132–135, 144, 166, 174–177, 179, 181–182, 190, 194–199, 206, 209–213
Rhizobium leguminosarum, 9–11, 15, 18, 60, 74–77, 80, 92, 96, 98, 109, 111, 114–118, 135, 143–145, 152, 153, 156, 157, 160, 177, 180, 190–199, 211
Rhizobium lupini, 11, 14, 15, 18, 144, 190, 194, 209
Rhizobium meliloti, 15, 72–75, 107, 109, 114–119, 143–145, 152, 156, 160, 169, 175–178, 191, 194, 196, 198–199, 209
Rhizobium phaseoli, 114, 144, 166, 190, 194–199, 209
Rhizobium strain identification
  cross-inoculation groups for, 190
  flagellin-specific antibodies for, 198–199
  lipopolysaccharide composition for, 191–194
  lipopolysaccharide-specific antibodies for, 194–198
  nodulation assay for, 190–191
  use of bacteriophages in, 199–201
Rhizobium trifolii, 5, 8–14, 18, 72, 104, 108–111, 114–122, 135, 143–144, 152–153, 156, 160, 165–183, 190–196, 199, 209
Rhizocoenosis, *see* Associative symbiosis

Salmonella typhimurium, 8
Serradella spp., 21, 36
Sorghum, 245–253
Soybean, 75–79, 106, 132–134, 165, 166, 177, 179, 191, 199, 206, 211, 286
Specificity in legume symbioses
  adsorption of rhizobia to root hairs, 176–178
  anomolous interactions in, 167
  clover / *R. trifolii* system, 168–175
    2-deoxyglucose in, 169, 175, 176
  "cowpea-type" / tropical legume systems, 167–168
  importance of, 166
  lectin binding vs. infectivity correlation, 205–209
  lectin recognition hypothesis of, 166–177
  nature of lectin receptor on host plant, 182–183, 209–211
  nature of lectin receptor on rhizobia, 179–181, 211–213
  polarity in rhizobia effect in, 182
  regulation of recognition in, 181–182
  schematic representation of, 170
Staphylococcus aureus, 273

Terramus unicinatum, 18
Trifolium spp., *see* Clover

Vetch, 18, 21, 60, 75, 113, 178
Vicia spp., see Vetch
Vigna spp., *see* Cowpea

Wheat, 245–249

An essential reference for botanists and plant scientists...

CO_2 METABOLISM AND PLANT PRODUCTIVITY

Edited by **R. H. Burris, Ph.D.,** Professor of Biochemistry, University of Wisconsin (Madison) and **C. C. Black, Ph.D.,** Professor of Botany and Biochemistry and Head, Department of Botany, University of Georgia.

For scientists studying ways to increase crop production this volume presents large amounts of new data from biochemical and physiological studies of CO_2 metabolism combined with evaluations of current work on plant metabolism. It is an essential reference source for botanists, biochemists, plant physiologists, plant breeders and agronomists.

The book examines complex interactions of photosynthesis and assesses, from several viewpoints, central problems concerning manipulation of CO_2 assimilation by environmental, chemical, and genetic means. A section is included on new methods that can supplement classical techniques in plant breeding programs to improve productivity.

This volume contains material not available elsewhere on photosynthesis, photorespiration, and dark respiration, with special consideration of the pentose phosphate and C_4 pathways, glycolate biosynthesis, glycine and serine metabolism, C_3 plants, and crassulacean acid metabolism (CAM).

Its thorough and authoritative coverage of the subject make it valuable reading for graduate students of plant physiology, plant biochemistry, crop production, and crop physiology as well as a reference for all workers and advanced students concerned with plant metabolism and crop improvement.

416 pages *Illustrated* *1976*